ANALYSI
QUANTITATIVE
DATA

SAGE
50
YEARS

ANALYSING QUANTITATIVE DATA

Variable-based and Case-based Approaches to Non-experimental Datasets

Raymond Kent

Los Angeles | London | New Delhi
Singapore | Washington DC | Boston

Los Angeles | London | New Delhi
Singapore | Washington DC

SAGE Publications Ltd
1 Oliver's Yard
55 City Road
London EC1Y 1SP

SAGE Publications Inc.
2455 Teller Road
Thousand Oaks, California 91320

SAGE Publications India Pvt Ltd
B 1/I 1 Mohan Cooperative Industrial Area
Mathura Road
New Delhi 110 044

SAGE Publications Asia-Pacific Pte Ltd
3 Church Street
#10-04 Samsung Hub
Singapore 049483

Editor: Jai Seaman
Assistant editor: Lily Mehrbod
Production editor: Victoria Nicholas
Copyeditor: Neville Hankins
Proofreader: Andy Baxter
Marketing manager: Sally Ransom
Cover design: Shaun Mercier
Typeset by: C&M Digitals (P) Ltd, Chennai, India
Printed and bound by
CPI Group (UK) Ltd, Croydon, CR0 4YY

MIX
Paper from responsible sources
FSC® C013604
FSC www.fsc.org

Library of Congress Control Number: 2014947234

British Library Cataloguing in Publication data

A catalogue record for this book is available from the British Library

ISBN 978-1-4462-7340-1
ISBN 978-1-4462-7341-8 (pbk)

At SAGE we take sustainability seriously. Most of our products are printed in the UK using FSC papers and boards. When we print overseas we ensure sustainable papers are used as measured by the Egmont grading system. We undertake an annual audit to monitor our sustainability.

CONTENTS

LIST OF FIGURES, TABLES AND BOXES

Figures

Tables

Boxes

ABOUT THE AUTHOR

Raymond Kent is now retired, but was Senior Lecturer in Marketing at the University of Stirling, UK. He has published on topics as diverse as product range policy, marketing communications with sensitive groups, private trading on the Internet, improving email responses and the use of fuzzy logic in data analysis. This is his eighth book; earlier books have been in the areas of the history of sociological research, survey data analysis and marketing research. After each book, he says it will be his last. We'll see ...

COMPANION WEBSITE

This book is supported by a brand new companion website (https://study.sagepub.com/kent). The website offers a wide range of free teaching and learning resources, including:

- chapter summaries;
- recommended reading lists;
- free access to SAGE journal articles;
- answers to exercises from the book;
- PowerPoint slides for each chapter.

In addition, there is:

- an overview of data analysis packages;
- an introduction to SPSS;
- weblinks to alternative datasets.

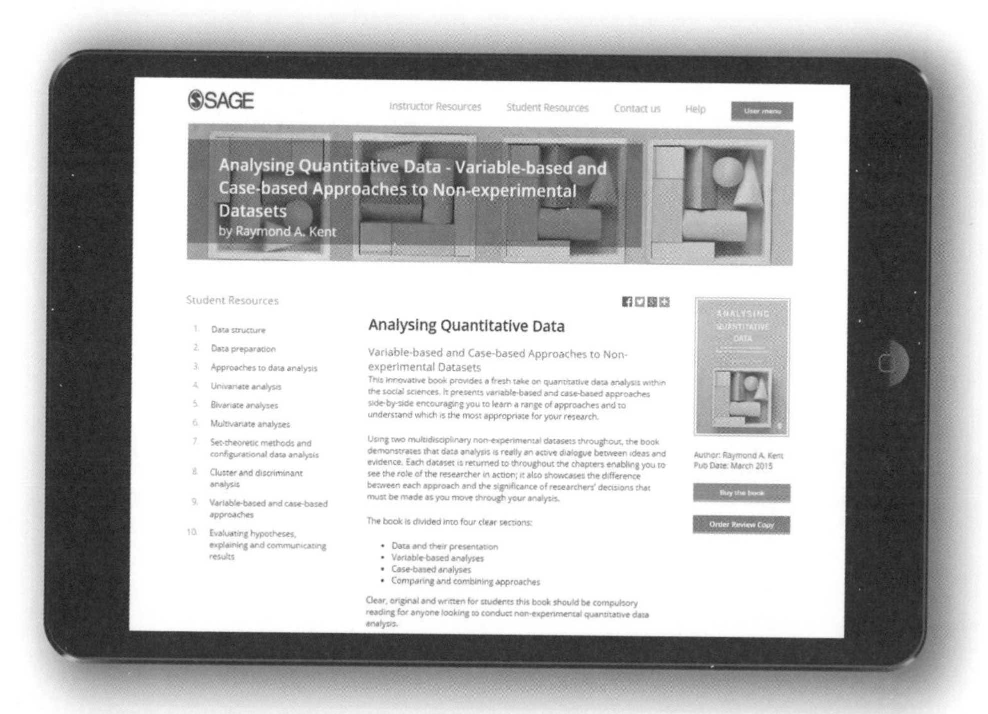

PREFACE

This text concentrates on the analysis of quantitative data rather than their generation, but takes a broad view of what 'data analysis' entails. It is more – much more – than simply 'doing statistics'. It means understanding the data – how they were constructed in the first place, what kind of data they are and the ways in which they are structured. It includes the various processes involved in preparing and transforming the data ready for analysis, creating a data matrix, and then going through the processes of describing, interpreting, relating, evaluating, explaining, applying and presenting the results. Any dataset, furthermore, needs to be approached holistically, that is, as a complete, self-contained entity, set in the context of the objectives for which the research that generated the dataset was designed to achieve and with a well-rounded view of what *all* the evidence is saying. This may mean combining hard and soft research data with informal individual experience, knowledge and intuition and seeing this all within the context of the 'bigger picture'.

Data analysis means getting the most out of a dataset, approaching it in several different ways so that the data tell the complete story. All too often, researchers go to a lot of trouble to construct their data, only to limit the analysis by throwing at such data a few statistics with which the researcher happens to be familiar or believes are appropriate to the data. Students in the social sciences normally undertake a course or module in research methods that includes an explanation of how a selected range of statistics is calculated. However, when approaching a dataset students will often limit themselves to asking 'OK, so what statistics should I use here?', instead of thinking about 'What are my research objectives?', 'What do I want my analysis to show or investigate?' or 'What are the different ways in which the analysis could be approached?'

Analysing data, furthermore, is seldom a one-off enterprise accomplished in a single session. Analysis is a dialogue between ideas and evidence: researchers move backwards and forwards between constructed data and the objectives for which the research was undertaken, often on different occasions or at different phases of the research.

After studying this book, readers should be able to:

- understand how datasets can be handled by being taken through, in a kind of 'master class', how a given dataset can be analysed in several different ways;
- question basic statistical concepts and compare the results of using different techniques;
- think about the choices they need to make in the analysis of non-experimental quantitative data by considering the dataset as a whole;
- understand that quantitative case-based methods offer an alternative to or an addition to the more standard variable-based methods.

The subtitle of this book indicates that its focus is on non-experimental datasets. Experimental datasets are generated from experimental designs in which researchers deliberately manipulate factors that are seen as potentially causal (often referred to as 'treatments' or 'interventions') and carefully observe the effects of such actions while controlling as far as possible for extraneous factors. Data analysis will focus on the nature of those designs and how results can be tested using statistical inference. Non-experimental datasets, by contrast, will be generated largely from various kinds of survey, but may also be a result of using electronic or observational data capture techniques. The kinds of manipulations and controls in experimental designs are seldom possible with survey and other non-experimental data, while samples tend to be much larger, making the role of statistical inference far less important. Statistical inference is still covered in this text, but as part of variable-based analyses, taking variables one, two and three or more at a time.

Within the context of non-experimental datasets, this text is designed as a stand-alone, complete introduction to both variable-based and case-based approaches to data analysis. To date, books tend to focus on one or the other, while comparisons between approaches are few. The book assumes no prior knowledge of statistics. However, details of statistical calculations or how to use particular pieces of software are put into boxes so that students or researchers can skip these if they are already familiar with the procedures involved. The approaches, methods and techniques are all illustrated with one main dataset based on research carried out at the Institute of Social Marketing at the University of Stirling, which investigates the role of alcohol marketing on the drinking behaviour of young people. The dataset is far from perfect; in fact there are many problems with it. This, however, turns out to be an advantage since many typical problems in the analysis of real-life data come to light and remedies or ways to handle them can be considered.

This book should be of interest to final-year undergraduates and postgraduates who are undertaking modules or courses in research methodology in the areas of sociology, business studies, marketing, health and education. In particular, it should be of interest to students who are doing projects, dissertations or theses and who are wondering what approaches to data analysis are possible and what data analysis strategies they should adopt. It should also be of interest to researchers who usually go to great lengths to construct their data, but often find that traditional, variable-based statistics produce disappointing or inconclusive results.

Part One of this book provides an overview of the nature of quantitative data, their structure, preparation and analysis. Chapter 1 introduces the alcohol marketing dataset, it considers what data are, how they are constructed, how they are structured and how errors can arise in the process of construction. Chapter 2 looks in some detail at how quantitative data need to be prepared ready for analysis, for example by performing a number of transformations on them. Chapter 3 explains the notions of datasets and data matrices, outlines the various elements that go into the data analysis process and explores some of the ethical considerations when constructing and analysing data.

Part Two then turns to an approach to data analysis that reviews data on a variable-by-variable basis, looking at the distributions of values and their frequencies across a set of cases. Chapter 4 shows how variables can be displayed and summarized one at a time and how inferences can be drawn should the data be based on a random sample. Chapter 5 then takes variables two at a time, showing the variety of relationships that is possible and how such relationships can be displayed, summarized and have inferences drawn from them. Finally, Chapter 6 introduces multivariate analysis which takes three or more variables at a time. Part Two will be familiar to those readers who have already studied traditional statistics. However, I have tried very hard to show what these procedures accomplish (and what they do not) without being, at this stage, overly evaluative or critical of the approach. All the techniques are illustrated using the survey analysis package IBM® SPSS® Statistics software. Explanations of how to use SPSS are put into boxes, so while this text is by no means a manual on how to use the software, focusing on the boxes will get you started on this program. SPSS has been around a long time (it was developed in the 1960s) and is available in most university and college labs. However, there are many other packages and these are reviewed briefly at the end of Chapter 4. The version of SPSS used in this text is version 19.0. At the time of writing, the latest version is 22.0. However, most users are likely to have earlier versions. The guidelines on the use of SPSS in the boxes in this book are unlikely to be affected by the version being used.

An alternative approach to variable-based data analysis, which is the focus of Part Three, is to review a dataset on a case-by-case basis, looking at the configurations of values for each case across a set of case characteristics. Chapter 7 on set-theoretic methods and configurational data analysis, and which uses fuzzy set analysis, is not likely to be familiar to readers. While the logic of this approach can be daunting, I have tried to explain and illustrate each step of the way. To those who are uneasy about 'numbers', the good news is that there is very little by way of calculation in this chapter. Furthermore, if the research is based (or likely to be based) on a relatively small sample or population (of between about 30 and 100 or so) then this approach is well worth looking at. This chapter introduces a particular freeware program for fuzzy set analysis called fsQCA, which stands for fuzzy set Qualitative Comparative Analysis.

Part Four compares the strengths and weaknesses of variable-based and case-based approaches, considers how they can be mixed or combined, and discusses how both can be used to evaluate hypotheses, establish causal relationships, explain the findings and communicate results to an audience.

I have used several learning tools to assist the reader. Each chapter begins with a list of learning objectives that outline what the reader can expect to learn in that chapter. The list will allow readers to monitor their understanding and progress during the chapter. An introduction then explains how the chapter is organized, provides some background to its content and links it to other chapters in the book. At the end of most major sections in each chapter there are key points and wider issues. These provide a quick summary of the key points and go on to consider any issues that might arise from the section by looking at the wider context. Boxed areas provide more detailed information on either the calculation of particular statistics or the operation of the software that is being used for illustration. Each chapter makes frequent reference to the alcohol marketing dataset to illustrate procedures and points being made and, at the end of each chapter, the implications of the chapter content for the alcohol marketing dataset are discussed. The chapter is then summarized, there are exercises and questions for discussion, and suggestions are made for further reading. These are annotated to help readers to decide what is worth reading as a follow-up to the chapters from their point of view. At the end of the book there is a glossary, all the references are collected together and there is a full index. In the text, a selection of key terms to be found in the glossary are highlighted in bold the first time they appear, and in selected places where it is felt this would be useful to the reader.

On matters of style, the word 'data' is treated as plural throughout. Where numbers or words used in the text refer to what appears or might appear in or are to be entered into software, they are in `Courier New` font. I have tried to develop a terminology that is consistent throughout the text so that words like 'case', 'variable', 'value', 'score', 'measure', and so on have the same meaning throughout. I have, of necessity, felt the need to introduce in many places terms that will be explained later in the text. I have indicated where this is so, and in addition the terms will be in the glossary. The material in the boxes can be skipped without detracting from the meaning of the text.

I would like to thank the people who have helped me to complete this text. First, my editor, Jai Seaman, for her encouragement, suggestions and critical comments. Second, the eight anonymous reviewers of my book proposal whose comments and suggestions have been incorporated as far as possible. Finally, I would like to thank Gerard Hastings at the University of Stirling for permission to use the alcohol marketing dataset, and John Hall, now retired, formerly of the Social Science Research Council Survey Unit, who has given permission to use the datasets on his very helpful website.

I

QUANTITATIVE DATA: STRUCTURE, PREPARATION AND ANALYSIS APPROACHES

1

DATA STRUCTURE

Learning objectives

In this chapter you will learn that:

- data are constructed rather than collected and result from a process of systematic record-keeping;
- records are created in a social, economic and political context and for purposes specific to individuals or groups within organizations;
- qualitative data consist of words, phrases, narrative, text and visual images, while quantitative data arise as numbers that result from the systematic capture of classified, ordered, ranked, counted or calibrated characteristics of a specified set of cases;
- all quantitative data have a structure that consists of cases, properties and values;
- the construction of data of any kind is likely to give rise to errors from various sources;
- the dataset used throughout this text consists of 61 properties for 920 cases, but they do not constitute a random sample and there are many potential sources of error in the data.

Introduction

All research involves analysing data at some point – but what do we mean by 'data'? What kinds of data are there? How are they constructed? How are quantitative data structured? This chapter provides an overview of the nature and characteristics of data in general and shows how quantitative data in particular are constructed and structured.

The procedures used by researchers to structure and analyse a dataset are illustrated throughout this text with a study carried out by the Institute of Social Marketing at the University of Stirling, which studies the impact of alcohol marketing on the drinking behaviour of young people aged between 12 and 14. The findings are based on a survey that involves an interview-administered questionnaire measuring awareness and involvement with alcohol marketing and a self-completion questionnaire measuring alcohol drinking and associated behaviours. The homes of all second-year pupils attending schools in three local authority areas in the west of Scotland were contacted, generating a sample of 920 respondents (Gordon et al., 2010a).

The key research hypotheses are that the more aware of and involved in alcohol marketing that young people are, the more likely they are to have consumed alcohol, and the more likely they are to think that they will drink alcohol in the next year. To measure awareness, respondents are asked if they have seen any adverts for alcohol in any of 15 channels, for example television, cinema, newspapers, websites or sponsorships. Responses are recorded into 'Yes', 'No' and 'Don't know'. To measure involvement in alcohol marketing, pupils are asked whether they have, for example, received free samples of alcohol products, free gifts showing alcohol brand logos or promotional mail or email.

Drinking behaviour is measured in four main ways. Drinking status is assessed by asking whether pupils have ever had a proper alcoholic drink, not just a sip. Future drinking intention is assessed by asking about the likelihood that they would drink alcohol during the next year – 'Definitely not', 'Probably not', 'Probably yes' and 'Definitely yes'. They are also given a 'Not sure' option. Initiation is measured by asking how old they were when they took their first drink, plus a **measure** of the number of alcoholic units last consumed. The study uses a range of control variables suggested in the literature, for example parental attitudes towards drinking and alcohol consumption, perceived parental drinking approval, sibling and peer drinking behaviour, liking of school and **rating** of school work. Demographic controls include gender, social grade (based on the occupation of the head of household), ethnicity and religion. The data from the alcohol marketing research are available online from the Sage website, https://study.sagepub.com/kent.

Data and their construction

Data are often thought of as 'the facts' – things that are known to be true. The dictionary tells us that the word is a plural noun (although commonly treated as singular) and derives from the Latin word that translates literally as 'things given'. Data are thus portrayed as a form of knowledge – sheer, plain, unvarnished, untainted by social values or ideology and, for the most part, unchallengeable. The assumption is that they exist independently of our research activities and that we can simply go out and discover or 'collect' them like so many tadpoles in a pond.

In reality, however, data are not collected or discovered, but constructed. They are generated as a result of the human activity of systematic record-keeping, for example in registers of births, marriages and deaths, hospital records, invoices, questionnaires, electronic meters, audio or video recordings. Record-keepers, furthermore, construct data for their own purposes. They have their own agendas and personal circumstances; they have careers to pursue, their own fears and hopes; they have bosses to impress or subordinates to guide or deploy. Data construction is a process, furthermore, that takes place in a social, moral, economic, political and historical context. There are, for example, colleagues or academic peers to consider, respondents or subjects to bear in mind, consumers, clients, funding or sponsoring agencies to take into account.

All this is not to say that data are just concocted – meaningless artefacts, subject to manipulation, doctoring or media spin. They are, however, constructed in a particular context for specific purposes. It has been argued that everyday reality (Berger and Luckmann, 1966), scientific facts (Latour and Woolgar, 1979) and many other things like gender, homosexual culture or ideas about illness are socially constructed. By being specific about *what* is being socially constructed, there is the implicit admission that not everything (like material objects) is a social construction and that there may be degrees of construction involved (Hacking, 1999). Social reality does, however, both constrain and facilitate data construction; so do the agreed (or disputed) practices and routines of scientific procedure.

Few data, furthermore, are perfect. Errors, to varying degrees, will almost certainly be made in the data construction process. Different researchers will often produce different results, apparently from researching the 'same' phenomena. Even government statistics are often based on questionnaire surveys, and there are many things that can go wrong with this process. Issues of error in data construction are taken up later in this chapter. Apart from the absence or presence of error, the quality of data will also vary in their comprehensiveness, the speed or timeliness with which they are delivered, and in the manner of their construction.

Data, in short, are not 'the facts' or 'things given'; they are social products. The records created are not reality itself; rather they are a result of researchers' attempts to observe or measure traces or evidence of phenomena situated within complex systems (Byrne, 2002). The records that researchers create come in very different forms. The historian likes to think of church registers, diaries of famous people, or transcripts of what was said by politicians as 'data'. A sociologist with an audio recorder studying women's emotional reactions to domestic violence, or participating in 'street corner society' and making notes of his or her experiences, likes to think that he or she is collecting 'data'. An anthropologist looking at some unusual, remote tribe of people considers that he or she is generating 'data' by making records describing their culture. The archaeologist uses physical traces or remains as evidence or data on past events, conditions or social behaviour. The manager of a business organization may think more in terms of sales data or information on balance sheets and profit and loss statements. The market researcher is more likely to see the results of a questionnaire survey or the record of a focus group discussion as 'data'.

Data may, in fact, consist of three rather different kinds of constructed record, for example:

- words, phrases or narrative captured in audio tape or digital recordings, interview transcripts or field notes; alternatively, text already recorded in minutes of meetings, reports, historical or literary documents, personnel records or newspaper clippings;
- images, for example paintings, sketches, drawings, photographic stills, DVD recordings, computer-generated images, posters, advertisements;
- numbers that result from the systematic capture of classified, ordered, ranked, counted or calibrated characteristics of a sample or population of cases, for example the number of males and females in an organization or the sizes of supermarkets in square metres of floor space.

Words, phrases, narrative, text and visual images (which are often combined, for example in posters) are usually regarded as 'qualitative' data. Data that arise as numbers are 'quantitative'. What is commonly described as 'qualitative research' will usually result in the construction of largely **qualitative data**, while quantitative research will focus mainly on generating **quantitative data**, but both types of research will usually be a mixture of both sorts of data. The focus in this text is on the analysis of quantitative data, but Part Three does consider mixed methods and how some forms of qualitative data can be quantified.

Data construction may take place either during the routine capture of information, for example on patients admitted to the accident and emergency department in a hospital, or they may be a result of research activity. Data construction in the latter context will include two key elements: the design of the research, which provides the context within which it is intended to construct data, and the actual capture of the data themselves.

The purpose of any research design is to ensure that the data constructed enable the researcher to address the objectives for which the research was undertaken, for example to answer research questions or to test research hypotheses. Writers of texts on research methods are apt to propose listings of different types of design: for example, there are qualitative designs, quantitative designs and mixed designs (e.g. Creswell, 2009); there are exploratory, descriptive and causal designs (e.g. McGivern, 2009). De Vaus (2002) suggests that all designs in the social sciences fall into one of four main groups: experimental, longitudinal, cross-sectional or case study.

Classifications of different types of research design such as those above imply alternative combinations of elements that are for the most part mutually exclusive. An actual piece of research, however, will usually utilize more than one type of design element. So, any design is usually specific to a particular enquiry and will be a unique combination of elements that involve mixing different types of research in the same project. A design may usefully be seen as a series of 'sub-designs', for example a design for the specification and selection of the entities that are to be the focus of the research, a design for the role, construction and measurement of selected characteristics of those entities, a design for the capture of data and proposals for their analysis.

A key element in any research design is the clarification of research objectives. These spell out what the research is designed to show or achieve. The more specific these are, the easier it is to design a piece of research that will construct relevant data and the easier it is to see what kinds of data analysis might be appropriate. Ideally, stated research objectives should consist of two key elements: a statement of the general research area, purpose or aim and more specific research questions or research hypotheses. The general research purpose may broadly be exploratory or verificational, for example it may be 'to explore, investigate or study the effect of playing background music on consumer behaviour in the retail environment' or 'to demonstrate or show that the playing of background music has a significant impact on consumer behaviour in the retail environment'. More specifically, a research question might be 'What is the effect of playing loud music on the amount spent?' or, phrased as a hypothesis, 'The faster the music, the less time customers spend in the retail environment'.

The actual capture of the data will require the use of one or more data capture instruments. For qualitative data, the creation of a record could be by way of manual or electronic notebooks, audio or video recorders, camcorders, still cameras or seeking commentary via open-ended questions in questionnaires, email, web pages, blogs, Facebook, and so on. For quantitative data, the most common way to capture data will be through the use of fixed choice responses in a questionnaire, but these may be of very different types. For example, they may be completed by respondents themselves or by interviewers on behalf of the respondent either in a face-to-face situation or over the telephone. Self-completed questionnaires may be delivered personally, by post or using the Internet. An alternative instrument that is commonly used but seldom explained is the diary. These get respondents to record instances of behaviour as and when they occur and may, for example, relate to records of personal contacts or media use – radio listening, for example, is commonly recorded in this way. Increasingly, however, quantitative data are captured electronically using bar scanners, set meters (for television viewing, for example), passive sensing devices, portable data entry terminals or the Internet.

The data from the alcohol marketing study are largely quantitative and are constructed using an academic cross-sectional survey research design. It was cross-sectional in the sense that the study was treated as a 'one-off' with measures taken as a single time period. The alternative would be a longitudinal design where measures are taken at intervals with the express purpose of measuring changes. Although the data used in this book are cross-sectional, in reality they are part of a wider study at the Institute of Social Marketing at the University of Stirling which is a two-wave cohort design, the first study carried out from October 2006 to March 2007 with a follow-up of the same respondents two years later. This nicely makes the point that real designs are combinations of elements. Data capture, too, was a combination of interviewer-completed and self-completed (but personally delivered) questionnaires.

Key points and wider issues

Data do not exist 'out there', waiting to be 'collected' or 'discovered'. Rather, they are constructed by individuals within a social, economic, political and moral matrix of possibilities and constraints. They are generated as a result of the human activity of systematic record-keeping using a range of data capture instruments. Data are of very different types. They may be qualitative or quantitative, or some mixture. There are, furthermore, different types of each. Qualitative data may be words, phrases, text, images or a mixture. Quantitative data, as is explained in the next section, may be categorical or calibrated according to some specified metric like years, euros or kilograms. Data come in different qualities (good, adequate and poor), they are all in various ways subject to error and may be judged in different ways, for example in terms of comprehensiveness, accuracy, timeliness, relevance, and so on, so what counts as 'good' data may not, in any case, be clear-cut.

Data construction may be a routine process or it may be a result of research activity, in which case the construction entails the design of the research and the purposeful capture of the data. A research design is specific to a 'piece' of research and is made up of a number of elements whose combination is usually unique to that research.

The wider implications of this view of the nature of data for data analysis are that, before any analysis takes place, the researcher should think about the quality of the data, the design of the research and the contexts in which the data were captured. There is little point, for example, in fussing over the finer points of statistical analysis if the data are of dubious quality to begin with or are inappropriate for the purposes of achieving the research objectives.

The structure of quantitative data

All quantitative data result from the systematic capture of classified, ordered, ranked, counted or calibrated characteristics of a sample or population of entities like individuals, groups, organizations, societies, nation-states, places or objects. They have a structure that consists of three elements: cases, properties and values. Cases are the entities that are the focus of inquiry; they are instances of conditions, behaviours, events or circumstances whose characteristics merit the researcher's attention; properties are the characteristics of those cases that the researcher has chosen to observe or measure and then record, while the values are what are actually recorded.

Cases

All scientists, whether of the physical or social variety, study cases. They look for patterns or causal mechanisms among a set of entities under investigation – entities that they deem to be 'instances' of the phenomena being studied. Cases may be 'micro' entities like individuals, families, social groups of individuals or households, or they may be 'macro' entities like organizations, governments or nation-states. Alternatively, cases might be places, geographical areas, objects or events. Usually, these entities have specified characteristics in common, for example they are 'alcoholics', 'single-person households', 'small businesses', 'private hospitals', 'faith schools' or 'democratic countries'. Cases may be legally recognized entities like a joint stock company, they may be researcher constructs, like a 'small to medium enterprise', or they may be a statistical artefact such as a health sector generated from a cluster analysis. Cases may be complex adaptive systems like nation-states; they may have open boundaries, emergent properties and transformational potential like social networks.

Cases are always located in time and geographical area. Time may be a moment of time or a period of time. They become sites, or potential sites, for the capture of data and while each case is inherently separate from other cases and unique in its multifarious characteristics, it is nevertheless treated as sufficiently 'similar' to other cases in its display of the characteristics being studied. Researchers define cases by outlining the selected characteristics that distinguish between members and non-members of a research population, but often they do so without offering a rationale for the time and places selected, for the characteristics selected or discussing what other characteristics may be important. Researchers are apt to use terms like 'case', 'number of cases' or 'case study', but often without any consideration of what a 'case' is, how it may be defined for any particular piece of research, why its creation or selection is relevant and appropriate to the objectives of the research, and why a single case, a small number of cases, a sample of cases or an entire population of cases selected for study is suitable.

Researchers who are constructing quantitative data tend to treat cases as 'real' objects and, furthermore, to accept established entities like households, neighbourhoods, small to medium enterprises, organizations, firms or cities as cases. These are seen to exist independently of any particular research effort; however, the researcher still needs to define them, for example what counts as a 'household' or a 'city'. Researchers seldom problematize the nature and boundaries of their cases; for example, what does the concept of 'a social network' mean in the context of this research and can there be degrees of membership of this category? Researchers quite frequently assume that entities like organizations or groups of individuals can 'behave' and 'think' like individuals. Companies in particular are often described as 'doing' things like putting up prices or reacting to competitors in particular ways. Universities or hospitals may even be seen to have attitudes and opinions. Organizations can now even be sued for 'corporate manslaughter'. Committees within the organization, perhaps as a result of a vote or some other process, 'decide' to do things or to take a view on an issue.

In survey research the cases are typically individuals. This is understandable since it is only an individual who can respond to survey questions; furthermore, the individual as an entity is easy to define – we know what an 'individual' is and what its boundaries are – but researchers still need to define what sorts of individual are members of the research population, for example 'current prisoners serving sentences of five or more years in the UK in December 2013'. The individual acts as the 'base' unit against which values are recorded and from aggregations of which higher order units can be constructed and described. Thus the characteristics of 'departments' in an organization might be derived from aggregations of characteristics of the individuals who are its members. If the researcher calculates an average age or proportion male then these are features of the higher order unit, not of any individual cases. If the researcher is interested in making comparisons between departments in an organization, then implicitly, if not explicitly, departments now become cases.

The problem with this aggregative approach is that it may tempt the researcher to avoid considering issues relating to holistic, 'global' properties of cases, for example the boundaries between departments in an organization, the 'reality' or 'entity-ness' of households, along with issues of emergent properties that are not just the sum of individual characteristics. There are all kinds of assumptions – conceptual, theoretical and meta-theoretical – that are involved here. In short, the process of case creation or 'casing' – to use Ragin's (1992) term – and the resultant 'case-ness' of the entity are not just philosophical issues, but have very practical implications for the process of data construction.

Any quantitative research will focus its attention on a specified set of cases. The set may be small or large; it may be a sample of cases, an entire population of cases or an incomplete attempt to engage the entire population. In any one particular piece of research there may be more than one type of case; indeed cases may be nested in hierarchical fashion – individuals within department, within organizations, within industries, within regions, for example. When researchers talk about the 'number of cases' in their research, they may be referring to a variety of different things: the total population of cases in the defined set (but whose precise number may not be known), the number of cases selected for the sample, the number of usable returns in a survey, the number for which values have been successfully recorded for a given property, or the number used in a particular calculation or statistical procedure.

Particularly where macro entities are involved, the number of cases available for study may be severely limited, for example there may only be 20 or 30 organizations of the type the researcher wished to study. Researchers may, alternatively, decide to restrict severely the number of cases to study so that each one can be investigated in detail. Cases may well be selected in ways that meet the purposes of the researcher, for example because they are ones to which researchers have access, that they consider to be the most important, that represent extremes in terms of the phenomena or outcomes being studied, that are typical or that correspond to the needs of particular research designs. Thus cases may be selected because they are most similar except in terms of the outcome and factors that theory suggests may be important in contributing to that outcome. Extraneous factors are thus 'controlled' as far as possible and differences in outcome may be attributed to the remaining factors that differentiate the cases. Alternatively, the strategy may be to select cases that are most different in the belief that contrasting cases will eliminate factors that are not linked to identical outcomes. Case selection may be an iterative process whereby new cases may be added or others dropped in the ongoing piece of research when new hypotheses arise that may be confirmed with more similar, or falsified with more different, cases.

Larger numbers of cases may be used in research, perhaps because the population of cases is quite large and researchers want to study them all, or because researchers want a sample that is, as far as possible, representative of the population of cases from which the same was drawn. This might enable researchers to draw conclusions about the population from evidence in the sample. Cases may be selected using randomized techniques which are the equivalent of drawing names out of a hat and are independent of human judgement, or, in a survey, interviewers may be asked to make a selection, but according to agreed rules, or they may be given quotas of types of respondent to fill.

For descriptive studies, it may be sufficient or appropriate to select only those cases that have experienced the phenomenon being studied, for example 'failing' schools. However, for research that goes on to examine relationships between properties, it will be usual to include both cases that have experienced the phenomenon and cases that have not. If, for example, a researcher wishes to study patterns of political mobilization within ethnic groups, then he or she will include countries that have and have not had this experience. However, does the wider population include all countries on the globe, or just those countries in which political mobilization is thought possible? If the latter, is that all countries in which there is ethnic diversity, countries where there is also ethnic inequality or countries where there is in addition a non-repressive political system?

Key points and wider issues

Cases are the entities under investigation in a particular piece of research. Quantitative data analysis will always relate to a particular set of cases, usually referred to as the 'sample' even if that set is not actually a sample but the total number in a population of cases. The set will normally be defined in terms of particular characteristics that define membership and all members will have those characteristics in common. Ideally, the rationale for the selection of such characteristics, for example in terms of their relevance to the inquiry, should always be explained. The number of cases used in any particular analysis will be known and will relate either to the entire set or to some subset. In some research, membership of a research population may be a matter of degree, for example an 'alcoholic', a 'workaholic', a 'bulimic' or an 'autistic'. Here, the researcher will be constrained to define a particular quantity, for example number of units of alcohol consumed per day or per week, as a cut-off point between inclusion and non-inclusion in the sample of research cases.

In some research projects there may be more than one set of cases, sometimes arranged hierarchically as sets within sets, or sets at different points of time. Sometimes, when a sample is taken, the precise number of cases in the population that the sample is meant to be a sample of is unknown or may or may not be estimated.

Properties

Each case will be a configuration of a potentially infinite set of characteristics. In practice, researchers undertaking quantitative research will focus on a limited set of properties – variously described as 'variables', 'set memberships', 'causes', 'effects', 'conditions' or 'outcomes' – that are taken as a basis for recording characteristics. Properties are the characteristics of cases that are included in the research sample or population and that the researcher has chosen to observe or measure and then record. Some of these properties will be common to all the cases, for example all the cases in a sample of individuals are female, aged between 18 and 40 and resident in the UK. These properties are sometimes called 'constants' and they define which cases are considered to be members or potential members of the research population. Alternatively, some properties will relate to characteristics that vary between cases, for example some nation-states in a research population of states have parliamentary democracies and others do not.

The type of characteristic to which properties may relate can be broadly classified into demographic, behavioural or cognitive. **Demographic properties** relate to features that researchers have chosen to characterize the nature or condition of a case like a person's age and sex, a household's size, an organization's legal status or a country's rate of net immigration. These qualities may be fixed or relatively fixed (like gender or organizational legal status), or slow to change (like age). Some may be subject to sudden changes interspersed with periods of stability, for example an individual's social, economic or marital status, or an organization's location of company head offices.

Behavioural properties relate to what cases did in the recent past, to what they usually or currently do, or to what they might do in the future. Typical **measures** for individual consumers in marketing, for example, relate to the purchase and use of products and brands like purchase/non-purchase of a product or brand over a specific time period, brand variant purchased, quantity/size of pack, price paid, source of purchase, other brands bought, nature of purchase, and use/consumption of the product. These measures may, in turn, be used to generate calculations of brand loyalty, brand switching behaviour and frequency of purchase. If the research is a product test or product concept test, consumers may be asked about future behaviour, for example the likelihood of trial of a new product and likely frequency of purchase.

Cognitive properties relate to mental processes that go on within individuals and include their attitudes, opinions, beliefs and images. These are notoriously difficult to assess. Attitude scaling, which is explained in the next section, focuses on how researchers have attempted to address this problem. There is an issue of whether or not aggregations of individuals or macro units can 'do', 'think' or 'believe' things; that will be something on which researchers will need to take a view and decide.

In terms of the various roles that demographic, behavioural or cognitive properties may play in research, we can distinguish properties being used as descriptors and properties being used either as potentially causal factors or as outcomes. Descriptors are properties that are studied one at a time in order to illustrate or summarize the key features of a set of cases. They are not being investigated for their potential relationship to other characteristics. Demographic properties in particular are often used to provide a framework for defining and describing the key characteristics of the cases that are providing the data in a piece of research; for example, a sample of online shoppers may be described in terms of the numbers of males and females, the age distribution and whether or not they have access to broadband. However, behavioural and cognitive properties may also be used for descriptive purposes. Where, in a piece of research, properties are being used solely in this fashion, then it may be called a 'descriptive' study.

Alternatively, properties may be used precisely for the purpose of investigating the nature of their relationships to other properties. In some research the purpose of the study may be to explore whether or not, or the extent to which, patterns exist; for example, that males are more likely than females to instigate divorce proceedings. Most researchers, however, are interested in examining whether some properties, variously called 'conditions, 'independent variables' or 'causes', have some influence or impact on other properties – 'outcomes', 'dependent variables' or 'effects'. The notion of causality is extremely complex and is considered in detail in Chapter 10.

Behavioural, cognitive and some demographic properties may be used in any of the three roles in research, as illustrated in Figure 1.1. Some demographic properties, however, are difficult to conceive as being 'effects', for example trying to 'explain' a person's gender or age! Some properties may be used in more than one role in a piece of research. Thus some demographics may be used for both structural and analytic purposes, for example using age to describe the sample of respondents and also using it to see how far it 'explains' variation in one or more of the other properties. Some characteristics may be used by researchers as both cause and effect in the same piece of research. Thus customer satisfaction may be seen both as a result of a customer's prior expectations about the product or service (it is an effect) and in turn as causing or influencing repeat purchase behaviour (it is also a cause).

	Role in research		
Attribute	Descriptor	Cause	Effect
Demographic			
Behavioural			
Cognitive			

Figure 1.1 Attributes and roles of properties in research

Properties are, in effect, researcher constructs: they are what the researcher has defined them to be. Either they are the deliberate creation of researchers who have decided how, where and when the assessment of properties are to take place, or researchers accept the constructions of other individuals, taking them as appropriate for their own research. In some instances the degree of 'construction' is limited, as in recording the gender of a respondent in a survey, although even here there may be the odd discrepancy between observed gender and self-reported gender. In other situations, recording a person's social class may be the result of a highly complex process. Such processes may be referred to as 'measurement', 'scaling' or creating 'operational definitions'. They are the means by which researchers create their 'yardsticks' for categorizing, counting or calibrating the values to be recorded. This may be achieved in one of four main ways:

- directly;
- indirectly;
- deriving from two or more separate measures;
- creating a multidimensional profile.

The values of some properties may be directly observable by the researcher, for example the number of individuals in a group. Sometimes a construct is redefined so that there is a one-to-one correspondence between the construct and what can be observed or recorded. Social class might, for example, be defined as perceived social class so that respondents in a survey can be asked what social class they think they are and answers are taken at face value as a 'true' record of respondents' perceptions.

This is fine if, as researchers, what we wish to measure is perceived health status, perceived likelihood of drinking alcohol in the next year, or self-defined social class; however, different individuals will define these in different ways. Furthermore, for various reasons, the respondent may give a 'wrong' answer, for example because he or she has incorrect recall, has misinformation, is exaggerating or fabricating. The respondent's answers are also likely to be affected by mood, situational factors, willingness or reluctance to impart feelings or information, the wording of the question, the way it was addressed, or the understanding of the question.

In these circumstances, researchers may seek more 'objective' measures – ones that are more readily observable or recordable and are more likely to be comparable across respondents. Researchers may, for example, take an indicator of the concept rather than the concept itself. Gross national product is commonly taken as an indicator of a country's wealth; repeat purchase may be taken as an indicator of brand loyalty. Indirect measurement assumes that there is a degree of correspondence between the concept and the indicator deployed, but recognizes that the indicator is not the concept itself, only a reflection of it. Such measurement depends on the presumed relationships between observations and the concept of interest.

With concepts as complex as health status, social class or academic ability, asking just one question of respondents or taking just one measure of a nation's wealth may be insufficient. Such concepts will tend to have several dimensions, aspects or facets. Each is then used to derive an overall measure. This may involve adding up recorded values and then taking an average, it might entail subtracting one value from another to derive differences, or it may mean using more complex statistical techniques. One of the most commonly used methods of derived measurement in the social sciences that is used to measure attitudes is the **summated rating scale**. A **rating** is an ordered classification of a grade given by a respondent in a survey, such as 'Excellent', 'Good', 'Fair', 'Poor', 'Very poor'. In order to be able to add up ratings for several aspects, a numerical value is assigned to each category, for example 5, 4, 3, 2 and 1. These can now be totalled to give an overall **score**.

Suppose 150 respondents in a survey are asked to rate their level of satisfaction with five aspects of a service from 'Very satisfied' to 'Very dissatisfied' and values are allocated as illustrated in Figure 1.2. Total scores can now be added up. The maximum score a customer can give is 5 on each aspect, totalling 25. The minimum total is 5. These totals can then be divided by five to give an average value for each case.

How satisfied were you with the performance of our staff on each aspect of service when you last telephoned us?

	Very satisfied	Fairly satisfied	Neither	Fairly dissatisfied	Very dissatisfied
Speed of getting through	5	4	3	2	1
Getting the right person	5	4	3	2	1
Politeness	5	4	3	2	1
Staff knowledge of products	5	4	3	2	1
Efficiency	5	4	3	2	1

Figure 1.2 A summated rating scale

A particular version of a summated rating scale to measure attitudes was developed by Likert in 1932. Likert scales are based on getting respondents to indicate their degree of agreement or disagreement with a series of statements about the object or focus of the attitude. Usually, these are on five-point ratings from 'Strongly agree', through 'Agree', 'Neither agree nor disagree', 'Disagree' to 'Strongly disagree'. Likert's main concern was with single dimensionality, that is, making sure that all the items would measure the 'same' thing. Accordingly, he recommended a series of steps:

1. A large list of attitude statements, both positive and negative, concerning the object of the attitude is generated, usually based on the results of qualitative research.
2. The response categories are given codes, typically 5 for 'Strongly agree' down to 1 for 'Strongly disagree' (these may need to be reversed for negative statements). The assigned codes are then treated as numerical values.

3. The list is tested on a screening sample of 100–200 respondents representative of the larger group to be studied and a total is derived for each respondent by adding up the values.
4. Statements that do not discriminate (i.e. everybody gives the same or similar answers), or that do not correlate with the total, are discarded. This is a procedure Likert called 'item analysis' and it avoids cluttering up the final scale with items that are either irrelevant or inconsistent with the other items.
5. The remaining statements, such as the ones in Figure 1.3, are then administered to the main sample of respondents, usually as part of a wider questionnaire survey. The items in Figure 1.3 were generated by 'converting' the items in Figure 1.2 into a set of Likert items.
6. Totals are derived for each respondent. These totals can be used in a variety of ways that are explained in Chapter 6.

Below is a series of statements that people have made about our Call Centre. Please indicate to what extent you agree or disagree with each statement by putting a circle around the appropriate number

	Strongly agree	Agree	Neither	Disagree	Strongly disagree
I get through very quickly	5	4	3	2	1
I always get the right person	5	4	3	2	1
The staff are **not** very polite	1	2	3	4	5
Staff know their products well	5	4	3	2	1
The staff are **not** very efficient	1	2	3	4	5

Figure 1.3 A Likert scale

There are a number of fairly fundamental problems with the Likert scale, and indeed all summated rating scales:

- The totals for each respondent may be derived from very different combinations of response. Thus a score of 15 may be derived either by neither agreeing nor disagreeing with all the items or by strongly agreeing with some and strongly disagreeing with others. Consequently, it is often a good idea also to analyse the patterns of each response on an item-by-item basis.
- The derived totals are in no sense absolute; they only show relative positions. There are no 'units' of agreement or disagreement, while often, as in this example, the minimum score is not 0 so that a respondent scoring 20 is not 'twice' as favourable as another scoring 10. All we can really say is that a score of 20 is 'higher' than a score of 10 or 15 or whatever.
- The screening sample and subsequent item analysis are often omitted by researchers who simply generate the statements, probably derived from or based on previous tests, and go straight to the main sample. This is in many

ways a pity, since leaving out scale refinement and purification will result in more ambiguous, less valid and less reliable instruments.

- The process of summating the ratings is potentially imposing a number system that forces metric characteristics (see 'values' in the next section) onto concepts that may not inherently possess these characteristics.
- Such scales assume that individuals lie along a single dimension from positive to negative.

The analysis of data from summated rating scales can be quite complex, yet is seldom discussed in books on research methodology or data analysis. It involves using a range of univariate, bivariate and, sometimes, multivariate techniques, which are considered in Chapters 4–6 of this book. For a specific discussion of an example of analysing such data, see Kent (2007: 323–8).

While derived measures create a single total for each case, multidimensional models, by contrast, allow for the possibility not only that there is more than one characteristic underlying a set of observations, but also that these cannot be summed or transposed into a derived score. One possibility is to generate a profile of each dimension which is described separately in order to present a more complete picture. Ratings can be used to calculate an average across cases separately for each item, so that, for Figure 1.3, for example, there would be an average score for *I get through very quickly* and another for *I always get the right person*, and so on. There would be no attempt to add up scores for the five items. A more common way of obtaining a profile is to use a semantic differential. These measures were developed by Osgood et al. (1957) and were designed originally to investigate the underlying structure of words, but have subsequently been adapted to measure, for example, images of organizations or the services they offer. They present characteristics as a series of opposites, which may be either bipolar, like 'sweet' through to 'sour', or monopolar, like 'sweet' through to not 'sweet'. Respondents may be asked to indicate, usually on a seven-value rating, where between the two

Please put an X at a point between the two extremes which indicates your view about the service you receive from the Club.

	7 6 5 4 3 2 1	
Fast to get through	:_:_:_:_:_:_:_:	Slow to get through
Get the right person	:_:_:_:_:_:_:_:	Get the wrong person
Staff are polite	:_:_:_:_:_:_:_:	Staff are impolite
Staff know products	:_:_:_:_:_:_:_:	Staff do not know products
Staff are efficient	:_:_:_:_:_:_:_:	Staff are inefficient

Figure 1.4 Profiling: a semantic differential

extremes their views lie, as illustrated in Figure 1.4. The ratings are then given numerical values of 1–7 and treated as if they are metric, allowing an average to be calculated separately for each item across the respondents. It is then possible, for example, to compare profiles of two or more organizations in a 'snake' diagram as in Figure 1.5.

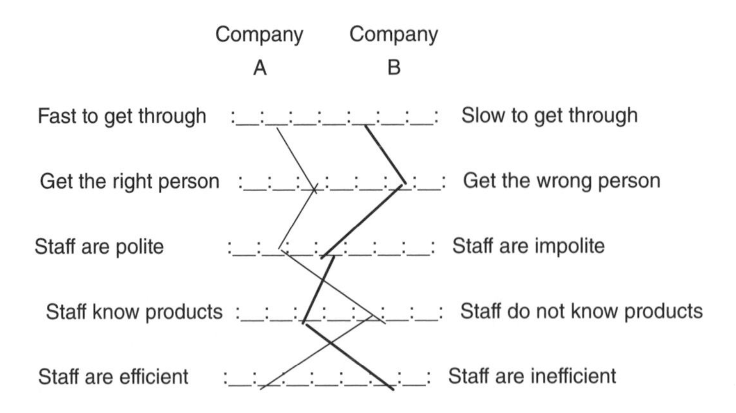

Figure 1.5 A snake diagram

An alternative to profiling is to locate each case as a single point in multidimensional space. What is known as multidimensional scaling (often referred to as MDS for short, or as perceptual mapping) refers in fact to a series of techniques that help the researcher to identify key characteristics underlying respondents' evaluations. Such techniques attempt to deduce the underlying dimensions from a series of similarity or preference judgements of objects, products, services, organizations, and so on made by respondents. MDS is explained in more detail in Chapter 6.

Key points and wider issues

Properties are the characteristics of cases that the researcher has chosen to observe or measure and then record. They may be demographic, behavioural or cognitive and they may play one or more roles in a research project as descriptors, as causes or as effects. Properties are to different degrees researcher constructs that are generated directly, indirectly, derived from two or more properties or treated multidimensionally. Properties do not always constitute 'reality' as such, but rather tend to reflect researcher attempts to locate traces of complex systems such as the degree of bureaucracy within an organization.

Most researchers will use the term 'variable' rather than 'property', to refer to characteristics of cases. However, the next section makes a distinction between properties as variables and properties as set memberships. The distinction is crucial to an understanding of the difference between variable-based and case-based approaches to data analysis. In the alcohol marketing study, the properties were originally used as variables, which are listed in Table 1.1. These are a mixture of demographic, behavioural and cognitive variables. Some are used purely as descriptors while others may in addition play a role as dependent or independent variables. Drink status (Have they ever had a proper alcoholic drink?) is an interesting variable because it could be seen either as an outcome (e.g. What factors are associated with whether or not pupils have already had a proper alcoholic drink?), or it could be interpreted as an independent variable (How does it impact on the likelihood that respondents think they will have an alcoholic drink in the next year?). Some of the variables are measured directly (like watched television in the last seven days), some indirectly (like social class) and some are derived (total importance of brands).

Table 1.1 The variables used in the alcohol marketing study

Variable	Values	Measure
Watched television in the last 7 days	Yes	Nominal
Read a newspaper in the last 7 days	No	
Read a magazine in the last 7 days	DK	
Listened to the radio in the last 7 days		
Used the Internet in the last 7 days		
Brand importance for chocolate or sweets	Very important	Ordered category
Brand importance for fizzy drinks	Quite important	
Brand importance for crisps	Neither	
Brand importance for trainers	Quite important	
Brand importance for clothes	Very important	
Brand importance for magazines	DK	
Brand importance for perfume/aftershave		
Brand importance for cigarettes		
Brand importance for alcohol		
Total importance of brands	Numeric	Discrete metric
Seen ads for alcohol on street, large posters/billboards, bus shelters or on sides of vehicles	Yes No/DK	Binary
Seen ads for alcohol in press		
Seen sponsorship of sports or sports teams by alcohol brands		
Seen sponsorship of music events, festivals, concerts or venues by alcohol brands		

(Continued)

Table 1.1 (Continued)

Seen sponsorship of TV programmes/films on TV/cinema by alcohol brands		
Seen signs or posters about alcohol in shops or on shop fronts		
Seen alcohol logos on clothing including football and other sports tops		
Seen special price offers for alcohol		
Seen promotional emails, chain emails or joke emails that mention alcohol brands		
Seen famous people in films, on TV or in music videos, that show a particular make or brand of alcohol		
Seen unusual bottle or can designs for alcohol		
Seen websites or pop-ups for alcohol brands that have brand logos or names on them		
Seen mobile phones or computer screensavers containing pictures of alcohol products		
Seen web home pages containing alcohol brand logos		
Seen any other ways that companies try to attract attention to alcohol		
Total number of channels seen	Numeric	Discrete metric
Channels seen recoded	None 1–5 channels 6 or more	Discrete metric
Received free samples of alcohol products	Yes	Binary
Received free gifts, showing alcohol brand logos	No	
Received special price offers for alcohol		
Received promotional mail, emails or joke, chain or wind-up emails mentioning alcohol brands		
Owned clothing, with an alcohol brand name or logo on it		
Looked at a website for alcohol brands		
Downloaded a mobile phone or computer screensaver containing an alcohol brand name or logo		
Used a web home page containing an alcohol brand name or logo		

Total involvement	Numeric	Discrete metric
How do you feel about alcohol ads as a whole?	Like ads a lot Like ads a little Neither Dislike ads a little Dislike ads a lot	Ordered category
How much do you like or dislike school?	Like a lot Like a little Neither Dislike a little Dislike a lot DK/unstated	Ordered category
Have you ever had a proper alcoholic drink?	Yes No	Binary
Do you think you will drink alcohol at any time during the next year?	Definitely not Probably not Not sure Probably yes Definitely yes	Ordered category
How old were you when you had your first proper alcoholic drink?	Numeric	Continuous metric
How often do you usually have an alcoholic drink?	Every day or almost About twice a week About once a week About once a fortnight About once a month Only a few times a year I never drink alcohol now	Ordered category
Total units of alcohol last consumed	Numeric	Continuous metric
Do any of your brothers or sisters drink alcohol?	Yes No, DK	Binary
Does your mother drink alcohol nowadays?	Yes No, not sure, no mother	Binary

(Continued)

Table 1.1 (Continued)

Does your father drink alcohol nowadays?	Yes No, not sure, no father	Binary
Which of these statements best describes your smoking behaviour?	Never smoked Tried once Never smoke now Smoke sometimes 1–6 cigarettes a week More than 6 a week	Ordered category
Gender	Male Female	Binary
Social class of chief income earner	A B C1 C2 D E DK/not stated	Ordered category
Do you regard yourself as belonging to any particular religion?	Christian Other religion None DK/unstated	Nominal

Values

Values are what researchers actually record as a result of the process of assessing properties. Such records may relate either to variables or to set memberships. The values recorded for variables arise from one or more of the activities of classifying, ordering, ranking, counting or calibrating the characteristics of cases. All variables at a minimum classify cases into one of two values, but there may be many or even an infinite number of possible values. The range of values deployed to record a case property either may consist of a defined number of categories that are mutually exclusive – they do not overlap – and are exhaustive of all the possibilities, or are a result of applying a metric. The former may be called 'categorical' or 'non-metric' variables, the simplest of which are **binary variables**. These consist of a record of the presence or absence of a property. Thus an organization may be commercial or non-commercial; a nation-state may be democratic or non-democratic; an individual may be married or not married. Some variables are naturally binary, for example a product is either on the shelf in a supermarket or it is not. Some variables with a limited number of possible values may be readily converted into binary sets, for example employment status. If the possibilities are 'employed full time', 'employed part time' or 'unemployed' then these can become either 'employed full time/not employed full time' or 'employed, either full or part time/unemployed'. However, there are nearly always complicating issues. Even the apparently simple distinction between married and not married becomes complicated by decisions about whether 'not married' includes divorced, separated, widowed or in a civil partnership. Despite these issues, much of our thinking is binary in nature. We often think that people are 'right' or 'wrong', that propositions are 'true' or 'false', that countries are 'democratic' or 'undemocratic'. Binary logic, furthermore, has been at the forefront of developments in electronic circuits, computer science and computer engineering, which are all based on binary language.

Where cases are classified not into the presence or absence of a characteristic, but into contrasting groups, then we have a **nominal variable**. Dichotomies, for example, consist of two categories that represent two contrasting groups like black/white, male/female, or they may be polar opposites like rich/poor, fast/slow, instrumental/expressive. Such variables may be better treated as two binary variables, 'rich/not rich' and 'poor/not poor', or better still as fuzzy sets with degrees of membership of these categories. Fuzzy sets are explained below. Strictly speaking, dichotomies are not binary in the sense that white, for example, is not the absence of black, and female is not the absence of male. Cases are either type A or type B rather than A or not A. If there really are only two categories, as with gender, then it does no harm to treat such variables as if they are binary. However, for a yes/no answer in a questionnaire, if it is binary, then 'no' really means 'not yes' and may, for example, include those who refused to answer, did not have an answer, or the question was inapplicable. In a dichotomy, 'no' means that the answer 'no' was given and these other possibilities are excluded.

Nominal variables are sometimes converted into binary variables so that, for example, the dichotomy A/B becomes A/not A and B/not B. A trichotomy becomes A/not A, B/not B and C/not C. Statisticians sometimes call these **dummy variables** and they are useful because they have particular properties and can be used in some statistical procedures where nominal variables are inappropriate.

A key feature of nominal variables is that where there are three or more categories, the order in which the values appear in a table makes no difference to any statistical calculations that may appropriately be applied to the data. The values do need to be listed in some sequence (which might, for example, be alphabetical), but it is not a graduated series from 'high' to 'low' or 'large' to 'small'. Some variables, however, define the relationships between values not just in terms of categories that are exhaustive and mutually exclusive, but the categories are also arranged in relationships of greater than or less than, although there is no metric that will indicate by how much. Thus product usage can be classified into 'Heavy', 'Medium', 'Light' and 'Non-user'; there is an implied order, but no measure of the actual usage involved. The various social classes, social grades or socio-economic groups used in various European countries are good examples of such **ordered category variables**. The individual items used to generate summated rating scales such as the Likert scale, which were explained in the previous section, are also common examples of ordered categories.

In ordered category variables there is usually a limited number of categories into which researchers may map a large number of cases. So, 200 students might be mapped onto five degrees of the extent to which they say they have been bullied at school. However, in other situations it may be possible to rank-order each respondent. In **ranked variables** each case being measured is given its own ranking. Thus 30 hospital patients may be ranked 1–30 on the basis of their summated ratings of a hospital radio programme, or schools are ranked according to the performance of their pupils in examinations. We would normally rank-order only a fairly limited number of people or objects. To rank 300 people 1–300 would be rather cumbersome. Alternatively, respondents in a survey may be asked to rank a number of items; for example, customers may be asked to rank seven brands 1–7 in terms of value for money. Respondents may find this tricky, so paired comparisons may be used. If, for example, seven brands of beer are to be ranked, then respondents are asked to say which of two brands they prefer, taking each combination of pairs, of which there will be $n(n - 1)/2$ pairs or 21 combinations. The results can be converted into a rank order by counting the number of times each brand is preferred.

Binary, nominal, ordered category and ranked variables all use sets of values that are usually labelled in words. However, in order to be able to enter these values into data analysis software like IBM SPSS (which is introduced in the next chapter), the labels need to be identified by numbers which are used as **codes** that 'stand for' each value. There are few rules that suggest how this should be achieved. They may be assigned arbitrarily and it does not matter if we assign 1 = male and 2 = female, or 1 = female and 2 = male, or even 26 = male and 39 = female. What we certainly cannot do, for example, is, if we take 1 = male and 2 = female and we have 60 males and 40 females, calculate the 'average sex' as 1.4! As we will see later, any self-respecting computer will happily perform this calculation for you: the trick is to realize that the result is total nonsense. At the ordered category level, again we can assign the numbers arbitrarily, but they must preserve the order.

Metric variables arise when either there is a metric like age in years that can be used to calibrate distances between recorded values, or the values are a result of counting the number of instances involved as a measure of size. Think about how we might measure the size of a car park. We could either measure the area in square metres, or count up the number of parking spaces it provides. The first procedure might give any value in square metres and fractions of a square metre up to however many decimal places are required. The second method will produce only whole numbers or integers. In statistical parlance, the first is usually called a **continuous variable** and the second a **discrete variable**. The values for metric variables are numeric rather than in words, as with categorical variables.

Table 1.1 lists the variables used in the alcohol marketing study, the values recorded and the types of measure. Notice that only two of the variables are listed as continuous metric – the age at which respondents first had their alcoholic drink and the total units of alcohol last consumed. It is quite common in survey research that most of the variables are binary, nominal, ordered category or discrete metric.

Variables may be seen as containers, and each case has a place in each container (one container for each property) either in one of two or more compartments or at a certain 'level' inside the container. Set memberships, by contrast, focus on whether or not (or the extent to which) cases 'belong' in a container. Cases, then, may be members of some sets, but not others. Thus a nation-state may be a member of the sets 'democratic', 'having strong trade unions' and 'low crime rate', but not of the sets 'unregulated press' or 'strict controls on the possession of guns by individuals'. The focus is then on which combinations of memberships characterize each case. Sets are based on notions of inclusion or exclusion; boundaries are defined in a way that creates containers into which cases may or may not be assigned.

Set memberships may be crisp or fuzzy. With **crisp sets**, cases are unambiguously members or not members of a set. Thus the UK is a member of the EU but not a member of the eurozone. Crisp sets are identical to binary variables in that they record the presence or absence of a characteristic, but they are allocated not codes but set membership values. For this reason, the latter are, in this text, indicated in square brackets. Full membership is always indicated with a value of [1] and non-membership with a value of [0]. Crisp sets have only these two values. Crisp sets are at the base of set theory and what has become known as Boolean logic. George Boole (1847) was a nineteenth-century mathematician and logician who developed an algebra suitable for properties with only two possible values. Set theory and Boolean logic are explained in more detail in Chapter 7 on configurational approaches to data analysis.

In reality, the world and large parts of social science phenomena do not come naturally in binary form. Membership of the category 'democratic country' or 'profitable organization' may be a matter of degree. **Fuzzy sets** record degrees of membership of a defined category by permitting membership values in the interval between [1] and [0]. They distinguish between cases that are 'more in' a set than out of it and are given values above [0.5], for example a value of [0.8] to indicate that a case is mostly in a set, and those more out of a set than in it are given values below [0.5]. The crossover point of [0.5] indicates cases that are neither in nor out of a set. It is the point of maximum ambiguity. This, Ragin (2000) emphasizes, should be conceptually defined according to the theory or theories being applied, or according to empirical evidence, research findings or researcher understanding of the cases involved. The researcher has to decide, for example, what being a 'heavy' viewer of television entails, for example in terms of hours viewing per day or per week and at what point a viewer is no longer in the set 'heavy viewer'. This should not be an arithmetical mean or average, which is driven by the particular dataset used, but an absolute value that is not affected by other values in the set. The assessment of crisp and fuzzy sets is explained in detail in Chapter 7.

Key points and wider issues

Values are what researchers actually record as a result of the process of assessing properties of cases. However, they may arise either as variables or as set memberships. The values of variables assess cases relative to one another; sets define memberships in absolute terms according to generally agreed external standards or based on a combination of theoretical knowledge and practical experience of cases. The values recorded for variables arise from one or more of the measurement activities of classifying, ordering, ranking, counting or calibrating the characteristics of cases. These activities result in variables that may be binary, nominal, ordered category, ranked, discrete metric or continuous metric. These types of variable can themselves be seen as a variable having ordered category characteristics of increasing complexity from binary up to continuous metric.

Sometimes the distinction between the different types of measure is blurred, or difficult to make or open to interpretation. Thus the numerical totals derived from summated rating scales should, strictly speaking, be used only to create ordered categories from high to low. The totals only indicate relative positions, so that while 15 is 'higher' than 12, by how much is unclear because there are no 'units' of measurement. There is no metric for measuring people's attitudes, opinions or beliefs. In practice, however, researchers commonly treat the results, particularly of Likert scales, as if they are metric and will calculate average scores and use the results in statistical procedures that require metric data. Such a practice is making the assumption of equivalence of distance between recorded values, so that the difference, for example between 'Strongly agree' and 'Agree', is the 'same as' the distance between 'Agree' and 'Neither' and between 'Disagree' and 'Strongly disagree'. This may be a reasonable assumption for the standard Likert scale and little error may result from acting as if the resulting scale is metric, but for other kinds of scale the assumption may be unwarranted or at least questionable.

The assessment of set membership may result in crisp or fuzzy sets. Crisp sets are the same as binary variables and record cases as either members or non-members of a set. Fuzzy sets allow degrees of membership and combine binary with metric characteristics with membership values between [1] and [0] with [0.5] as the crossover point.

Error in data construction

The construction of data of any kind is likely to involve errors from various sources. These might include, for example:

- inappropriate specification of cases;
- biased selection of cases;
- random sampling error;
- poor data capture techniques;
- non-response;
- response error;
- interviewer error;
- measurement error.

The appropriateness of the type or types of case specified for a piece of research is often taken for granted rather than argued and justified. Ideally, not only must cases share sufficient background characteristics for comparisons to be made between them, but also those characteristics must be relevant to the topic under investigation. So, selecting university students or 'housewives' to study attitudes of hostility towards allowing female or gay bishops in the Church of England may not be appropriate.

Once the characteristics that define the research population of cases have been specified, the selection of cases to be used in the research may be made in a variety of different ways, but should, as far as possible, avoid the over- or under-representation of types of case. This may arise because, for example, the sampling frame used to select cases from omits certain kinds of case, or non-random methods of case selection have been used, for example interviewers have been asked to select respondents. Any **biases** in selection procedures will not be reduced by increasing the size of the sample.

Even in carefully conducted random samples, there will always be potential for fluctuations from sample to sample such that values of properties recorded for a particular sample may not reflect the actual values on the population from which the sample was drawn. The probability of getting sampling errors of given sizes can be calculated using **statistical inference**, which is explained at various points in Chapters 4–6.

Error in the construction of data can arise as a result of poor data capture techniques, for example in questionnaire design. There are many things that can go wrong both in the design of individual questions and in the overall design of the questionnaire. Some of these problems arise because the researcher has not followed the guidelines for good questionnaire design, in particular the stages for questionnaire construction, for question wording, routeing and sequencing. Any of these problems will result in errors of various kinds and their extent is unlikely to be known. It has been shown many times over that the responses people give to questions is notoriously sensitive to question wording. However, answers are also affected by the response options people are given in fixed choice questions, by whether or not there is a middle category in a rating, by whether or not there is a 'don't know' filter, or by the ordering of the questions, the ordering of the responses or their position on the page. All the researcher can do is to minimize the likelihood of errors arising from poor questionnaire design through design improvements.

A source of error in virtually all survey research is non-response. It is seldom that all individuals who are selected as potential respondents are successfully contacted, and it is seldom that all those contacted agree to co-operate. Non-contacts are unlikely to be representative of the total population of cases. Married women with young children, for example, are more likely to be at home during the day on weekdays than are men, married women without children, or single women. The probability of finding somebody at home is also greater for low-income families and for rural families. Call-backs during the evening or at weekends may minimize this source of bias, but it will never be eliminated.

The contact rate takes the number of eligible cases contacted as a proportion of the total number of eligible cases approached. Interviewers may be compared or monitored in terms of their contact rates. Potential respondents who have been contacted may still refuse co-operation for a whole variety of reasons including inconvenience, the subject matter, fear of a sales pitch or negative reaction to the interviewer. The refusal rate generally takes the number of refusals as a proportion of the number of eligible cases contacted. Once again, refusals are unlikely to be representative, for example they may be higher among women, non-whites, the less educated, the less well off and the elderly. The detection of refusal bias usually relies on checking differences between those who agreed to the initial contact and those who agreed only after later follow-ups on the assumption that these are likely to be more representative of refusals.

Most researchers report a response rate for their study and this will normally combine the ideas of a contact rate and a refusal rate. However, in terms of its actual calculation a bewildering array of alternatives is possible. Normally, it is the number of completed questionnaires divided by the number of individuals approached. Sometimes the number found to be ineligible is excluded from the latter. Yet others will argue that the same applies to non-contacts, terminations and rejects. The result will be dramatically different calculations of the response rate. Whichever of these is reported, however, what is important as far as error in data construction is concerned is the extent to which those not responding – for whatever reason – are in any way systematically different from those who successfully completed. Whether or not this is likely to be the case will depend substantially on whether or not there were call-backs and at what times of the day and days of the week individuals were approached.

Apart from non-contacts, refusals and those found to be ineligible, there will, in addition, usually be item non-response where individuals agree to participate in the survey but refuse to answer certain questions. A refusal to answer is not always easy to distinguish from a 'don't know', but both need to be distinguished from items that are not responded to because they have been routed out as inappropriate for that respondent. All, however, are instances of 'missing values', which are considered in Chapter 2.

Researchers faced with unacceptable non-response rates have a number of options:

- simply report the response rate as part of the findings;
- try to reduce the number of non-respondents;
- allow substitution;
- assess the impact of non-response;
- compensate for the problem.

Many researchers choose to report survey results based only on data derived from those responding and simply report the response rate as part of the results. This shows that the researcher is unaware of the implications of non-response, believes them to be negligible or has chosen to ignore them. Non-response may not itself be a problem unless the researcher ends up with too few cases to analyse. What is important is whether those not responding are in any significant ways different from those who do.

The number of non-respondents can usually be reduced through improvements in the data collection strategy. This might entail increasing the number of call-backs, using more skilled interviewers or offering some incentive to potential respondents. The effort to increase the rate of return becomes more difficult, however, as the rate of return improves and costs will rise considerably. Allowing substitution can sometimes be a sensible strategy in a sample survey provided the substitutes are selected in the same way as the original sample. This will not reduce bias from non-response, but it is a useful means of maintaining the intended or needed sample size. For censuses, substitution is, of course, not an option.

Assessing the impact of non-response implies an analysis of response rates, contact rates, and so on, plus an investigation of potential differences between respondents and non-respondents, and some model of how these relate to total survey error. There are various ways of checking for non-response bias. Researchers sometimes take late returns in a postal survey as an indication of the kind of people who are non-responders. These are then checked against earlier returns. In an interview survey, supervisors may be sent to refusers to try to obtain some basic information. Interviewers can also be sent to non-responders in a postal survey. Another technique is to compare the demographic characteristics of the sample (age, sex, social class, and so on) with those of the population. If this is known, the comparison is relatively straightforward, although deciding on how 'similar' they should be to be acceptable is not clear-cut. If differences are discovered then, again, this can simply be reported along with suitable caveats applied to the results. Alternatively, the researcher may try to compensate for the problem by using a weighted adjustment of responses. A weight is a multiplying factor applied to some or all of the responses given in a survey in order to eliminate or reduce the impact of bias caused by types of case that are over- or under-represented in the sample. Thus if there are too few women aged 20–24 in a sample survey compared with the proportions in this age group known to exist in the population of cases, such that only 50 out of a required 60 are in the achieved sample, the number who, for example, said 'Yes' to a question will be multiplied by a weighting that is calculated by dividing the target sample number by the actual sample number, in the example 60/50 or 1.2.

Even if individuals are responding, there may be differences between respondents' reported answers and actual or 'true' values. Response errors arising through dishonesty, forgetfulness, faulty memories, unwillingness or misunderstanding of the questions being asked are notoriously difficult to measure. Research on response error, furthermore, is limited due to the difficulty of obtaining some kind of external validation. In interview surveys, whether face to face or by telephone, interviewers may themselves misunderstand questions or the instructions for filling them in, and may be dishonest, inaccurate, make mistakes or ask questions in a non-standard fashion. Interviewer training, along with field supervision and control, can, to a large extent, reduce the likelihood of such errors, but they will never be entirely eliminated, and there is always the potential for systematic differences between the results obtained by different interviewers.

Errors arising from non-response, erroneous responses or interviewer mistakes are specific to questionnaire survey research. Errors from the inappropriate specification of cases, from biased case selection, from random sampling error or poor data capture techniques may arise in all kinds of research. Errors of different kinds will affect the record of variables or set memberships for each property in various ways and to different extents.

What researchers do in practice is, separately for each property, to focus on likely measurement error – discrepancies between the values recorded and the actual or 'true' values. The size of such error is usually unknown since the true value is unknown, but evidence from various sources can be gathered in order to estimate or evaluate the likelihood of such errors. Researchers focus on two aspects of such discrepancies: **reliability** and **validity**.

A measure is said to be reliable to the extent that it produces consistent results if repeat measures are taken. We expect bathroom scales to give us the same reading if we step on them several times in quick succession. If we cannot rely on the responses that a questionnaire item produces, then any analysis based on the question will be suspect. For a single-item question or a multi-item question (generating a derived measure) the measures can be retaken at a later date and the responses compared. Such test–retests give an indication of measure stability over time, but there are fairly key problems with this way of assessing reliability:

- it may not be practical to re-administer a question at a later date;
- it may be difficult to distinguish between real change and lack of reliability;
- the administration of the first test may affect people's answers the second time around;
- how long to wait between tests;
- what differences between measures count as 'significant'.

An alternative is to give respondents two different but equivalent measures on the same occasion. The extent to which the two measures covary can then be taken as a measure of reliability. The problem here is that it may be difficult to obtain truly equivalent tests. Even when they are possible, the result may be long and repetitive questionnaires. Another version of this equivalent measures test is the split-half test. This randomly splits the values on a single variable into two sets. A score for each case is then calculated for each half of the measure. If the measure is reliable, each half should give the same or similar results and across the cases the scores should correlate. The problem with this method is that there are several different ways in which the values can be split, each giving different results.

Where the measure taken is a multi-item scale, for example a summated rating scale, it is possible to review the internal consistency of the items. Internal consistency is a matter of the extent to which the items used to measure a concept 'hang' together. Ideally, all the items used in the scale should reflect some single underlying dimension; statistically this means that they should correlate one with another (the concept of **correlation** is taken up in detail in Chapter 5). An increasingly popular measure for establishing internal consistency is a coefficient developed in 1951 by Cronbach that he called alpha. **Cronbach's coefficient alpha** takes the average correlation among the items and adjusts for the number of items. Reliable scales are ones with high average correlation and a relatively large number of items. The coefficient varies between zero for no reliability to one for maximum reliability. The result approximates taking all possible split halves, computing the correlation for each split and taking the average. It is therefore a superior measure to taking a single split-half measure. However, there has been some discussion over the interpretation of the results. This discussion is summarized in Box 1.1.

Box 1.1 The interpretation of Cronbach's coefficient alpha

Alpha has effectively become *the* measure of choice for establishing the reliability of multi-item scales. Its availability at the click of a mouse button in survey analysis programs like SPSS has almost certainly meant that it is commonly reported by researchers, but often with little understanding of what it means and what its limitations are.

Despite its wide use, there is little guidance in the literature (and none from Cronbach himself) as to what constitutes an 'acceptable' or 'sufficient' value for alpha to achieve. Most users of the statistic cite Nunnally's (1978) recommendation that a value of 0.7 should be achieved. The coefficient is then treated as a kind of 'test'; if alpha for any scale is greater than 0.7 then it is deemed to be sufficiently reliable. However, all those authors making recommendations about acceptable

(Continued)

(Continued)

levels of alpha, including Nunnally, indicate that the desired degree of reliability is a function of the purpose of the research, for example whether it is exploratory or applied. Nunnally himself in 1978 suggested that for preliminary research 'reliabilities of 0.70 or higher will suffice'. For 'basic' research, he suggests that 'increasing reliabilities much beyond 0.80 is often wasteful of time and funds' (Nunnally, 1978: 245). In contrast, for applied research, 0.80 'is not nearly high enough'. Where important decisions depend on the outcome of the measurement process a reliability of 0.90 'is the minimum that should be tolerated'.

None of Nunnally's recommendations, however, have an empirical basis, a theoretical justification or an analytical rationale. Rather they seem to reflect either experience or intuition. Interestingly, Nunnally had changed his own recommendations from his 1967 edition of *Psychometric Theory*, which recommended that the minimally acceptable reliability for preliminary research should be in the range of 0.5 to 0.6.

Peterson (1994) reports the results of a study to ascertain the values of alpha actually obtained in articles and papers based on empirical work. From a sample of over 800 marketing and psychology-related journals, conference proceedings and some unpublished manuscripts, he reviewed all alpha coefficients found in each study, resulting in 4,286 coefficients covering a 33-year period. Reported coefficients ranged from 0.6 to 0.99 with a mean of 0.77. About 75 per cent were 0.7 or greater and 50 per cent were 0.8 or greater. Peterson found that reported alphas were not greatly affected by research design characteristics, such as sample size, type of sample, number of scale categories, type of scale, mode of administration or type of research. One exception to this is that during scale development items are often eliminated if their presence restricts the value of alpha. Not surprisingly, the alpha coefficients reported were significantly related to the number of items eliminated.

It is important to remember that alpha measures only internal consistency; if error factors associated with the passage of time are of concern to the researcher, then it will not be the most appropriate statistic. However, since alpha approximates the mean of all possible split-half reliabilities, it can be seen as a superior measure of scale equivalence. It is not, however, as is commonly supposed, an indication of unidimensionality. Alpha can, in fact, be quite high despite the presence of several different dimensions (Cortina, 1993).

When interpreting alpha coefficients, it is often forgotten, furthermore, that the values achieved are in part a function of the number of items. Thus for a three-item scale with an alpha of 0.80, the average inter-item correlation is 0.57. For a ten-item scale with an alpha of 0.80 it is only 0.28. What needs to be kept in mind is that in evaluating, say, a 40-item scale, alpha will be relatively large simply because of the number of items, and the number of items is not exactly a great measure of scale quality. When many items are pooled, internal consistency estimates are inevitably large and invariant, and therefore of limited value.

Alpha, in short, should be used with some caution. It is appropriate only when the researcher requires a measure of internal consistency, and is helpful only then if the number of items used is fairly limited. The value of alpha to be taken as 'acceptable' must be related to the purpose of the research, and even then only used as an indication rather than a 'test' to be passed with a fixed value. Furthermore, if researchers are concerned about dimensionality, then procedures like factor analysis are probably more appropriate. For a more recent review of Cronbach's alpha see Lee and Hooley (2005).

A record of a value for a case is said to be valid to the extent that it measures what it is intended to measure: in other words, the measure and the concept must be properly matched. Validity relates to the use that is made of the measure, so stepping on bathroom scales may be a valid measure of weight, but not of overall health. A reliable measure, furthermore, is not necessarily valid. Our bathroom scales may consistently over- or under-estimate our real weight (although they will still measure change). A valid measure, on the other hand, will, of course, be reliable.

There is no conclusive way of establishing validity. Sometimes it is possible to compare the results of a new measure with a more well-established one. The results of a device for measuring blood pressure at home might be compared with the results from a doctor's more sophisticated equipment. This assumes, of course, that our GP's results are valid. For many concepts in the social sciences there are few or no well-established measures. In this situation, researchers might focus on what is called 'content' or 'face' validity. The key to assessing content validity lies in reviewing the procedures that researchers use to develop the instrument. Is there a clear definition of the concept, perhaps relating this to how the concept has been defined in the past? Do the items that have been used to measure the concept adequately cover all the different aspects of the construct? Have the items been pruned and refined so that, for example, items that do not discriminate between respondents or cases are excluded, and any items that overlap to too great an extent with other items are avoided?

Another way of establishing validity is to assess the extent to which the measures produce the kind of results that would be expected on the basis of experience or well-established theories. Thus a measure of alienation might be expected to associate inversely with social class – the lower the class, the higher the alienation. If the two measures do indeed behave in this way, then this is evidence of what is often called construct validity. If they do not, the researcher may be unclear whether either or both measures are faulty or whether the relationship between the two measures is contrary to theoretical expectations. Conversely, the expectation might be that the two measures are unconnected and therefore should not correlate. If it turns out that they are indeed not correlated, then this is sometimes called discriminant validity.

Other more complex methods of measuring construct validity have been developed, such as multitrait–multimethod validity (Campbell and Fiske, 1959), pattern matching (Cook and Campbell, 1979) or factor analysis (see Chapter 6). Evidence of validity has to be argued for and may be gathered in a number of ways. No single way is likely to provide full evidence and even a combination of approaches is unlikely to constitute final proof.

Key points and wider issues

In constructing data, there is the potential for error from a range of sources including inappropriate specification of cases, biased selection of cases, random sampling error, poor data capture techniques, non-response, response error, interviewer error and measurement error. In reporting the results of social survey research, most researchers tend to focus on random sampling error. This is largely because they can calculate the probability of given sizes of such error based on assumptions about values, differences or covariations in the population of cases from which the sample was drawn, the size of the sample and the variability of measures taken for a given property. These calculations are explained in Chapters 4–6. However, random sampling error is likely to be only a small proportion of overall error and will, furthermore, only exist if a random sample has been taken. Assael and Keon (1982), for example, estimated that random sampling error accounted for less than 5 per cent of total survey error. Some researchers will present evidence of measure reliability; less often, evidence of validity will be suggested. There is a considerable literature on ways of calculating error of various kinds (for a review see Lessler and Kalsbeek, 1992). However, the calculations can get quite complex and most formulae assume metric data, taking the 'mean square error' as the key dependent variable which is explained by a range of sources of **bias** plus random error.

In practice, researchers are more likely to focus on ways of minimizing the likelihood of error arising in the first place by adopting strategies and procedures to control its occurrence.

Error of various kinds can always be reduced by spending more money, for example on more interviewer training and supervision, on sophisticated random sampling techniques, on pilot testing or on getting a higher response rate. However, the reduction in error has to be traded off against the extra cost involved. Furthermore, errors are often interrelated so that attempts to reduce one kind of error may actually increase another; for example, minimizing the non-response errors by persuading more reluctant respondents may well increase response error. Non-sampling errors tend to be pervasive, not well behaved and do not decrease – indeed may increase – with the size of the sample. It is sometimes even difficult to see whether they cause under- or over-estimation of population characteristics. There is, in addition, the paradox that the more efficient the sample design is in controlling sampling fluctuations, the more important in proportion become bias and non-sampling errors.

The implications of this chapter for the alcohol marketing dataset

The alcohol marketing study is based on the construction of quantitative data using a combination of interviewer-completed and self-completed (but personally delivered) questionnaires in a cross-sectional survey research design. The resulting dataset is far from being error free and there are many ambiguities. The researchers explain that they sent an information pack 'to the homes of all second year (12–14 years) pupils attending schools in three local authority areas in the west of Scotland' (Gordon et al., 2010a). This, they say, 'generated a sample of 920 respondents'. Although they call this set of cases a 'sample', they do not mention what population the sample is meant to be a sample of. They do not say how many packs were sent out, so there is no idea of what the response rate was. Accordingly, it is probably safest to regard the 920 cases as an incomplete census of all the pupils in the selected schools. It is almost certainly not a random sample, or indeed not really a sample of any kind. The date when these packs were sent is not stated, but from other sources it was probably between October 2006 and March 2007. The authors do not discuss why the 12–14-year age group is the appropriate one to choose for this research or why the west of Scotland was chosen for its location.

In the original dataset, over 1,600 variables were entered for each of the 920 cases. A lot of variables were created because, for example, each brand of alcohol drunk last time was listed as either drunk or not drunk and recorded as a separate binary variable. For this text, two datasets were created from the original, one for a selection of 61 properties recorded as variables and another for a selection of those properties recorded as set memberships. Table 1.1, presented earlier in this chapter, lists the variables, the range of values used for each variable and the type of measure. Table 1.2 lists the set memberships, whether they are crisp or fuzzy and whether they are being used as conditions or outcomes.

Table 1.2 Properties used as set membership in the alcohol marketing study

Set	Type	Role
Aware of many channels for advertising alcohol	Fuzzy	Condition
Involved in alcohol marketing	Fuzzy	Condition
Like alcohol ads a lot	Fuzzy	Condition
Like school a lot	Fuzzy	Condition
Has brothers or sisters who drink alcohol	Crisp	Condition
Intention to drink alcohol in the next year	Fuzzy	Outcome
Drink status	Crisp	Outcome

The properties of the 920 cases are a mixture of demographic, behavioural and cognitive characteristics which play a range of different roles as descriptors, as potentially causal factors or as outcomes. All 61 properties may be used as descriptors and some, like drink status, may be seen either as an outcome or as a potentially causal factor. Many of the properties are measured directly, for example 'Read a newspaper in the last seven days'. Some, like total importance of brands, are derived. Social class is measured indirectly from the occupation of the chief income earner (CIE) in the household.

Potentially there are many sources of error in the dataset, particularly from non-response (we do not know how many households refused their consent) and from response error (there is considerable scope for exaggeration or minimization of drinking behaviours). Some errors, for example from poor questionnaire design, can be addressed through recoding or transforming properties in ways that are considered in Chapter 2.

Chapter summary

This chapter has shown that data are a deliberate and thoughtful construction by researchers or other individuals rather than simply 'collected' and they result from a process of systematic record-keeping. Records are created in a social, economic and political context and for purposes specific to individuals within organizations. Data may be qualitative or quantitative; the former consist of words, phrases, narrative, text and visual images, while quantitative data arise as numbers that result from the systematic capture of classified, ordered, ranked, counted or calibrated characteristics of a specified set of cases. All quantitative data have a structure that consists of cases, properties and values. Cases are the entities under investigation in a particular piece of research. The number used in any particular analysis will be known and will relate either to the population of cases or to some subset of them. In some research projects, there may be more than one set of cases, sometimes arranged hierarchically as sets within sets, or sets at different points of time. Properties are the characteristics of cases that the researcher has chosen to observe or measure and then record. They may be demographic, behavioural or cognitive and they may play one or more roles in a research project as descriptors, causes or effects. Values are what researchers actually record as a result of the process of assessing properties. Such records may relate either to variables or to set memberships. The values of variables assess cases relative to one another; sets define memberships or degrees of membership in absolute terms according to generally agreed external standards or based on a combination of theoretical knowledge and practical experience of cases.

The values of variables may be recorded into different types of measure which have in this text been classified into binary, nominal, ordered category, ranked, discrete metric and continuous metric. Properties that relate to set memberships may be crisp or fuzzy. The distinction between types of measure is not always clear-cut and may be open to interpretation. The creation of measures, furthermore, is subject to many kinds of error in data construction. Errors can often be reduced by devoting extra resources to their minimization, but usually at extra cost in time and money.

Exercises and questions for discussion

1. Are all data 'manufactured' in some way or are there some data that we can accept as 'given'?
2. If a social researcher wanted to measure the extent to which individuals are 'religious', suggest how this could be achieved in a way that is (a) direct, (b) indirect, (c) derived or (d) multidimensional.

3. Make a list of variables that (a) are naturally binary, (b) can sensibly be made binary and (c) would be unwise to convert into binary.
4. What type of measure would you use for each of the following?

   (i) Degree of satisfaction or dissatisfaction with the services offered by the local social services department.
   (ii) Attitudes towards the BBC's Radio 2.
   (iii) The degree of local support for the creation of a 'free' school in an area of urban deprivation.

5. Examine Table 1.1 and consider which variables are demographic, which ones are behavioural and which ones are cognitive. Also consider which ones have been measured directly, which ones indirectly and which ones are derived.
6. Explain the type of measure indicated in Table 1.1 for each of the variables in the alcohol marketing study.

Further reading

DeVellis, R.F. (2011) *Scale Development: Theory and Applications*, 3rd edn. London: Sage.
A classic text on scale development, presenting complex concepts in a way that helps students to understand the logic underlying the creation, use and evaluation of measurement instruments and to develop a more intuitive feel for how scales work.

Diamantopoulos, A. and Schlegelmilch, B. (1997) *Taking the Fear Out of Data Analysis*. London: Dryden Press. Republished by Cengage Learning, 2000.
This book does what it says on the tin, all written in a jokey style. Part I considers what are data, the process of sampling and measurement. Succinct and well worth a read.

Gordon, R., Harris, F., Mackintosh, A.M. and Moodie, C. (2010a) 'Assessing the cumulative impact of alcohol marketing on young people's drinking: cross-section data findings', *Addiction Research and Theory*, Early Online, 1–10, Informa UK.
This article presents the initial results of the survey on alcohol marketing which is used throughout this text.

Kent, R. (2007) *Marketing Research: Approaches, Methods and Applications in Europe*. London: Thomson Learning (now Cengage, Andover).
This is an earlier text by the author, but it focuses on marketing research. Chapter 5 covers much of the material in this chapter.

Ragin, C. and Becker, H. (eds) (1992) *What Is a Case? Exploring the Foundations of Social Inquiry*. Cambridge: Cambridge University Press.
This is a seminal book on the role of cases in social research. In particular, have a look at the introductory chapter by Ragin, which reviews the many different ways in which the concept of 'case' has been used, and Chapter 10, also by Ragin, on the process of 'casing' in social inquiry. Also have a look at the chapter by Abbott, 'What do cases do? Some notes on activity in sociological analysis'.

Suggested answers to the exercises and questions for discussion can be found at the end of this text, pp. 293–321, and on the companion website, (https://study.sagepub.com/kent), which also give links to relevant free online Sage journal articles, PowerPoint slides, an overview of data analysis packages, an introduction to SPSS and weblinks to alternative datasets.

2

DATA PREPARATION

Learning objectives

In this chapter you will learn:

- how datasets are prepared, ready for the next stages of data analysis;
- that all data need to be checked, edited, coded and assembled before any further processes can take place;
- that the transformation of some of the properties in a dataset may involve one or more of a number of activities, including, for variables, regrouping values on a nominal or ordered category measure to create fewer categories, creating class intervals from metric measures, computing totals or other scores from combinations of several variables, treating groups of variables as a multiple response question, upgrading or downgrading measures, handling missing values and 'Don't know' responses, or coding open-ended questions;
- that, for set memberships, transformations may entail creating crisp sets or fuzzy sets from existing variables;
- how survey analysis software like SPSS can be used for assembling data and assisting in data transformations;
- that many of the codes used in the original alcohol marketing dataset were illogical or inconsistent and many data transformations were needed before analysis could begin.

Introduction

Before quantitative data that have been constructed by researchers can be analysed using techniques appropriate for pursuing the objectives of the research, they need to be prepared in various ways to make them ready for analysis. In their raw form, captured data will consist of stacks of completed paper questionnaires or diaries, entries into an online questionnaire or records made by researchers themselves. Before statistical techniques can be applied to the data, they will need to undergo many of the various processes listed in Figure 2.1.

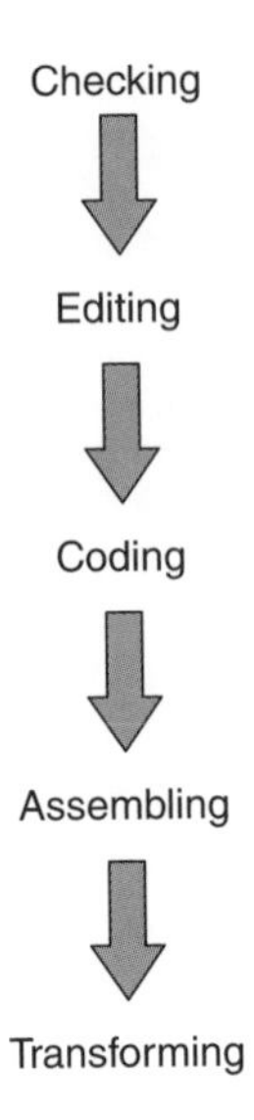

Figure 2.1 The data preparation process

Checking and editing

Most quantitative data in the social sciences will have been captured using some form of questionnaire, whether paper or electronic, so the first step involves checking these for usability as they are received. Questionnaires returned by interviewers or by respondents may be unusable for a number of reasons, for example:

- an unacceptable number of the questions that are appropriate for a given respondent have not been answered;
- the pattern of responses is such that it indicates that the respondent did not either understand or follow instructions – for example, questions that require a single response may have been given two or more responses;
- one or more pages are physically missing;
- the questionnaire has been answered by somebody who is not a member of the survey population;
- the questionnaire was received too late to include in the analysis.

The number of returned but unusable questionnaires will generally be quite small, so discarding these will not usually be a problem. If the number of discards is large, the researcher will need to check whether they are in any obvious way different from those that are usable. Either way the number of discards should be declared in the report of the research.

Editing involves verifying response consistency and accuracy, making necessary corrections and deciding whether some or all parts of a questionnaire should be discarded. Some of these checks include:

- logical checks, for example a 16 year old claiming to have a PhD, a male claiming to have had an epidural at the birth of his last child, or a respondent

may have answered a series of questions about his or her usage of a particular product, but other responses indicate that he or she does not possess one or has never used one;
- range checks, for example a code of 8 is entered when there are only six response categories for that question;
- response set checks, for example somebody has 'strongly agreed' with all the items on a **Likert scale**.

Where a question fails a logical check, then the pattern of responses in the rest of the questionnaire may be scrutinized to see what is the most likely explanation for the apparent inconsistency. Range check failures may be referred back to the original respondent. Response set checks may indicate that the respondent is simply being frivolous and the questionnaire may be discarded.

If questionnaires are checked and edited as they come in, it may still be possible to remedy fieldwork deficiencies before they turn into a major problem. If problems are traced to particular interviewers, for example, then they can be replaced or asked to undergo further training. It may be possible to re-contact respondents to seek clarification or completion before the last date on which data can be processed. If this is not feasible, the researcher can treat these as **missing values**, that is, treat the questions involved as unanswered. This may be suitable if the unsatisfactory responses are not key properties and the number of questions concerned is quite small. If this is not the case, values may be imputed. How this can be done is discussed later in the chapter. The alternative is to discard the questionnaire.

Coding

As was explained in Chapter 1, data analysis software usually requires that all the values to be entered are either already numerical (as in age = 23) or they are given a number that is a code that 'stands for' values that are in words. Binary variables will normally be coded either as 1 or 0 or as 1 or 2. The categories for nominal and ordered category variables will generally be numbered 1, 2, 3, 4, and so on. Note that it makes sense for ordered categories to give the highest code number to the highest or most positive value as in Figures 1.2 and 1.3 in Chapter 1. Metric data already have numerical values that can be entered directly, for example the number of units of alcohol consumed last week as 10.7. Some, perhaps all, of the categorical responses on a questionnaire will have been pre-coded, that is they are already numbered on the questionnaire. If not, they need to be coded afterwards by the researcher.

Qualitative responses to open-ended questions will normally be classified into categories, which are then coded. The categories developed should meet the minimum requirements for a binary or nominal measure, namely, they should be exhaustive, mutually exclusive and refer to a single dimension. If, however, most of the spaces left for text in the questionnaire have been left empty, it may not be worthwhile doing this. Some pre-coded questions may have an 'Other, please specify' category, in which case some further coding may be worthwhile.

If a question is unanswered, the researcher, when entering data into a survey analysis program, can record a missing value or enter a code for, for example, 'Not applicable' or 'Refused to answer'. For multiple response questions where the respondent can indicate more than one category as applicable, each response category will need to be treated as a separate variable, and will usually be coded as 1 if the category is ticked and 0 or 2 if not. The treatment of open-ended and multiple response questions is considered in more detail later in this chapter.

In large-scale projects, and particularly when data entry is to be performed by a number of people or by subcontractors, researchers will often develop a **codebook**, which lists all the variable names (which are short, one-word identifiers), the variable labels (which are more extended descriptions of the variables and which appear as table or chart headings), the response categories used and the code numbers assigned. This means that any researcher can work on the dataset irrespective of whether or not they were involved in the project in its formative stages. Codebooks, however, are not always needed. Survey analysis packages like SPSS record all this information as part of the data matrix.

Assembling

Data assembly means gathering together all the checked, edited and coded questionnaires, diaries or other forms of record, and entering the values for each variable for each case into data analysis software. This is usually achieved in a framework of rows and columns for storing the data called a **data matrix**. Data matrices are explained in more detail in Chapter 3. The range of software for assembling data is briefly reviewed at the end of Chapter 4. Some researchers like to assemble data first into a spreadsheet like Excel before exporting to a survey analysis package. Some survey analysis software packages allow the researcher to enter the data by clicking the appropriate box against the answer given on an electronic version of the questionnaire. In the background, the software creates the data matrix. The researcher does not need to **pre-code** response categories in the questionnaire nor engage in post-coding when the questionnaires have been completed. With online surveys, the data matrix is automatically built up as respondents submit their completed online questionnaires. Box 2.1 explains how to enter data into SPSS, the package that will be used throughout this text.

Mistakes can, of course, occur in the data entry process. Any entry that is outside the range of codes that have been allocated to a given variable will quickly show up in a table. Provided the questionnaires have been numbered, it is a simple matter to check the number of the respondent from the data matrix where the wrong codes have been entered and find what the code should have been from the questionnaire. In some packages any entry that is outside of a specified range will be flagged as the data are being keyed in. To detect erroneous codes that are inside the specified range, data may be subjected to double-entry data validation. In effect this means that the data are entered twice, usually by two different people, and any discrepancies in the two entries are flagged up by the computer and can be checked against the original questionnaire.

Box 2.1 Entering data into IBM SPSS Statistics

This software is one of the most widely used survey analysis computer programs and focuses exclusively on variable-based statistical analysis. It has gone through many versions, the latest being version 22.0. This text uses version 19.0. For more information and a free download of a demo version visit www.spss.com/spss.

You can almost certainly obtain access to SPSS by logging on to your own university or college network applications. The first window you will see is the `Data Editor` window (Figure 2.2). Before getting to the `Data Editor`, however, you need to tell SPSS what you want to do – open an existing data source, type in new data, and so on. If you are entering data for the first time, check the `Type in data` radio button. The `Data Editor` offers a data matrix whose rows represent cases (no row should contain data on more than one case) while the columns list the variables. The cells created by the intersection of rows and columns will contain the values of the variables for each case. No cell can contain more than one value.

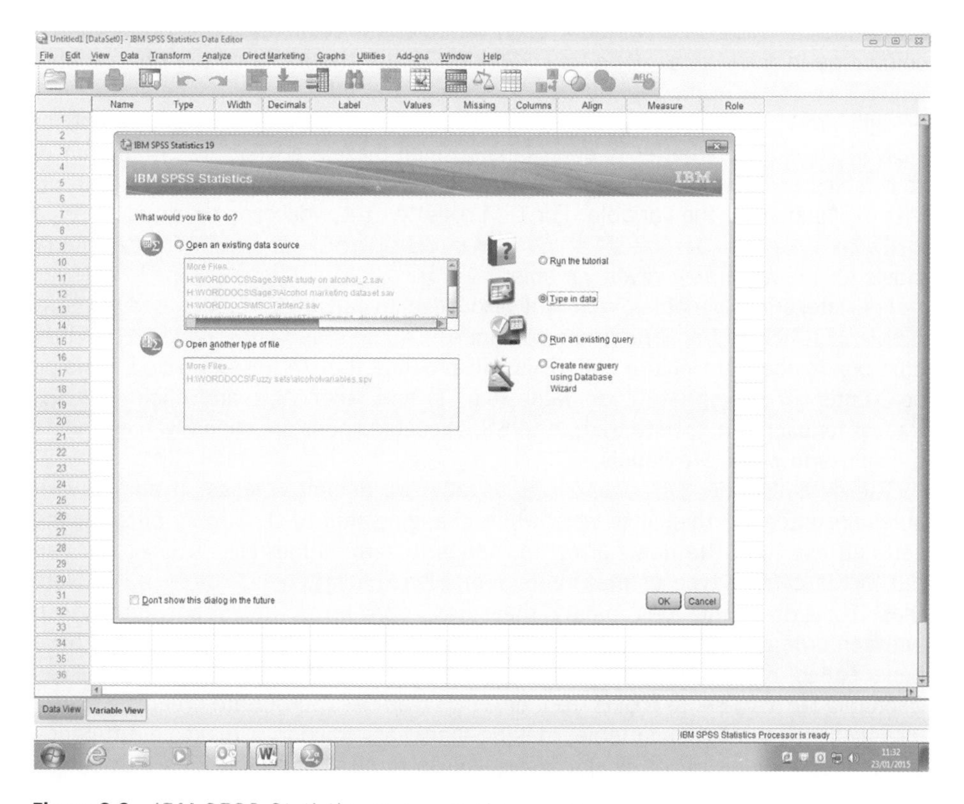

Figure 2.2 IBM SPSS Statistics `Data Editor` screen

(Continued)

(Continued)

The alcohol marketing dataset was introduced in Chapter 1. The full dataset consists of 61 properties and 920 cases, and is available at https://study.sagepub.com/kent. As an exercise in data entry, instead of attempting to enter over 40,000 values, try entering just the nine key variables for the first 12 cases that are illustrated in the next chapter in Figure 3.1.

The key **dependent variables** relate to alcohol drinking behaviour and included here are Drinkstatus (whether or not they had ever had a proper alcoholic drink), Intentions (whether they think they will drink alcohol at any time in the next year) and Initiation (how old they were when they had their first proper alcoholic drink). Three key **independent variables** have been picked out: Totalaware (the number of alcohol marketing channels seen), Totalinvolve (the number of marketing involvements) and Likeads (how they feel about alcohol ads as a whole). Finally, there are two demographics: Gender (male or female) and Socialclass (A, B, C1, C2, D, E).

Before entering any data, it is advisable first to name the variables (if you do not, you will be supplied with exciting names like var00001 and var00002). These names must begin with a letter and must not end with a full stop/period. There must be no spaces and the names chosen should not be one of the key words that SPSS uses as special computing terms, for example and, not, eq, by, all.

To enter variable names, click on the Variable View tab at the bottom left of the Data Editor window. Each variable now occupies a row rather than a column as in the Data Editor window. Enter the name of the first variable Drinkstatus in the top left box. As soon as you hit Enter or the down arrow or right arrow, the remaining boxes will be filled with default settings, except for Label. It is always better to enter labels, since these are what are printed out in your tables and graphs. Labels can be the wording of the questions asked or a further explanation of the variable. For Drinkstatus, you can, for example, type in Have you ever had a proper alcohol drink? You can put in labels for the remaining seven variables.

For categorical variables, you will also need to put in Values and Value Labels. Click on the appropriate cell under Values and click again on the little blue box to the right of the cell. This will produce the Value Labels dialog box. Enter an appropriate code value (e.g. 1) and label Yes and click on Add. Repeat for each value. Note that, in SPSS, allocated codes are called 'values', while the values in words are 'labels'.

The default under Decimals is usually two decimal places. If all the variables are integers, then it is worthwhile changing this to 0. Simply click on the cell and use the little down arrow to reduce to zero. Under Measure, you can put in the correct type of measure – Nominal, Ordinal or Scale. Note that Nominal includes binary measures, Ordinal does not distinguish between ordered category and ranked measures, and Scale refers to what have been called metric measures in Chapter 1. The default setting is Scale. Changing Measure to Nominal or Ordinal as appropriate creates a useful icon against each listed variable, making them easy to spot; it makes a difference to some operations in SPSS and forces you to think about what kind of measure is attained by each variable.

To copy any variable information to another variable, like value labels, just use `Edit/Copy` and `Paste`. SPSS does not have an automatic timed backup facility. You need to save your work regularly as you go along. Use the `File|Save` sequence as usual for Windows applications. The first time you go to save, you will be given the `Save As` dialog box. Make sure this indicates the drive you want. `File|Exit` will get you out of SPSS and back to the `Program Manager` or Windows desktop. SPSS will ask you if you want to save before exiting if unsaved changes have been made. Always save any changes to your data, but saving output is less important because it can quickly be recreated. The completed `Variable View` is shown in Figure 2.3.

*Data entry.sav [DataSet1] - IBM SPSS Statistics Data Editor

File Edit View Data Transform Analyze Graphs Utilities Add-ons Window Help

	Name	Type	Width	Decimals	Label	Values	Missing	Columns	Align	Measure	Role
1	Drinkstatus	Numeric	8	0	Have you ever had a proper alcoholic drink?	{0, No}...	None	8	Right	Nominal	Input
2	Intentions	Numeric	1	0	Do you think you will drink alcohol at any time during the next year?	{1, Defi...	None	6	Right	Ordinal	Input
3	Initiation	Numeric	2	0	How old were you when you had youor first proper alcoholic drink?	None	None	6	Right	Scale	Input
4	Totalaware	Numeric	8	0	Total number of channels seen	None	None	8	Right	Scale	Input
5	Totalinvolve	Numeric	8	0	Total involvement	None	None	7	Right	Scale	Input
6	Likeads	Numeric	1	0	How do you feel about alcohol ads as a whole?	{1, I lik...	None	5	Right	Ordinal	Input
7	Gender	Numeric	8	0	Sex of respondent	{0, Fe...	None	7	Right	Nominal	Input
8	Socialclass	Numeric	1	0	Social Class of CIE	{1, A}...	7	7	Right	Ordinal	Input
9											
10											
11											
12											
13											
14											
15											
16											
17											
18											
19											
20											
21											
22											
23											
24											
25											
26											
27											
28											
29											
30											
31											
32											
33											

Figure 2.3 The completed `Variable View`

Key points and wider issues

The careful checking, editing, coding and assembly of data should never be neglected. If poor-quality data are entered into the analysis, then no matter how sophisticated the statistical techniques applied, a poor or untrustworthy analysis will result. In this context the phrase 'Garbage In, Garbage Out' (or GIGO) is often mentioned. Checking, editing, coding, assembly and entry of data into a data matrix will commonly account for a substantial amount of time that an analyst will spend on the data.

Transforming

Before beginning data analysis, the researcher may wish to transform some of the variables in a number of ways that might include:

- regrouping values on a nominal or ordered category measure to create fewer categories;
- creating class intervals from metric measures;
- computing totals or other scores from combinations of several values of variables;
- treating groups of variables as a single multiple response question;
- upgrading or downgrading measures;
- handling missing values and 'Don't know' responses;
- coding open-ended questions;
- creating crisp or fuzzy set memberships from nominal, ordered category, ranked or metric measures.

Most of these tasks can be accomplished by using SPSS procedures, which are explained in Boxes 2.2–2.6 in this chapter.

Regrouping values

Where there are several or many values in a nominal or ordered category variable, particularly if the number of cases in the dataset is fewer than 300 or so, or if the frequencies in some of the categories are very small, it may make sense to add together the frequencies in adjacent categories if the variable is ordered, or in a way that 'makes sense' if it is nominal. In the alcohol marketing survey, respondents were asked what they felt about alcohol adverts on the whole. Figure 2.4, for example, shows that only 13 out of a total of 920 respondents from the alcohol marketing survey responded 'I like alcohol adverts a lot'. It seems sensible to add these to the next category, 'I like alcohol adverts a little'.

How do you feel about alcohol ads as a whole?

		Frequency	Percent
Valid	I like alcohol adverts a lot	13	1.4
	I like alcohol adverts a little	90	9.8
	I neither like nor dislike alcohol adverts	386	42.0
	I dislike alcohol adverts a little	150	16.3
	I dislike alcohol adverts a lot	264	28.7
	DK	17	1.8
	Total	920	100.0

Figure 2.4 Liking of alcohol adverts

Likeads3

		Frequency	Percent	Valid Percent
Valid	I like alcohol adverts	103	11.2	11.4
	I neither like nor dislike alcohol adverts	386	42.0	42.7
	I dislike alcohol adverts	414	45.0	45.8
	Total	903	98.2	100.0
Missing	System	17	1.8	
Total		920	100.0	

Figure 2.5 Liking of alcohol adverts collapsed to three categories

To keep the value set balanced, the disliking of adverts a little and a lot can also be added together. The 17 who responded 'Don't know' can be treated as missing values (the handling of missing values is considered later in this chapter). The resulting table is shown in Figure 2.5. Note that the `Valid Percent` is based on the 903 non-missing responses. Box 2.2 shows you how to do this in SPSS.

Box 2.2 Regrouping values in SPSS

If you need to transform a variable by regrouping categories, then it is the `Recode` procedure that you need. From the `Value Labels` box (which you can obtain from the `Variable View` screen) you can see that the codes allocated for the responses to how they felt about alcohol adverts as a whole are as shown in Figure 2.6. We need to add together codes 1 and 2, codes 4 and 5, and treat code 6 as a missing value. In SPSS, from the `Menu bar`, select `Transform|Recode Into Different Variables`. From the list of variables, select `Likeads` ('How do you feel about alcohol adverts on the whole?') and transfer to the `Input Variable -> Output Variable` box. Now click on `Old and New Values`. We need codes 1 and 2 to become 1, so in the `Old Value` dialog area on the left click on the first `Range` radio button and enter `1` then `through` and `2`. In the `New Value` dialog area on the right enter `1` in the `Value` box and click on `Add`. This instruction will now be entered into the `Old -> New` box. We want to change code 3 to 2, so click on the `Value` radio button under `Old Value` and enter `3`. Now enter `2` under `New Value` and click on `Add`. We now want codes 4 and 5 to be 3. Click on the `Range` radio button and enter `4 through 5`. Under `New Value` enter `3` and click on `Add`. Click on `Continue`. Under `Old Value`, add code 6 and click on `System-missing` under `New Value`. Give the `Output Variable` a name in the `Name` box, for example `Likeads3`, and click on `Change` then `OK`. The new variable will appear as the last column.

(Continued)

(Continued)

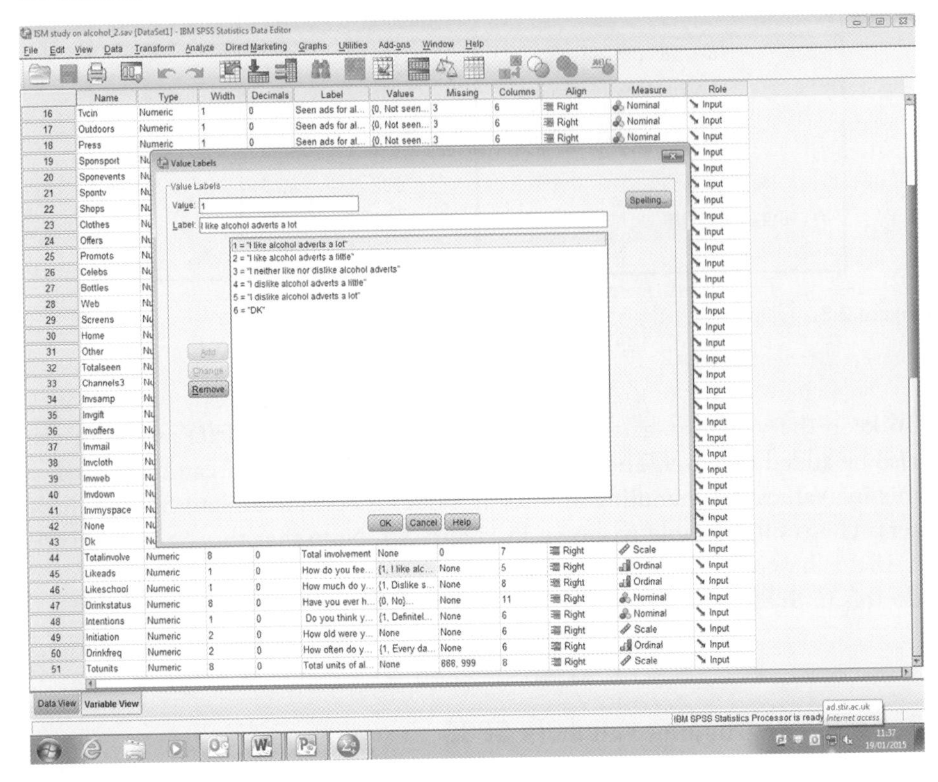

Figure 2.6 Liking of alcohol adverts

To add value labels for the new variable, change to the `Variable View`. Click on the right corner of the `Values` cell in the appropriate row and obtain the `Value Labels` dialog box. Enter `1` in `Value` and `Important` under `Value Label` and click on `Add`. Now enter `2` in `Value` and `Neither` under `Value Label` and click on `Add`. Finally, enter `3` in `Value` and `Unimportant` under `Value Label` and click on `Add`. Now click on `Continue` and `OK`. You can now check this out using the `Analyze|Descriptive Statistics|Frequencies` procedure.

Creating class intervals

These are used to group together ranges of values on metric measures to enable the researcher to get an overview of the distribution. The intervals must be non-overlapping and as far as possible of the same width. In the alcohol marketing data, young people were asked whether or not they had seen adverts for alcohol in any of 16 different channels, and a new variable, `Total number of channels seen (Totalseen)`, was created. Figure 2.7 shows the frequencies for each number of channels. Thus 33 claim to have seen no adverts for alcohol on any of these listed channels, one had seen them on all channels, but the majority, just over

50 per cent, said they had seen them on between four and seven channels. The creation of class intervals may be approached in a number of ways. From the `Cumulative Percent` column, you can see that nearly 50 per cent (49.3 per cent) had seen such adverts on up to five channels. If the researcher wanted to create a two-value measure, then the 920 cases could be split into those who had seen alcohol adverts on up to five channels and those who had seen six or more channels, as shown in Figure 2.8. This is not a binary measure in the sense that `Six or more channels seen` is not the 'absence' of membership of the category `Up to five channels seen`. It is really a two-value nominal measure – a dichotomy. Furthermore, the latter category includes the 33 who said they had not seen any. These do not sit well in any classification of the number of channels in which alcohol adverts were 'seen'. It would make sense to keep the 33 as a separate category in the three-value measure in Figure 2.8. Another option is to treat the 33 as 'missing values' and exclude them from the table. The treatment of missing values is considered later in this section.

The two- and three-value solutions above can be considered not so much as 'intervals' as ordered category solutions. To keep the measure metric, the number of channels in which adverts for alcohol were seen could be grouped

		Frequency	Percent	Cumulative Percent
Valid	0	33	3.6	3.6
	1	40	4.3	7.9
	2	56	6.1	14.0
	3	84	9.1	23.2
	4	104	11.3	34.5
	5	137	14.9	49.3
	6	121	13.2	62.5
	7	106	11.5	74.0
	8	83	9.0	83.0
	9	71	7.7	90.8
	10	43	4.7	95.4
	11	24	2.6	98.0
	12	3	.3	98.4
	13	11	1.2	99.6
	14	3	.3	99.9
	16	1	.1	100.0
	Total	920	100.0	

Figure 2.7 Number of channels on which adverts for alcohol have been seen

		Frequency	Percent
Valid	Up to 5 channels seen	454	49.3
	Six channels or more seen	466	50.7
	Total	920	100.0

		Frequency	Percent
Valid	None	33	3.6
	Seen 1-5 channels	421	45.8
	Seen 6 or more channels	466	50.7
	Total	920	100.0

Figure 2.8 Figure 2.7 recoded into two- and three-value measures

into several intervals of equal width, for example, of three channels, 0–2, 3–5, 6–8, 9–11, 12–14 and 15–16. The last interval is only two channels: nothing is ever perfect and compromises have to be made. The usefulness of doing this is limited when there are few values to be grouped. However, if there were 100 or so (as there would be with age in years for individuals), then grouping into class intervals (of perhaps 10 years) and obtaining a frequency distribution would enable the researcher to overview the entire distribution pattern in a simple table.

The number, width and placing of the intervals are matters for researcher judgement and may be subject to trial and error, with the researcher trying out different groupings to see to what extent this may affect the results. To view a distribution a useful rule of thumb is to create between about 5 and 15 intervals. If there are **outliers** – values that are substantially different from the general body of values – then there may need to be open-ended classes at either or both ends of the table. This is quite a common way of dealing with extreme values, but it does mean that the width of the open-ended intervals is unknown. Creating class intervals in SPSS is explained in Box 2.3.

Box 2.3 Creating class intervals in SPSS

To create class intervals in SPSS you need the `Recode` procedure again. In `Old and New Values` enter the ranges 0–5 and 6–16, giving these the new codes of 1 and 2 for the two-value solution, and 0–2, 3–5, 6–8, and so on for the metric class interval solution, again giving each interval a new code. Note that 'Up to 5 channels seen' and 'Six or more channels seen' are the new researcher-defined values, but what SPSS is calling `New Values` are the codes that 'stand for' the new categories.

Computing totals

It is sometimes helpful to add together the values recorded either for two or more variables within cases or for a single variable across cases – provided, however, they are, or can reasonably be assumed to be, metric values. In the alcohol marketing data,

	Very important	Quite important	Neither important nor unimportant	Quite unimportant	Very unimportant	DK
	5	4	3	2	1	0
Chocolate or sweets	58	244	285	208	117	8
Fizzy drinks	66	274	249	220	104	7
Crisps	47	234	287	232	103	17
Trainers	319	357	131	74	34	5
Clothes	330	362	121	78	24	5
Magazines	57	242	282	200	115	24
Perfume/Aftershave	197	413	153	100	48	9
Cigarettes	33	79	153	124	377	154
Alcohol	36	117	160	125	336	146

Figure 2.9 The importance of choosing well-known brands

Total importance of brands

		Frequency	Percent	Cumulative Percent
Valid	0	2	.2	.2
	8	1	.1	.3
	9	11	1.2	1.5
	10	4	.4	2.0
	11	3	.3	2.3
	12	5	.5	2.8
	13	5	.5	3.4
	14	7	.8	4.1
	15	13	1.4	5.5
	16	13	1.4	7.0
	17	15	1.6	8.6
	18	32	3.5	12.1
	19	17	1.8	13.9
	20	22	2.4	16.3
	21	44	4.8	21.1
	22	37	4.0	25.1
	23	51	5.5	30.7
	24	53	5.8	36.4
	25	46	5.0	41.4
	26	46	5.0	46.4
	27	72	7.8	54.2
	28	54	5.9	60.1
	29	52	5.7	65.8
	30	67	7.3	73.0
	31	46	5.0	78.0
	32	38	4.1	82.2
	33	30	3.3	85.4
	34	32	3.5	88.9
	35	25	2.7	91.6
	36	18	2.0	93.6
	37	16	1.7	95.3
	38	12	1.3	96.6
	39	11	1.2	97.8
	40	8	.9	98.7
	41	7	.8	99.5
	42	2	.2	99.7
	43	1	.1	99.8
	44	1	.1	99.9
	45	1	.1	100.0
	Total	920	100.0	

Figure 2.10 Total scores in brand importance

respondents were asked how important they thought it was to choose a popular well-known brand when deciding to buy each of several different products. The SPSS results are shown in Figure 2.9. If the codes allocated (5 for `Very important` and so on) can reasonably be treated as calibrations of the degrees of importance, then the nine items can be treated as a **summated rating scale** and totals calculated for each of the respondents who gave a rating. The maximum total score is 45 (the person thinks choosing popular known brands is `Very important` for all items). The minimum depends on what the researcher decides to do with the 'Don't know' answers. If given a value of 0 (as in Figure 2.10) this will mean a minimum score of 0 if somebody responds 'Don't know' to each item. Indeed, an individual respondent will achieve a lower total score if any responses are 'Don't know'. Alternatively, these answers could be treated as a missing value and ignored in the calculation.

Figure 2.10 shows the frequency of each total score calculated by SPSS. How to do this in SPSS is explained in Box 2.4. Thus two respondents had a score of 0, so they indicated 'Don't know' for each item. One person scored the maximum of 45, but the majority had scores somewhere in the middle. There are various ways in which a distribution like this can be summarized, including creating class intervals as in the previous section, but other ways will be explored in Chapter 4.

Box 2.4 Computing totals in SPSS

To get SPSS to compute the totals that were used to create Figure 2.10 select `Transform` then `Compute Variable`. You will obtain the `Compute Variable` dialog box (Figure 2.11). Notice that there are lots of functions that we could perform on the variables – but all we want to do is add the nine variables together, so highlight `Chocolate or sweets` and put this into the `Numeric Expression` box by clicking on the arrow. Now click on the + button and bring over the next variable, then on + again, and so on until you have the nine variables added together. Enter a variable name, something like `Totbrand`, in the `Target Variable` box and click on `OK`. A new variable will appear in your data matrix, giving the total scores for each case. You can now use `Recode` to group the responses into, say, high-, medium- and low-score categories.

Multiple response questions

There are often questions in a survey that allow respondents to pick more than one answer. The very first question on the alcohol survey, for example, asks respondents whether they have watched television, read a newspaper, read a magazine, listened to the radio or used the Internet in the last seven days. Respondents can reply 'Yes' to as many of these that apply. For the purpose of analysis, each medium will need to be treated as a separate variable, each one of which is either selected or not selected (so, it is binary). The five items then need to be analysed together. The results are shown in Figure 2.12. This indicates that there were 3,418 yeses. The 909 who indicated that they watched television constituted 26.6 per cent of the 3,418 yeses and 98.9 per cent of the 920 cases. Box 2.5 explains how to do this in SPSS.

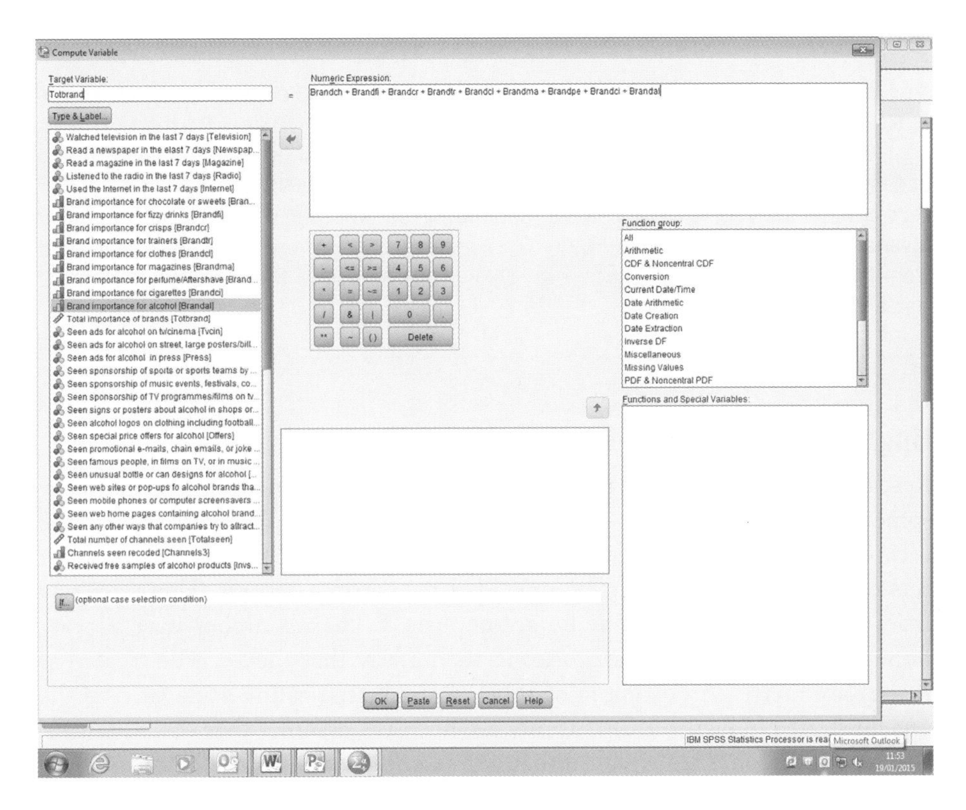

Figure 2.11 The `Compute Variable` dialog box

		Responses		Percent of Cases
		N	Percent	
In the last 7 days	Watched television	909	26.6%	98.9%
	Read a newspaper	526	15.4%	57.2%
	Read a magazine	558	16.3%	60.7%
	Listened to the radio	669	19.6%	72.8%
	Used the Internet	756	22.1%	82.3%
Total		3418	100.0%	371.9%

Figure 2.12 SPSS multiple response

Box 2.5 Multiple response items in SPSS

To treat the five items relating to media consumption as a multiple response question, select `Analyze|Multiple Response|Define sets`. Bring the five variables across to the `Variables in Set` box. Since the code of 1 was entered for those who said 'Yes', enter `1` in the `Counted Value` box. Make sure the `Dichotomies` radio button is clicked under `Variables Are Coded`

(Continued)

(Continued)

As. You will also need to give the new variable a name, like Media. Click on the Add button to add the name to the Multiple Response Sets box, then on Close. The new variable, however, does not appear in the data matrix. To access it, click on Analyze|Multiple Response and either Frequencies or Crosstabs depending on whether you want **univariate** or **bivariate analysis**. To produce Figure 2.12, select Frequencies. Move Media from the Multiple Response Sets box across to the Table(s) for box and click on OK.

Upgrading or downgrading measures

Researchers sometimes upgrade the complexity of the measures achieved by some of their variables in order to apply the more sophisticated statistical techniques that thereby become available. The most usual transformation is for sets of ordered categories to be upgraded to metric measures. There are two main ways in which this may be accomplished. The researcher may allocate numeric codes to ordinal categories in such a way that they can be treated as if they are metric. An example of this is a **summated rating scale** such as the one illustrated in Figure 1.2. This process, however, assumes that the 'distances' between each value are equal so that, for example, the distance between 'Very important' and 'Quite important' is the 'same' as the distance between 'Quite important' and 'Neither important nor unimportant'. Such an assumption may seem reasonable in this example. It also tends to be reasonable for Likert items (see Chapter 1) where respondents are being asked about their degree of agreement or disagreement with a number of statements. However, for levels of satisfaction, where a set of categories like 'Very satisfied', 'Fairly satisfied' and 'Dissatisfied' have been used, this assumption is suspect since the 'distance' between 'Dissatisfied' and 'Fairly satisfied' is probably much greater than the 'distance' between 'Fairly satisfied' and 'Very satisfied'. In any event, it would be unwise to treat average scores in any absolute sense. Thus an average score of, say, 4.0 for males and 2.0 for females does not mean that males are 'twice' as satisfied, or that they are two 'units of satisfaction' above the females. All we can say is that the average of the males is higher than that of the females. However, for measuring change, for example from one week to the next, then changes in the average scores *are* likely to reflect real changes in people's level of interest.

The other way to create metric measures is to define categories of an ordered set in terms of counting numbers of instances as a measure of size. Thus a distinction between 'Small', 'Medium' and 'Large' organizations is only an ordinal distinction. However, if the researcher defined 'Small' organizations as having fewer than 50 employees, 'Medium' as having between 50 and 200 employees, and 'Large' as having over 200 employees then a discrete metric variable has been created, the 'metric' in this case being size measured by the number of employees. With a larger number of categories, more precisely defined, with upper and lower limits, it becomes possible to calculate an average size. This procedure is fine provided there is accurate information, for example, in the situation above, on the number of employees in each organization of interest. By creating (or assuming the creation of) metric measures, the researcher can now, for example, add up and then calculate average scores, calculate standard deviations and use the variables in ways that will be explained in Chapters 4–6.

There are some circumstances when a researcher may downgrade a measure and treat it as if it were at a less complex level. Thus a metric variable may be treated as ranked by ignoring the distances between categories. A class test out of 100 may be used to create ranks of first, second, third, and so on. This may be undertaken by the researcher either because he or she feels that the assumptions of the original metric are unwarranted, or because the variable concerned is to be correlated with another **ranked variable** and the researcher wants to apply a statistic that requires two ranked variables. Metric variables can be ranked in SPSS using the procedure `Transform|Rank Cases`.

Another example of downgrading is when a researcher wishes to crosstabulate a **nominal** with an **ordered category variable**. An appropriate **coefficient** may be chosen that treats both variables as nominal, thereby ignoring the ordering of the categories in one of the variables. A more extreme example is when a researcher takes a continuous metric variable like age and groups respondents into a binary measure of 'old' and 'young' or into an ordered category measure of 'old', 'middle aged' and 'young'. This may be undertaken if the researcher wishes to crosstabulate age with another binary, nominal or ordered category variable, for example 'purchased' and 'did not purchase' a newspaper in the last seven days. The age split would normally be done in a way that creates two (or three or more as required) roughly equal groups. The SPSS procedure `Transform|Compute` can be used to create a new variable grouped in this way.

Handling missing values and 'Don't know' responses

In any survey, not all respondents will answer all the questions. This is less likely to be the result of individual refusal to answer some of the questions (although this does happen), or people accidentally omitting to consider some of the questions, than a result of questionnaire design whereby not all the questions are relevant to all the respondents. The result is that values will be missing from some of the cells in the data matrix.

Where a question *would* be appropriate to a given respondent, but an answer is not recorded, then such missing values may be referred to as 'item non-response'. Most researchers are inclined just to accept that there will be item non-response for some of the variables and will simply exclude them from the analysis. This is fine when the number of cases entered into the data matrix is large or at least sufficient for the kinds of analyses that are required. However, there is always the danger that this approach may reduce the number of cases used in a particular analysis to such an extent that meaningful analysis is not possible. There is, however, a bewildering array of techniques that have been suggested in the literature for ways of dealing with this situation. Most of these involve filling the gaps caused by missing values by finding an actual replacement value. The process is sometimes called 'explicit imputation' and the idea is to select a replacement value that is as similar as possible to the missing value. Where variables are metric, one remedial technique, for example, is to substitute the mean value for the missing value. For **categorical variables** one technique that is sometimes used is to give the questionnaire with the missing value the same value as the questionnaire immediately preceding it.

Most of the techniques assume, however, that question items not responded to are done so at random. This can be quite difficult to determine. Furthermore, when the amount of item non-response is small – less than about 5 per cent – then applying any of the methods is unlikely to make any significant difference to the interpretation of the data. Ideally, of course, researchers should, in reporting their findings, communicate the nature and amount of item non-response in the dataset and describe the procedures used to remedy or cope with it. How missing values are handled in SPSS is explained in Box 2.6.

Box 2.6 Missing values in SPSS

SPSS makes a distinction between two kinds of missing value: system missing values and user-defined missing values. The former result when the person entering the data has no value to enter for a particular variable (for whatever reason) for a particular case. In this situation the data analyst will just skip the cell and SPSS will enter a full stop in that cell to indicate that no value has been recorded. For most non-graphical outputs, SPSS will list in a separate `Case Processing Summary` the number of valid and the number of missing cases. In some tables, as in Figure 2.5 in an earlier section, valid and missing cases are shown in the printed output table itself. Percentages are then calculated both for the total number of cases entered into the data matrix and for the total of non-missing cases for that variable – what SPSS calls the `Valid Percent`.

User-defined missing values are ones that have been entered into the data matrix, but the researcher decides to exclude them from the analysis. To create them for any particular variable, from the `Variable View` select the little blue box in the `Missing` column against the variable you want and obtain the `Missing Values` dialog box. This enables you either to pick out particular codes to be treated as missing values by clicking on the `Discrete missing values` radio button and entering up to three codes, or to select a range of missing values.

It would be a sensible policy to reserve system missing values in effect for questions that are not applicable to the respondent in question and to give a special code for those where responses are missing for other reasons. The combination of system-defined and user-defined missing values can mean that, for some tables or calculations, the number of cases used is considerably less than the number of cases entered into the data matrix. Furthermore, it will mean that the number of cases included will vary from table to table or statistical analysis. If the number of cases in the data matrix is quite small to begin with, this can have serious implications for the analysis.

'Don't know' answers are one type of non-committal reply that a respondent may give along with undecided or neutral responses in a balanced rating with a middle point. These responses may be built into the design of the questionnaire with explicit options for a non-committal response. In Figure 2.9, there are, for example, separate categories for 'Neither important nor unimportant' and for 'Don't know': 285 or over 30 per cent gave a neutral response for the importance of choosing well-known brands for chocolate or sweets, and 154 (over 16 per cent) indicated 'Don't know' for cigarettes. In some questionnaires the 'Don't know' answers are included in the 'Neither' category.

An understanding of the pattern of such replies is important for formulating research methodology, particularly questionnaire design, item phrasing or the sampling plan, or for interpreting the results when there are many 'Don't know' responses. Non-committal replies have been interpreted very differently by researchers. These interpretations fit into two broad patterns:

- 'Don't know' responses are a valid indicator of the absence of attitudes, beliefs, opinions or knowledge.
- 'Don't know' replies are inaccurate reflections of existing cognitive states.

The first interpretation provides a rationale for including explicit non-committal response categories in the questionnaire. It also implies that such responses should be excluded from the analysis (treating them as user-defined missing values in SPSS, see Box 2.6), even if this means that the number of cases on which the analysis is based is thereby reduced. If there are a lot of respondents in this category, then it is possible that the question to which people are being asked to respond is not well thought through and there may well be an argument for excluding the question from the analysis altogether.

The second interpretation has been used to set in motion various efforts to minimize 'Don't know' responses on the basis that only committed responses will reflect a respondent's true mental state. Such efforts will include providing ordered category measures that have no middle position or non-committal option, or that have interviewers probe each non-committal reply until a committal response has been obtained.

If there are relatively few non-committal responses then leaving them out of the analysis may well be the best course of action to take, particularly if the number of remaining cases is still adequate for the statistical analyses being proposed. There will certainly be a case, however, for including them in any preliminary analysis of the variables. A decision can then be taken about whether they are to be excluded from subsequent analyses.

Survey research findings are certainly not invariant to decisions about what to do with non-committal responses. Treating such responses as randomly distributed missing data points when in fact some responses are a genuine result of ambivalence or uncertainty may introduce **bias** into the data. The same would be true if responses are included as neutral positions when they are in fact an indicator of no opinion or refusal to answer. A first step in any analysis would be to investigate the extent to which non-committal responses are a function of demographic, behavioural or other cognitive variables. Some studies, for example, have reported an inverse relation between education and non-committal responses (the better educated are less prone to give them), but it has to be said that other research has found exactly the reverse. Durand and Lambert (1988) found that such responses vary systematically with socio-demographic characteristics and with involvement with the topic area.

Open-ended questions

Most surveys contain one or more open-ended questions where responses are recorded as words, phrases, sentences or even more extended text. To be used in quantitative data analysis, the responses need to be categorized and each category given a code. The result should be either a binary or a nominal measure such that the values are exhaustive and mutually exclusive, or a **fuzzy set** giving degrees of membership of a defined category.

The approach to coding can be split into two situations. In the first situation, the open-ended question is being used to capture factual information, since listing all the options for responses in a closed question would take up too much space. Where respondents can give their answer in numerical form, for example putting in their age, then no additional coding is necessary. The actual age can simply be put into the data matrix. Where responses are in words, like brand purchased last time, then coding will involve creating a list of all the possible answers, assigning a code to each and recording a code for each respondent's answer. It may be necessary to develop coding rules which specify codes to be allocated when the answer does not fit any of the obvious categories. For example, if respondents are asked 'Not counting yourself, how many other people were you with?' then most will give a clear number, but some may say '30–40' or 'a lot'. In this situation, one rule might be to give the mid-point of a range of values, so the answer '30–40' will be coded as 35.

Where open-ended questions are being used not to capture factual information but to record respondent opinions, attitudes, views, knowledge, and so on, then creating a sensible code frame is the most important part of the analysis. By definition this is likely to get quite complex – if it were easy then the question could no doubt be pre-coded! The aim is to formulate a set of categories that accurately represents the answers and where each category includes an appreciable number of responses. Ideally, the set of categories should be exhaustive, mutually exclusive and minimize the loss of information. Furthermore, they should be meaningful, consistent and relatively straightforward to apply. There may also need to be separate codes for 'No response', 'Not applicable' and 'Don't know'. Where the information is very detailed there may need to be many codes.

Developing a frame may require several 'passes' over the data. It is probably a good idea to have all the comments collected and typed out, but this may not be possible. A method of constant comparison is probably best. Begin by looking at a few of the comments and see whether they should be put into separate categories. Then look at a few more and see if some can be put into the same category or whether more categories will need to be developed. When too many categories begin to emerge, look for similarities so that some categories can be brought together. If there are a large number of responses then it may not be sensible to look through all of them to develop the fame, but take a sample. Thus if there are 500 cases, a sample of 50–100 should enable the frame to be finalized. It also helps if more than one person develops a code frame separately; they should then work together on a final code. This maximizes the **validity** and **reliability** of the process.

It helps if the researcher sets up the objectives for which the code frame is to be used before beginning the process. Thus if the objective is to look for positive and negative statements about a situation or a product then answers will be coded along this dimension, perhaps with categories of very positive, vaguely positive, mixed, vaguely negative and very negative. Sometimes answers to open-ended questions can be coded in several ways according to different dimensions. Thus a study of injuries following an earthquake could look at the way injuries occurred, the parts of the body affected, where the injury occurred, what the person was doing at the time, and so on. Each of these aspects may need to be recorded separately in a different variable.

At one time researchers had to code all open-ended questions before data entry could begin. With modern survey analysis packages like SPSS, however, this may be done after all the pre-coded questions have been entered. This is a big advantage because researchers are not always sure how responses to open-ended questions should be coded until they have started analysis of the data. In short, it is sometimes better to delay coding of open-ended responses until they are needed for analysis.

Key points and wider issues

Before engaging in the description of a dataset, or even following an initial overview of the distribution for each variable one at a time or each case one at a time, the researcher may wish to transform variables in a number of ways: for example, regrouping values on a nominal or ordered category measure to create fewer categories, creating class intervals from metric variables, computing totals or other scores from combinations of several variables, treating groups of variables as a multiple response question, upgrading or downgrading measures, handling missing values and non-committal responses, or coding open-ended questions. Some transformations might involve creating crisp or fuzzy set memberships as was explained in Chapter 1.

(Continued)

(Continued)

Data transformation is an important part of the data analysis process. There are no 'right' or 'wrong' ways of engaging in data transformation and there are usually several different ways in which it can be done. Perhaps the best strategy is what is sometimes called 'sensitivity analysis', whereby transformations may be tried in different ways to see how sensitive the results are to such processes. This is particularly true for how missing cases and 'Don't know' answers are handled.

Implications of this chapter for the alcohol marketing data

Many of the codings in the original dataset were illogical or inconsistent; for example, for some questions relating to whether or not they had done particular activities, respondents were given the choice between 'Yes', 'No' and 'Don't know', while for others it was just 'Yes' and 'No'. In a question asking respondents to indicate how often they had come across adverts for a range of different products, 'Very often' was coded 1 and 'Never' coded as 6, with 'Don't know' as 7. Besides being counter-intuitive (the higher the score, the less often), this way of coding makes it impossible to use the codes as metric values for summation since 'Don't know' has the highest value. Accordingly, a number of data transformations were needed before analysis could begin. It was also necessary to create new variables additional to those in the questionnaire, for example the number of channels on which respondents had seen adverts for alcohol.

Chapter summary

Before researchers can proceed with the next stages of data analysis, the data need to be prepared by checking questionnaires or other instruments of data capture for usability, editing responses for legibility, completeness and consistency, coding any responses that are not pre-coded, and assembling the data together by entering all the values for all the variables for all the cases into a data matrix. Data entry into the survey analysis package SPSS was explained in some detail.

Before analysis of the data can begin, some of the variables may need to be transformed in various ways and decisions may have to be made about how to handle missing values. The careful preparation of data ready for analysis should never be neglected. If poor-quality data are entered into the analysis, then no matter how sophisticated the statistical techniques applied, a poor or untrustworthy analysis will result. Handled with care, data preparation can substantially enhance the quality and usefulness of data analysis: paying inadequate attention to it can seriously compromise the validity of the results.

Exercises and questions for discussion

1. To what extent can treating codes allocated to ordered category measures as if they are numeric values be justified in the data analysis process?
2. When transforming variables, researchers make many decisions for which there are no 'rules' or even rough guidelines. What impact might these decisions have on the validity of the data?
3. What are the key circumstances in which missing values might be a severe problem for the data analyst?
4. Open IBM SPSS on whatever system you are using and enter the nine key variables for the first 12 cases for the alcohol marketing dataset that are illustrated in the next chapter in Figure 3.1. The procedures for doing so are explained in Box 2.1.
5. Figure 2.10 shows the total scores for the importance of well-known brands in choosing products. Try creating class intervals in various different ways using SPSS. The procedures for doing so are explained in Boxes 2.2 and 2.3.
6. Go to the website www.surveyresearch.weebly.com. Here you will find lots of interesting information about social surveys created by John Hall, previously Senior Research Fellow at the UK Social Science Research Council (1970–6) and Principal Lecturer in Sociology and Unit Director at the Survey Research Unit, Polytechnic of North London (1976–92). Download the `Trinians` dataset. Select `Survey Unit, Social Science Research Council`, then `Surveys by SSRC Survey Unit` and then the `'Trinians'` survey. Read the background to the survey, download the article in *Folio* and the questionnaire. Finally download and save the dataset from trinians.sav onto your SPSS file. Not all the questions in the questionnaire appear as variables and they are not all in the same order as in the questionnaire, but the question numbers are clearly marked. Check out the values being used from the `Values` column. Under `Measure`, they are all indicated as `Scale`. This is the default if researchers do not change any of these. Go down the variables and change to `Ordinal` or `Nominal` as appropriate (left click on `Scale` and the other two options will appear).

Further reading

Cragun, R. (2013) 'Using SPSS and PASW', Wikibooks. Available at http://en.wikibooks.org/wiki/Using_SPSS_and_PASW.

A useful (and free) wiki book giving a series of mini SPSS tutorials. For this chapter, have a look through the section on basic operations.

De Vaus, D. (2002) *Analyzing Social Science Data: 50 Key Problems and Data Analysis*. London: Sage.

Parts One and Two give quite detailed answers to frequently asked questions about data preparation in the form of 15 problems like 'How to code answers with multiple answers'.

Diamantopoulos, A. and Schlegelmilch, B. (1997) *Taking the Fear Out of Data Analysis*. London: Dryden Press. Republished by Cengage Learning, 2000.

Chapter 4 is about data preparation and transformation, but beware that the image of a data matrix is rather dated and does not look like an SPSS matrix.

Suggested answers to the exercises and questions for discussion can be found at the end of this text, pp. 293–321, and on the companion website, (https://study.sagepub.com/kent), which also give links to relevant free online Sage journal articles, PowerPoint slides, an overview of data analysis packages, an introduction to SPSS and weblinks to alternative datasets.

3

APPROACHES TO DATA ANALYSIS

Learning objectives

In this chapter you will learn that:

- a record of all the values for all the cases for all the properties in a research project constitutes a dataset, which will usually be laid out in the form of a data matrix;
- a data matrix is the beginning point of any data analyses, but some analyses focus on the columns, looking at the distributions of values of properties as variables across the cases, while others focus on the rows, reviewing the configurations of values as set memberships across the properties;
- the analysis of any kinds of data at a minimum includes a description of the dataset, but is likely also to involve elements of interpreting, relating, evaluating, explaining, applying and presenting results to an audience;
- the processes of data construction and data analysis raise a number of ethical issues;
- to date, all the publications relating to the alcohol marketing dataset have focused on analyses that are variable-based rather than case-based and, furthermore, serious ethical issues arise from the fact that the study involved interviewing young people aged between 12 and 14.

Introduction

Once data have been constructed and prepared in ways described in Chapter 2, they can be analysed. The beginning point of any analysis of quantitative data is the construction of a data matrix – a process that could be considered as part of data preparation, but is better explained as part of the analysis. This chapter explains the structure and creation of a data matrix and then turns to an outline of the overall process of data analysis, which is broadly similar whatever the types of data involved. There are two main approaches to data analysis which form the main theme of this text, namely variable-based approaches and case-based ones.

Data construction, their preparation and their analysis give rise to a range of ethical issues which are considered at the end of this chapter.

Datasets and data matrices

Data, as we have seen, consist of systematic records. Quantitative data that are to be used for data analysis will, in addition, have a structure that consists of a set of cases, properties and recorded values. A record of all the values for all the cases for all the properties in a research project is a **dataset**. This will usually be laid out in the form of a **data matrix**. This is a framework of rows and columns for storing numeric data in a table. It is usually accomplished electronically rather than on a large sheet of paper. Data matrices are rectangular and, by convention, the cases are allocated to the rows, the variables or set memberships are allocated to the columns, and the values are entered into the cells created by the intersection of rows and columns. Data analysis software works best when all the values in a data matrix are either numerical (as in age = 57) or given a number that acts as a code for values that are in words. Binary, nominal and ordered category variables need to be coded in this way. Figure 3.1, for example, shows part of a data matrix for the first 12 cases in the alcohol marketing survey. Thus `Drinkstatus` has been coded into 1 or 0 for the answers 'Yes' and 'No' to the question about whether or not respondents have ever had a proper alcoholic drink.

*Data entry.sav [DataSet2] - IBM SPSS Statistics Data Editor

File Edit View Data Transform Analyze Direct Marketing Graphs Utilities Add-ons Window Help

	Drinkstatus	Intentions	Initiation	Totalseen	Totalinvolve	Likeads	Gender	Socialclass	var	var
1	0	1	.	14	5	5	1	2		
2	0	2	.	4	1	2	2	3		
3	1	2	13	4	0	3	1	5		
4	0	1	.	1	0	5	2	6		
5	1	2	13	8	0	3	2	2		
6	0	3	.	11	1	3	2	2		
7	0	2	.	10	1	3	2	5		
8	0	2	.	11	0	3	2	3		
9	1	3	12	16	7	2	2	4		
10	1	4	12	3	0	3	2	2		
11	1	3	11	6	1	3	1	2		
12	0	1	.	7	0	3	2	2		
13										
14										
15										
16										
17										
18										
19										
20										
21										
22										
23										

Figure 3.1 The first 12 cases for the alcohol marketing study

In an interview survey, each value down a column will have come from a different respondent and different interviewers are likely to have contributed to each column. For the most part, each property will correspond with a particular question on the questionnaire and the numeric value, code or set membership value will be entered into the data matrix. In some situations, however, one question may give rise to several variables. This will be so, for example, for all multiple response questions where respondents can pick more than one response category or, indeed, as many categories as apply to them. Each category will be ticked or not ticked, so each gives rise to a separate binary variable. The order in which the properties are placed in the matrix will tend to follow the sequence in which they occur in the questionnaire. The order is unimportant for purposes of analysis, but such a sequence usually makes sense for data entry, otherwise it will be necessary to move about the questionnaire to find each value to be entered.

The rows in the data matrix are identified with the cases, one row for each case. In an interview survey this means one row per questionnaire. A series of values or codes are entered for each respondent and, in an interview survey for example, each row will be the work of a single interviewer. Again the order of the rows does not matter for purposes of analysis. Questionnaires are often entered in the order in which they are received. They will usually be given a serial number that will correspond with the row number. This means that if there are any queries about a particular questionnaire, it can be located in the data matrix.

The size of a data matrix is the product of the number of respondents and the number of properties. Where there are many respondents and many properties, such as in a large-scale survey, the data matrix will be very large. Thus an enquiry that has 200 respondents to a questionnaire survey with 100 questions will produce a matrix that contains 20,000 cells. Data matrices will have different shapes depending on the nature of the research. Intensive research will have relatively few respondents but many properties; extensive research will have many respondents and few properties. An opinion poll is a good example of the latter where a large sample (of 1,000 or so) is asked a few questions about voting intentions and party support. Usually, data matrices are rectangular; all the rows (cases) span across the same number of properties, while the columns (properties) span down the same number of cases. However, sometimes the matrix may have 'ragged' edges, with some of the rows spanning different numbers of properties, or columns spanning different numbers of cases. The former might arise where sets of properties do not apply to all the cases and the latter where groups of cases are not all subjected to the same measurements.

Most datasets (and hence data matrices) are cross-sectional and single level, that is they relate to a single moment or period of time and to one type of case. However, some may be longitudinal in that measures have been taken at two or more moments of time or the whole study is repeated at a later date. Longitudinal data are vital if the research is to address issues of change, trends or sequences.

There are three main kinds of non-experimental longitudinal data: repeat cross-sectional, panel studies and cohort studies. Repeat cross-sectional studies use a different sample of cases for each time period, so each case is measured only once. Panels consist of respondents who have agreed to provide repeat measurements, for example of television viewing, but panel members may have been substituted or replaced to keep the panel balanced and representative. The focus of both repeat cross-sectional and panel studies is on changes in properties over time. Cohort studies focus on exactly the same cases over time and changes in the cases themselves, even if there are some cases missing at later dates. There are, of course, implications for the structure of the data matrix. For all three there will be two or more measures for each variable for each time period, but in cohort studies, there will be no new cases, in panel studies there will be a few new cases and in cross-sectional studies there will be a new set of cases for each period. For cross-sectional studies, it is likely that each study will be entered into a different data matrix and comparisons between time periods will be made purely in terms of distributions of values across cases, for example changes in average attitude scores between studies. For panel studies and cohort studies, there may well be a mixture of such analyses and analyses based on case-by-case comparisons.

The cross-sectional alcohol marketing dataset that is used throughout this text is part of a wider longitudinal study that the researchers call a two-stage cohort study (Gordon et al., 2010b) in which the initial 920 young people were followed up two years later to see whether, for example, awareness of and involvement with alcohol at age 13 is predictive of whether or not those who were non-drinkers in the first study became drinkers in the follow-up or drinkers became more frequent drinkers. In all, 522 of the original 920 were successfully followed up. No new cases were introduced.

Multilevel datasets treat cases as nested within higher order units, for example individuals within organizations within regions. The simplest are two-level datasets, for example children in schools who are within different classes. Here, the behaviour and attitudes of the children are no longer independent of the class they happen to be in. The variation in individual continuous metric values may be greater between classes than it is within classes, for example. Two-level variable-based models can be handled with traditional statistics, but introducing further levels makes the analysis fiendishly complex. SPSS has a procedure called `Mixed Models` that can be used for both hierarchical and longitudinal effects.

Datasets can expand and grow into very large datasets which may be linked together with other datasets in a data warehouse. Researchers sometimes talk about 'big data', a term used to describe the huge volumes of data generated by traditional business activities and from new sources such as social media. Typical big data include information from store point-of-sales terminals, bank ATMs, Facebook posts and YouTube videos. Such data may subjected to what may be referred to as data mining, in which companies use sophisticated software to look for hidden patterns, trends and other insights.

Key points and wider issues

A data matrix is a framework of rows and columns for storing numeric data in a table. Cases are allocated to the rows, properties are allocated to the columns, and values are recorded for each case for each property in the cells created by the intersection of rows and columns. The matrix is usually constructed electronically by entering values into a data analysis package like SPSS as was explained in Chapter 2. Matrices may be large or small, they are generally rectangular, but may be cross-sectional or longitudinal and they may be single level or multilevel.

The construction of a data matrix is an essential preliminary to data analysis. However, some analyses are performed by analysing the properties as variables and looking at the distributions of values down the columns, going across the cases. Such variable-based analyses are described in Part Two of this book. It is quite possible to perform distributional analyses on set membership data, but there is no particular advantage in doing so and it is seldom undertaken.

Other analyses are performed by analysing the cases across the rows and looking at the configurations of values across each property for each case. Usually this is undertaken on properties as set memberships. This makes it possible to determine which configurations (combinations of set memberships) are most frequent, which ones do or do not exist, and, more importantly, which configurations may be sufficient for an outcome always or nearly always to arise. Case-based analyses are the focus of Part Three. Software that is designed for configurational analysis cannot use properties recorded as variables unless they are converted into set memberships.

Data analysis

Data analysis is not just about performing statistical calculations on numerical variables; it is about making sense of a dataset as a whole and thinking about a range of alternative ways of approaching its analysis, taking a well-rounded view of what all the evidence is saying.

This may mean combining evidence from the dataset itself perhaps with results from other datasets, with past research, with informal individual experience, knowledge and intuition. Analysis, furthermore, is seldom a one-off enterprise. Researchers seldom do it all at once; rather they move backwards and forwards between constructed data and the objectives for which the research was undertaken, often on different occasions or at different phases of the research. Analysis becomes a dialogue between ideas and evidence. Ideas may be vaguer than theories and will include our speculations about the world around us, tentative generalizations based on past experience or understandings, our explanations, even musings about how or why things happen. Similarly, evidence is a broader concept than data: evidence is often tentative, incomplete, inconclusive, circumstantial and frequently needs to be assembled or fitted together jigsaw-like with other clues, traces or everyday details of the entities that are under investigation to create a more complete image, projection or representation of social phenomena. Ideas and evidence interact; ideas enable researchers to derive implications from the evidence and, in turn, researchers use evidence to extend, revise, support or test ideas. The end product of this dialogue is the establishment of links between theory and data once ideas have been formalized and evidence is constructed into systematic records.

Although data construction and data analysis are separate processes, they are nevertheless interdependent. Both need to reflect the objectives of the research, while the analysis procedures used must bear in mind how the data were constructed, in particular the measurement and scaling procedures used, the likely sources of error and the structure of the overall dataset in terms of cases, properties and values. The fact that data construction and data analysis are separate procedures, however, means that good-quality data can be used for a variety of purposes using a number of different data analysis techniques. At the same time, poorly constructed data will mean that whatever the sophistication of the techniques of analysis used, the results will be suspect. This is sometimes referred to as the phenomenon of GIGO – Garbage In, Garbage Out.

Data analysis may be thought of as consisting of a number of elements, not all of which are necessarily present or accomplished in the same order, but elements that are essentially the same, whether the data are quantitative or qualitative, or whether the properties are being treated as variables or set memberships. These elements are illustrated in Figure 3.2.

In their raw form, quantitative data that have been captured will consist of stacks of completed paper questionnaires or diaries, entries into an online questionnaire or records of various kinds accessed from secondary sources or constructed by researchers themselves. Before data can be analysed using techniques appropriate for quantitative data, they will need to be prepared or transformed in various ways explained in the previous chapter to make them ready for analysis.

Data description entails producing an account of a dataset either by reviewing the distribution of values down each column of the data matrix, one property at a time, or by considering the combinations of values across the properties, one case at a time. Variable-based analyses, which are explained in detail in Part Two of this book, will display the distributions of variable values in charts or graphs so that it is possible to 'eyeball' the data at a glance, or such analyses will summarize or reduce those data by generating calculations that encapsulate the essence of the story the data are telling. For metric variables, researchers may, for example, calculate an average and a standard deviation to describe the central tendency and dispersion of the values for a variable, or, for categorical variables, a percentage or proportion of the total accounted for by each category. Such univariate analyses are explained in Chapter 4. For case-based analyses, which are explained in detail in Part Three, the researcher may, for example, summarize the number of times particular combinations of variable values or set memberships occur.

While data summaries stay close to the original data and encompass the physical, mechanical, functional part of handling the data, interpretation moves on to the conceptual, thinking part. To interpret is to reflect on the meaning of,

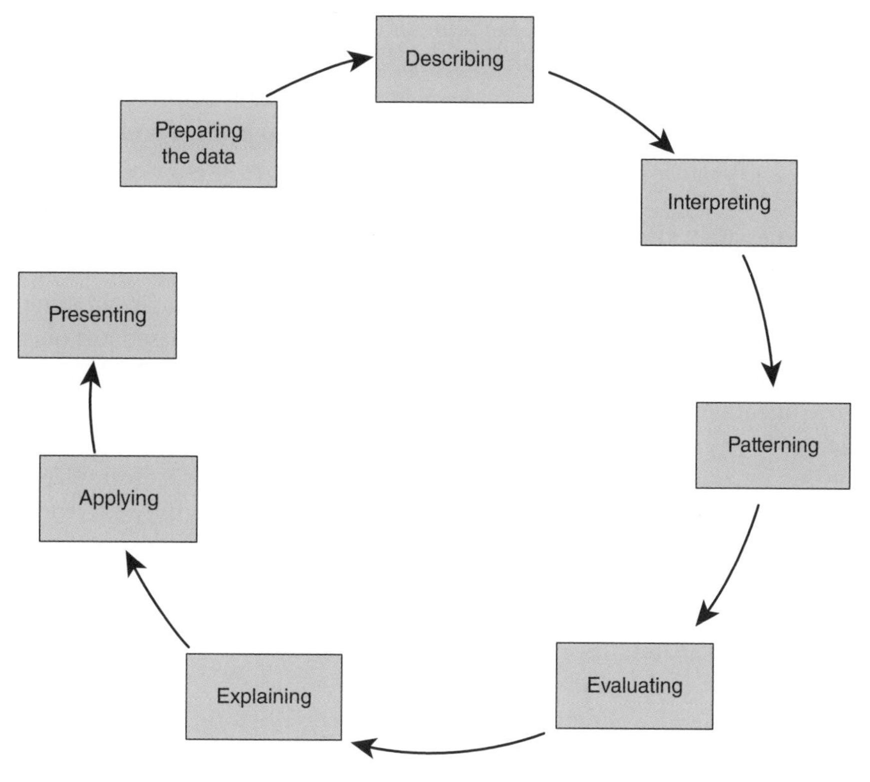

Figure 3.2 The processes of data analysis

to elucidate, unfold, show the purpose of, or translate into more familiar or intelligible forms. Interpretation can be creative, inventive or imaginative and will be less cautious methodologically than data summary. It still means staying with properties one at a time or cases one at a time, but considering the relevance or importance of the data, perhaps in the context of other research findings, of theory, of a particular audience or of a client. It may mean considering the validity and reliability of measurements taken, reviewing the likely accuracy of estimates of population values based on samples or testing the statistical significance of statements that relate to a single variable. These processes are explained in Chapter 4. It may, alternatively, mean reviewing the diversity and frequency of particular combinations of values or set memberships in a dataset and how that relates to theory, or attempting to understand why some configurations, though possible and plausible, have no empirical existence in the dataset. Chapter 7 looks at set-theoretic methods and configurational analysis.

Discovering or examining patterns in a dataset may entail reviewing changes, trends or sequences of events or measuring the extent of relationships between properties in terms of the manner in which the values of properties in a dataset differ, co-vary, cluster or combine together. Like data summary, this process will stick closely to the actual data in the dataset. Relating values as variables will entail measuring the size of differences, the degree of covariation, correlation or clustering and, when based on random samples, the statistical significance of discovered relationships. The variety of relationships between variables is outlined in Chapter 5 and taken up again in Chapter 6. Relating values as set memberships means considering which combinations of set memberships are associated with specified outcomes and assessing the extent to which particular configurations are consistent with notions of causal sufficiency or necessity. These issues are taken up in Chapter 7.

Evaluating discovered patterns means assessing how well they 'stack up' against research objectives, whether exploratory or verificational, whether as research questions or as research hypotheses. Neither variable-based nor quantitative case-based data analysis procedures themselves demonstrate the nature or degree of influence of one or more properties on any other properties. Even which properties count as conditions or causes and which ones are considered to be outcomes or effects are for the researcher to determine or hypothesize. The existence of causal connections can never be observed; it can only be rendered plausible from evidence in the dataset. The analysis of causality is very complex and the topic is taken up in more detail in Chapter 10.

To explain is to persuade an audience that understanding is being offered. A 'successful' explanation means that an audience subsequently understands that which it did not understand previously. To explain is to offer a rhetoric to readers, listeners, clients, colleagues or academic peers. Rhetoric is the theory and practice of eloquence; the art of using language to persuade others. A range of scientific rhetorics is available to quantitative researchers and these will include the rhetorics of causality, experimental design, statistical variable-based rhetoric and the rhetoric of quantitative case-based approaches. These rhetorics are explained in Chapter 10.

The application of the results of data analysis may mean reviewing the implications of findings for extant theory, for future research, for clients, for approaches to public policy, for medical practice or for commercial activity. The presentation of findings to live audiences or in journals or books may not normally be seen as part of 'data analysis', but if data analysis is seen in the broad sense of how data relate to research problems, then arousing the interest of audiences and persuading them of the veracity of the conclusions is part of the process.

For analysing **qualitative data** (as text and images) the same processes of data analysis may be applied. Such data still need to be prepared. They need to be assembled, for example, from voice recordings, still or moving images, field notes or previously recorded qualitative data. They may be entered into one of a range of software packages, usually referred to as computer-assisted qualitative data analysis software like NVivo or MaxQDA. Data description entails reviewing transcripts, tapes, photos or DVD clips on a case-by-case basis, accepting what individuals are saying at face value and producing a summary – perhaps including quotations from respondent comments as displays – of what people said, much as a journalist would for a magazine or newspaper. Interpretation for qualitative data might entail going beyond what respondents wrote or said to elucidate individual meanings, contexts and processes. Relating means considering the interweaving of the key factors involved in the focus of the study and looking for patterns, trends and changes. Evaluating involves reflecting on the believability or robustness of the conclusions drawn from the data.

Qualitative researchers might explain social phenomena by offering or deploying the rhetorics of case studies, qualitative comparative analysis, narrative analysis, phenomenology, ethnomethodology, and so on. Similarly, there are issues in applying and presenting the results from qualitative data analysis.

Key points and wider issues

Data analysis is the process whereby researchers take the raw data that have been entered into the data matrix and create information that can be used to address the objectives for which the research was undertaken. The processes that that analysis entails will include at least preparing and describing the data. Some 'descriptive' analyses may stop at that, but to create information that is useful to policy makers or to clients or will help academics to understand social phenomena, it is usually necessary to go beyond just giving an account of a dataset so that data are, in addition, interpreted, related, evaluated, explained, applied and presented. This is true whatever the kinds of data, but which particular activities are involved within each of these tends to differ according to whether the data are quantitative or qualitative and whether the approach is variable-based or case-based. Most research will be a mixture of different kinds of data and approaches to their analysis will often be mixed as well. The idea of mixed methods is taken up in Chapter 9.

The processes outlined in Figure 3.2 are not really 'stages', 'phases' or 'steps'. They are often not carried out in a linear fashion; they will often overlap, they may be undertaken simultaneously, or sequentially, but in different sequences. Yet, at the same time, they are not entirely non-linear. There is a 'logic' to some sequences so that data summary may be necessary before relating or interpreting can take place, although researchers may then return to description for its enhancement or emendation at a later stage. Researchers individually or in teams may have their own preferences for 'phases' of research; some of these process may be left out in part or in their entirety. Researchers may proceed from description to application or miss out description to go straight to relating or evaluating. There may be tendencies for researchers to go through particular sequences or phases in their data analysis, but it is only a tendency, not a rule that says that these are the 'correct' steps.

A word or warning here. Some researchers talk about 'secondary analysis', but this is open to many interpretations. It may mean the reanalysis of a dataset created by other researchers and may be subjected to the various processes outlined above. It may, alternatively, refer to the analysis of data accessed from, for example, government statistics. These will consist of tables of figures that cannot be reanalysed using variable-based or case-based procedures unless there is access to the original micro-data, that is, data that refer to each individual case. If these data are not available then secondary analysis really means the re-presentation of existing tabled data in a way that provides supporting evidence towards a researcher's research objectives.

Ethical issues

Ethics are moral principles or standards that guide the ways in which individuals treat their fellow human beings in situations where they might cause actual or potential harm, whether economic, physical or mental. Ethics in social research are concerned with professional standards of conduct and with the use of techniques in ways that avoid harm to respondents, to clients or to other parties. Ethical standards are important in a research context so that those involved in the research appreciate what is and what is not acceptable behaviour.

Social researchers often depend on the goodwill and participation by the public for the effective practice of their profession. At the same time members of society are becoming increasingly aware of their rights and sensitive about invasions of their privacy. Any individual, company or agency that violates the implicit trust of participants in a study makes it more difficult and more costly for all social researchers to approach and recruit future participants. Good ethical standards are good business. In consequence, various associations whose members are involved in social research have developed codes of conduct to guide the behaviour of their members.

The main ethical issues that arise in the conduct of social research concern:

- privacy;
- confidentiality;
- deception;
- imposition;
- integrity;
- misrepresentation.

If a researcher telephones a respondent to obtain an interview late on a Sunday evening, or if a researcher observes a customer's behaviour without the customer's knowledge, are these instances of invasion of privacy? It might be argued that since anybody can observe behaviour in public places, then the latter example is not unethical, particularly since no harm is involved. In the former case there may be mental harm if the outcome is the annoyance of the householder.

The issue of confidentiality might affect participants in research and, if the research is commissioned, the clients. If respondents in a survey are told or reassured that their replies will be treated with confidentiality, then it will be unethical for researchers to pass on this information to other parties, for example by selling mailing lists. If a client does not wish to be identified to respondents, then it would be unethical for interviewers to pass on this information. If the researcher is working on behalf of a particular client, then confidential information about the business should not be passed on to competitors.

Deception may come in many forms. Misleading a respondent into thinking that an interview will take 5 minutes when the researcher knows it will take 20 is unethical. Covertly numbering questionnaires that are meant to be 'anonymous' so that the researcher can determine who has and who has not returned them could also be seen as deception.

The rights of respondents to be able to refuse to grant an interview need to be respected and they should not be pressurized. There should be no adverse effects that result from participating in research, like receiving unsolicited sales material or price rises that result from questions about what maximum prices people would accept for a product.

Integrity includes both the technical and administrative integrity of the research so that the results are not 'doctored' or 'massaged' in any way or tied up in jargon just to baffle the audience, readers or clients. Lack of integrity shades into misrepresentation in which research results are presented in a way likely to mislead readers or clients. This might include deliberately withholding information, misusing statistics or ignoring relevant data. One form of dishonesty that is of concern to academics is plagiarism. This entails the passing off of somebody else's ideas as your own. Its most detectable form is the 'lifting' of text from published sources without acknowledgement of that source. The availability of published work online has made this process only too easy. However, special software has been developed that can detect the copying and pasting of text in this manner. This software is increasingly applied to student dissertations and theses, which are being submitted in electronic form.

Authors of textbooks are usually given guidelines by publishers, for example that authors must seek permission for the reproduction of any tables, charts, illustrations, statistics, quotations or adaptations from any text or material that is not completely the authors' own work. The only exceptions to this are quotations that fall into a 'fair dealing' category; this includes extracts that are used solely for the purposes of criticism or review and should be of modest amounts. What counts as 'modest' depends on the nature of the original work.

Many of the ethical issues that arise in the conduct of social research are covered by data protection legislation. The 1995 European Data Protection Directive is concerned with the rights of individuals who are asked for or who provide information about themselves. In particular it seeks to protect their rights to privacy by establishing the core principles of transparency and consent. Transparency means ensuring that individuals have a clear understanding about why the data are being collected or requested and how they will be used. Individuals must, furthermore, give their consent to data collection and be given the opportunity to opt out of any subsequent uses of the data.

The implementation of the EU Directive has varied across countries, which causes problems for international research – it causes confusion, adds to costs and undermines the free flow of research services. In the UK it is enshrined in the 1998 Data Protection Act. The Act requires that all those bodies who collect personal data (the data controllers) notify the Data Protection Commissioner that they intend to hold personal data and they will need to meet fairly stringent conditions. The Law covers the actions of all companies in the EU and any transfer of data to other countries outside of the EU. Transferring completely anonymous data is acceptable, but personal data, even a sampling frame, must have the same rules observed by end users as if they were in the EU.

In terms of the construction and analysis of datasets, which have been the main focus of this chapter, it is, for example, only too easy to frame categories of answer to a question in such a way as to manipulate the results. Thus a customer satisfaction survey could give respondents more 'satisfied' categories than 'dissatisfied' ones, thereby biasing results in a positive direction.

Implications of this chapter for the alcohol marketing dataset

To date, all the publications relating to this project have focused on analyses that are variable-based rather than case-based. Part Two of this text takes you through the implications of variable-based analyses and Part Three offers novel case-based approaches to the dataset, including set-theoretic methods, fuzzy set analysis, cluster analysis and discriminant analysis. All the components of data analysis outlined in this chapter will be displayed throughout the remainder of this text.

The fact that this study involves interviewing young people aged between 12 and 14 raises particular ethical issues of transparency and consent. Certainly the consent of parents was sought with a consent form for them to sign, while an information sheet attempted to make the objectives of the study transparent. Certainly in the publications of the results, no individuals and no schools are identified.

Chapter summary

A record of all the values for all the cases for all the properties in a research project constitutes a dataset, which will usually be laid out in the form of a data matrix, a framework of rows and columns for storing numeric data in a table. All data analyses begin with such a data matrix, but variable-based analyses focus on the columns, looking at the distributions of property values across the cases, whereas case-based analyses focus on the rows, reviewing the combinations of values across the properties. The analysis of any kind of data, whether qualitative or quantitative, whether variable-based or case-based, at a minimum includes data preparation and a description of the properties or of the cases in a dataset, but is also likely to include elements of interpreting, pattern seeking, evaluating, explaining, applying and presenting results to an audience. The analysis of any dataset needs to be approached holistically, that is as a complete, self-contained entity, set in the context of the objectives for which the research that generated the dataset was designed to achieve and with a well-rounded view of what all the evidence is saying. Data analysis means getting the most out of a dataset, approaching it in several different ways so that the data tell the complete story. These processes raise a number of ethical issues, for example of privacy, confidentiality, deception, imposition, integrity or misrepresentation.

Exercises and questions for discussion

1. Is data analysis any more than choosing the right kinds of statistics to apply to a dataset?
2. Access the alcohol marketing dataset which is available at https://study.sagepub.com/kent and check out the values that have been entered for the codes for the nine variables in Figure 3.1 for the non-metric measures. From Figure 3.1, try to summarize each variable by looking down each column. Try to summarize each case by looking across the variables.
3. What, do you think, are the key ethical issues raised by the alcohol marketing study?

Further reading

Diamantopoulos, A. and Schlegelmilch, B. (1997) *Taking the Fear Out of Data Analysis*. London: Dryden Press. Republished by Cengage Learning, 2000.

Read Chapter 6, which emphasizes the connection between research objectives and what the authors call analysis objectives. Beware, however, that the latter, apparently, amount to a choice between description, estimation and hypothesis testing. As is argued in Chapter 1 in this text, any research is likely to involve all these elements – and more, as we will see.

Kent, R. (2007) *Marketing Research: Approaches, Methods and Applications in Europe*. London: Thomson Learning.

Very few texts discuss the nature of data analysis in general. However, have a look at pages 296–9, which include some of my earlier thoughts on this topic, but these comments are largely within the context of deciding on appropriate variable-based techniques of analysis.

Suggested answers to the exercises and questions for discussion can be found at the end of this text, pp. 293–321, and on the companion website, (https://study.sagepub.com/kent), which also give links to relevant free online Sage journal articles, PowerPoint slides, an overview of data analysis packages, an introduction to SPSS and weblinks to alternative datasets.

II

VARIABLE-BASED ANALYSES

4

UNIVARIATE ANALYSIS

Learning objectives

In this chapter you will learn about:

- displaying categorical and metric variables in tables, charts and graphs;
- calculating statistical summaries for categorical and metric variables one at a time;
- making estimates derived from samples;
- testing univariate hypotheses for statistical significance;
- using SPSS for univariate analysis;
- how univariate analysis can be applied to the alcohol marketing dataset. In particular, you will discover that a majority of respondents say that they have never had a proper alcoholic drink and have no intention of doing so in the next year. However, there is high awareness of the marketing of alcohol, but little actual involvement.

Introduction

We saw in Chapter 3 that the analysis of a data matrix may be approached by focusing either on the distribution of values down the columns or on the configurations of values across the rows. The former uses properties as variables and will be referred to as 'variable-based' analysis. This form of analysis is the traditional approach and is the focus of most books on data analysis or statistics and also the focus of Part Two in this text. Chapter 7 introduces the configurational approach, which is seen as one of a number of approaches that focus on the cases rather than the variables and is an approach referred to here as 'case-based'.

Chapter 3 explained that any analysis of data will include the activities of describing, interpreting, relating, evaluating, explaining, applying and presenting a range of different kinds of data. It is also iterative: researchers go backwards and forwards between the data and the research objectives. A first move in this iterative process for variable-based analyses is commonly – and sensibly – to undertake univariate analysis, which is the focus of this chapter. Univariate analysis means producing a description of a dataset by reviewing the distribution of values down each column of the data matrix one variable at a time and might include displaying each variable in tables, charts or graphs and then summarizing it by generating calculations that encapsulate the essence of the story the data are telling. The analysis might also include interpreting the likely accuracy of estimates made from a random sample relating to a particular variable or testing the statistical significance of statements made about a single variable for a population of cases. At this point, the researcher may, in addition, reflect upon the implications of this univariate analysis for the research objectives, for theory development, for audiences or clients and for the next steps to be undertaken in the analysis of the data.

Univariate data display: categorical variables

Univariate data display means presenting the distribution of values inside variables one at a time in tables, charts or graphs so that it is possible for researchers to 'eyeball' the total distribution on each. As a first, descriptive, stage in data analysis it is likely that the focus will be on producing tables for each variable; as a final stage in the presentation of findings to clients, colleagues or other audiences, it is more likely that tables, charts and graphs will be used selectively to illustrate key or interesting results.

Categorical variables consist of two or more categories that encompass all the observations or records made for a dataset and are non-overlapping. However, they possess no metric with which to measure distances between recorded values. Chapter 1 explained that they may be sub-divided into binary, nominal, ordered category or ranked measures. The main techniques that are available for their display one at a time include frequency tables, simple bar charts and pie charts.

Frequency tables

A table is any layout of two or more values in rows, in columns, or in rows and columns combined. For binary, nominal and ordered category variables, the researcher can display the frequencies and percentages of the various categories. For binary variables, there will be only two categories giving, for example, the number of males and females in the alcohol marketing study, as shown in Table 4.1.

When reporting the findings from binary variables in the text that accompanies a table, it is really only necessary to state one of the two percentages (or proportions). If, for example, 47.1 per cent are male then, by definition, 52.9 per cent are female. For such binary data it could be argued that it is not really necessary to tabulate the results at all. For nominal variables there will often be

Table 4.1 Gender of participants in the alcohol marketing study (format for publication)

	Frequency	Per cent
Male	433	47.1
Female	487	52.9
Total	920	100.0

three or more categories. Precisely because the measure is nominal it does not matter in what order the categories are placed. However, if the number of categories is large it may make sense to list or group them in a particular way, for example alphabetically or by frequency. Where the categories have been coded, then SPSS will, as a default, put them in the order of the code numbers assigned with the lowest code number at the top, but these can be reversed or the categories can be ordered in terms of size or frequency.

For ordered category variables, the values should be in the intended order of magnitude, usually with the highest, largest, fastest, and so on, at the top. Figure 4.1, which is in the format presented by SPSS, for example, shows the degree of importance attached to choosing a popular, well-known brand when deciding to buy alcohol. 'Very important' in the dataset has been given a code of 5, 'Important' a code of 4, and so on, so to get SPSS to put high importance at the top of the list of values, the order of the values was reversed. How to do this is explained in Box 4.1. Note also, for example, that there were 336 who indicated that this was 'Very unimportant', which constituted 36.5 per cent of the 920 cases, but 43.4 per cent of the 774 who gave an importance rating. `Valid Percent`, in short, is the percentage of non-missing values. `Cumulative Percent` adds up the `Valid Percent` in order so that, for example, a total of 19.8 per cent thought choosing a well-known brand either very or quite important.

		Frequency	Percent	Valid Percent	Cumulative Percent
Valid	Very important	36	3.9	4.7	4.7
	Quite important	117	12.7	15.1	19.8
	Neither important nor unimportant	160	17.4	20.7	40.4
	Quite unimportant	125	13.6	16.1	56.6
	Very unimportant	336	36.5	43.4	100.0
	Total	774	84.1	100.0	
Missing	DK	146	15.9		
Total		920	100.0		

Figure 4.1 Importance of choosing a popular, 'well-known' brand when deciding to buy alcohol

Figure 4.1 illustrates a number of points about table layout:

- the response categories, the frequencies and the percentages are usually in columns;
- the columns are labelled at the top so that we know whether the figures represent frequencies, percentages or proportions;
- totals are given at the bottom of the columns;
- there is a table or figure number and a title, which is usually at the top of the table;
- there may be a source of the data at the bottom of the table if these data were derived from other than the results of the research being reported.

In terms of interpretation, some researchers might find it 'surprising' that nearly 60 per cent of those responding said that branding is either quite unimportant or very unimportant and under 20 per cent (19.8 per cent of those responding) said that it was important. This might run counter to accepted wisdom about the effect of marketing on brand choice; it might also, of course, reflect a reluctance to admit to such effects among those responding to a questionnaire.

Some tables might contain two or more adjacent univariate distributions, as in Figure 4.2. Such **multi-variable tables** do not relate or interlace the variables. Thus we do not know from Figure 4.2, for example, how many males have given which answers.

Looking at one-way tables enables the researcher to do two main things. First, the researcher can see whether any variables are failing to discriminate between respondents. If all, or most, respondents give the same answer, then, while it may be an 'interesting' finding, such a variable will be of limited value for further analysis. Second, and this is clearly related to the first point, if any categories are empty or very low in frequency, then the researcher will need to consider whether the values can be sensibly collapsed or recoded into a smaller number of categories (see Figures 2.4 and 2.5 for an example).

		Count
Sex of respondent	Male	433
	Female	487
Brand importance for alcohol	Very unimportant	336
	Quite unimportant	125
	Neither important nor unimportant	160
	Quite important	117
	Very important	36

Figure 4.2 Gender and brand importance

Simple bar charts and pie charts

'Chart' and 'graph' are terms that are often used interchangeably to refer to any form of graphical display; however, 'graph' is more sensibly reserved for metric variables. Charts for categorical variables are limited largely to bar charts and pie charts. In **bar charts** each category is depicted by a bar, the length of which represents the frequency or percentage of observations falling into each category. All bars should have the same width, and a scale of frequencies or percentages should be provided. What each bar represents should be clearly labelled or given a legend. A simple bar chart is illustrated in Figure 4.3, which shows the distribution of answers on the importance of choosing a well-known brand of alcohol, based on the same data as in Figure 4.1. Note that SPSS has, once again, put the bars in the order of the allocated codes so that it reads in the reverse order from Figure 4.1. However, there is an option to transpose the co-ordinate system that will construct the bars horizontally and put 'Very important' at the top.

In a **pie chart** the relative frequencies are represented in proportion to the size of each category by a slice of the circle. The bar chart, however, is normally preferred to the pie chart because the human eye can more accurately judge length comparisons against a fixed scale than angular measures. However, pie charts are nicer to look at and they clearly show that the total for all slices adds up to 100

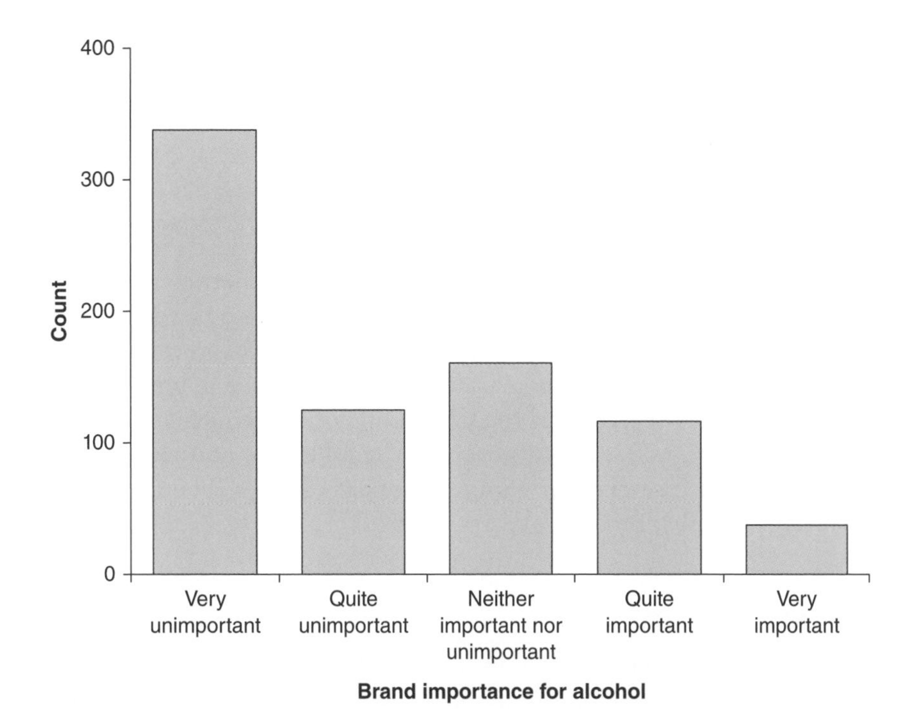

Figure 4.3 Importance of choosing a popular, 'well-known' brand when deciding to buy alcohol

per cent. Neither bar charts nor pie charts, however, are particularly useful for binary data. For nominal variables they can certainly help to display the data where there are between about 4 and 12 or so categories. Bar charts in particular can preserve the order of the categories for ordered category variables, but the sense of ordering may be lost in pie charts.

Univariate data display: metric variables

Metric variables possess a unit of measurement and may result from a process of counting the number of times a characteristic or an event occurs or from a process of calibration that gives rise to a continuous variable. The main techniques that are available for their display one at a time include frequency tables, metric tables, histograms and line graphs.

Frequency tables

Unlike categorical variables, metric variables, whether discrete or continuous, will usually consist of many values. However, these can be grouped into class intervals and then used to produce frequency tables or bar charts. Of course, when metric variables are deployed in this way, the information about distances is being ignored or thrown away. So why would any researcher want to do that? Usually, it is because the researcher wishes to look at the relationship between a categorical and a metric variable. In this situation, the researcher will group the metric variable into class intervals and then use **crosstabulation** with the categorical variable (see Chapter 5).

Metric tables

Tables are sometimes used to display not frequencies of metric values or class intervals, but actual metric quantities. If you look at a set of company accounts, you will see tables that express turnover, profits, revenue, costs in terms of pounds sterling or euros, or sales may be expressed in tons, kilograms, litres or units sold. The figures in the tables will not be frequencies, but actual metric values like £5,687,000 turnover. Metric tables are normally used only when there is one case or a very limited number of cases (like three or four companies).

Histograms and line graphs

A **histogram** is a graphical display for metric variables in which the length or height of the bars represents the frequency with which each interval occurs and the width of the bars the size of the **class intervals**. Figure 4.4 shows the frequencies of total scores for the importance of branding. The majority of the scores are between about 20 and 32 and relatively few at the extreme ends. SPSS has in this case chosen a class interval of one since the total scores are discrete rather than continuous.

Notice that there are two cases that are separate from the rest, having a total score of 0. These cases can be seen as **outliers** – values that are substantially different from the general body of values. These can have a marked impact on the analysis of the data. Researchers should always examine their data for their presence and consider whether or not they should be deleted from the analysis. Whether it is sensible to do so will depend on the objectives of the research and on the characteristics of each outlier; for example, could they be individuals who have misunderstood the question or what they are being asked to do?

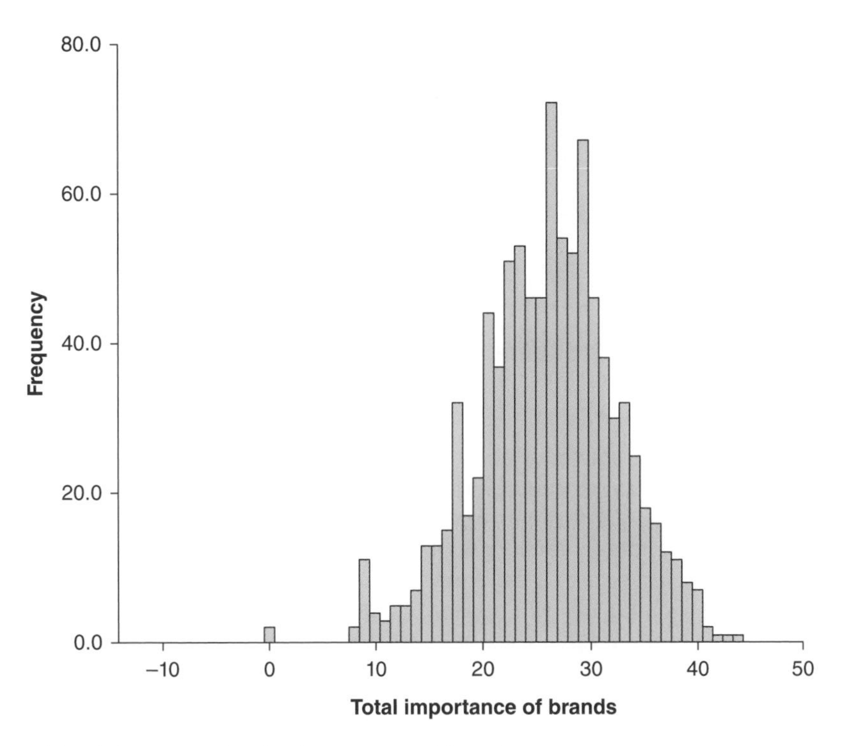

Figure 4.4 Total scores for the importance of brands

A **line graph** can be used to represent the frequencies or percentages of metric variables, as in Figure 4.5, but they are not particularly helpful when used in this way since it does not give much feel for the shape of the distribution. They are of more use as time graphs where the X-axis is used for elapsed time or for dates and the Y-axis for the quantities of a variable that are changing over time.

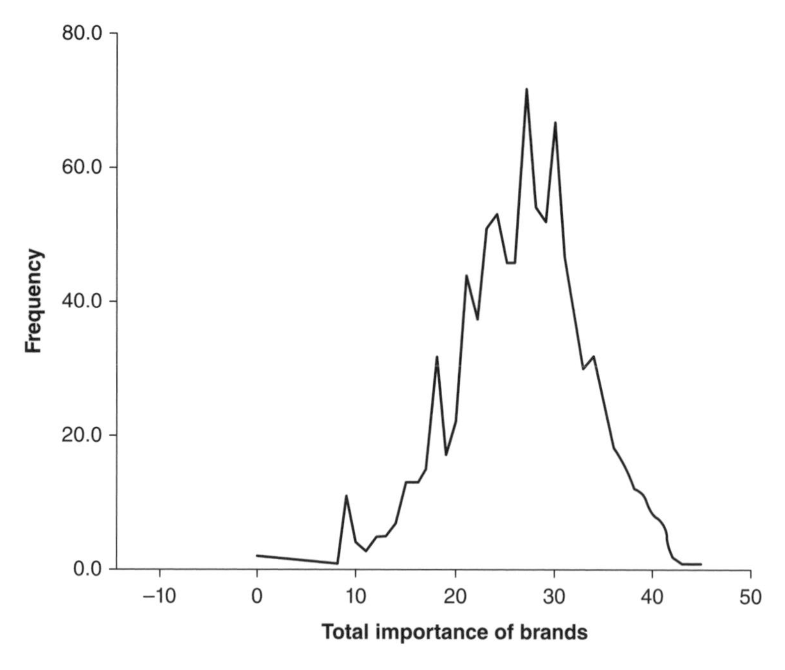

Figure 4.5 Total scores for the importance of brands

Box 4.1 Producing frequency tables, charts and histograms using SPSS

To obtain univariate frequency tables for categorical variables, you will need the SPSS `Frequencies` procedure. This is in the `Analyze | Descriptive Statistics` drop-down menu from the menu bar at the top. So, access the alcohol marketing data, click on `Analyze`, then move the pointer to `Descriptive Statistics` and then to `Frequencies` and left click. The `Frequencies` dialog box will appear. All variables are listed in the left box. To obtain a frequency count for any variable, simply transfer it to the `Variable(s)` box by highlighting it, then clicking on the direction button in the middle. To highlight blocks of adjacent variables or all the variables, hold down the shift key when highlighting the first variable, scroll down to the last variable in the block and click again. Figure 4.1 illustrates the default table produced by SPSS. Note that this is different from Table 4.1, which is in a format usually used for publication. To change the order of the categories presented in a table, from the `Frequencies` box select `Format` and then select the `Descending values` radio button, then click on `Continue` and `OK`.

Where there are no missing cases the `Percent` and `Valid Percent` are the same. You can edit the table to remove these columns if you wish. Just double-click on the table. This puts you into editing mode and gives you the `Formatting Toolbar`. Highlight the figures in the column (drag the pointer down the figures) and hit `Delete`. The column will disappear and the table will close up. You can do the same with the Cumulative Percent column. To get out of Edit mode just left click outside the table area. With the table highlighted (single click on the table – there will be a frame around it and a red arrow to the left) you can select Edit and Copy and then Paste it into any other application like Word or PowerPoint. Use the Paste Special procedures in PowerPoint or Word, otherwise these programs will try to edit the tables.

The Frequencies procedure will produce a separate table for each variable entered into the Variables box. To produce a multi-variable table, select Analyze | Tables | Custom Tables. Drag your first variable into the Rows box, then drag the next and subsequent variables to the foot of the lowest table shown. Alternatively, if the variables to be tabled are listed next to one another, highlight them all (by holding down the shift key) before dragging across. If the response categories for a number of variables are all the same and you want a table that sets out the responses as a matrix, create a multi-variable table as above, then under Category Position, select Row Labels in Columns. The result is shown in Figure 4.6 for the nine items relating to brand importance.

	Very important	Quite important	Neither important nor unimportant	Quite unimportant	Very unimportant
	Count	Count	Count	Count	Count
Brand importance for chocolate or sweets	58	244	285	208	117
Brand importance for fizzy drinks	66	274	249	220	104
Brand importance for crisps	47	234	287	232	103
Brand importance for trainers	319	357	131	74	34
Brand importance for clothes	330	362	121	78	24
Brand importance for magazines	57	242	282	200	115
Brand importance for perfume/Aftershave	197	413	153	100	48
Brand importance for cigarettes	33	79	153	124	377
Brand importance for alcohol	36	117	160	125	336

Figure 4.6 Brand importance responses

To obtain bar charts and pie charts, click on Charts in the Frequencies dialog box. Simply click on Bar Chart and indicate whether you want the axis label to display frequencies or percentages, click on Continue and then OK. This will give you a basic default bar chart in addition to the frequencies table. To obtain other kinds of bar chart like 'stacked' or 'clustered' bar charts or three-dimensional charts, you will need to select the Graphs drop-down menu. You can then choose between the Chart Builder, which gives you a kind of chart wizard, or the dialog boxes that were in previous versions of SPSS and which are now called Legacy

(Continued)

(Continued)

Dialogs to give you the Bar Charts dialog box. If you choose Chart Builder, drag the kind of chart you want into the Chart preview box, then drag the variable you want across to the X-Axis. The Y-Axis will change to Count. Click on OK.

Once you have obtained your chart you can edit it by double-clicking in the chart area. This will give you the Chart Editor. You can change the colours and a number of other chart features from the editor. Close it when you have finished. If you single-click on the chart area, you highlight it and it can be copied into other applications.

The Chart Builder can be used to produce both histograms and line graphs for metric variables. You can impose a **normal curve** on the histogram by clicking in the Display normal curve box in the Element Properties dialog box and clicking on Apply. The normal curve is explained later in this chapter.

Key points and wider issues

When a researcher is looking at a dataset that is ready for analysis, the first thing he or she will normally do is to produce tables of frequencies for each variable. This should give the researcher a feel for the distribution of each variable. From this it will become clear whether any further transformations of the variables in ways that were explained in Chapter 2 need to be applied. In particular, metric variables may need to be transformed into class intervals before they can be tabulated. Researchers will commonly use a selection of univariate tables, graphs and charts to describe the key features of the survey population or sample and selected 'interesting' results when they are presenting their analysis in a report or to an audience. When using SPSS, beware that the order of the categories presented in tables or charts will be determined by the codes allocated, beginning with the lowest code. The order can be reversed, but ensure that this has been done on all tables and charts.

Data summaries: categorical variables

The display of variables one at a time will give researchers an overview of the distribution of each. However, it is often useful to calculate a single summary measure that indicates a key characteristic or feature of a distribution. The calculation of univariate summary measures for categorical variables is, however, fairly limited. For binary variables, the percentage or proportion of cases that possess a characteristic can be seen as a summary measure. For nominal variables, the measure usually mentioned is the mode or modal category. This is the most commonly occurring value – the category with the highest frequency. It is a kind of 'average' that expresses the typical observation in a distribution of category frequencies, but the notion of 'centrality' is very weak, particularly since the categories may be arranged in any order and the category with the highest number is not in any sense in the 'middle' of a distribution. In this context, the mode is better thought of as the 'best guess' for a distribution rather than as a measure of central tendency. For example, in Figure 4.7, the modal category is 'Christian' for declared religion, which has the highest frequency. If you were asked to guess what is the stated religion for any particular individual, then in the absence of other information, you would have the least chance of being wrong by guessing the modal category. How good the modal category is as a guess depends on what proportion of the total set of cases is contained in the mode. For the variable religion, it is 62.4 per cent, so we are more likely to be right than wrong. The proportion in the modal category needs to be high to be good as a summary of the variable distribution. As we will see later in the chapter, the likelihood of being right multiplied by the likelihood of being wrong $[p(1 - p)]$ is an index of the degree of variation in a distribution. Maximum variation is when the modal category contains 50 per cent of the cases, leaving 50 per cent not in the modal category. Minimum variation is, of course, when all the cases are in the same category.

For ordered category variables, we could also indicate the modal category, but this ignores the ordinality in the data. Since ordered categories can, by definition, be arranged in order, the middle of that arrangement can be important as a summary. The middle or median category is the category that contains the 'middle' case. The middle case in a set of cases arranged in order occupies the $(n + 1)/2$ position, for example the fourth case up (or down) in a set of seven. Where the number of ordered categories is few (like high/medium/low, or a five-point Likert item), then unless the distribution is very unbalanced, the median category will always be the middle category. While this may be true, it may not be particularly insightful. Both the mode and the median are likely to be of more use when, for example, comparing the mode or the median for subgroups, like the mode or median category separately for males and females. They are also likely to be of more use when applied to metric variables, as explained below. For fully ranked measures, the median is always, by definition, the middle rank since each rank has a frequency of 1. Again true, but not particularly insightful.

		Frequency	Percent
Valid	Christian	574	62.4
	Other religion	46	5.0
	None	294	32.0
	DK/unstated	6	.7
	Total	920	100.0

Figure 4.7 Stated religion

Data summaries: metric variables

For metric variables, there is a large assortment of statistics for calculating univariate summary measures. These statistical techniques may be put into three main groups:

- central tendency;
- dispersion;
- distribution shape.

Central tendency

Central tendency focuses on the extent to which the metric values for a given variable are concentrated about a central figure. SPSS offers three measures of central tendency:

- the mean;
- the median;
- the mode.

Figure 4.8 shows the mean, median and mode for the total scores for the importance of brands. The **mean** is commonly what we understand by 'average'. It is calculated by adding all the values together to create what SPSS calls the sum, which is 24,446, and dividing by the number of values included (in this example, 920 cases), giving a mean of 26.57. The mean takes account of every value in the distribution. While this is normally seen as an advantage, it may be unduly affected by extreme values or **outliers**.

Another tricky feature of the mean arises when the measure is discrete rather than continuous. Being discrete, all the values will be integers, but the mean may be a fraction, like the average family having 2.4 children. Remember that the mean is appropriate only for metric data, but SPSS is very obliging and will, for example, happily take the variable 'liking of alcohol ads', which is ordinal, and calculate an average liking of 3.67. This is nonsense, of course (it is taking the assigned codes of 1–5 as metric values), but you cannot expect SPSS to know this.

The **median** for a metric variable is the middle point in an ascending or descending series of values with 50 per cent of the cases above that value and 50 per cent below it. It is that value which splits the observations into two halves.

N	Valid	920
	Missing	0
Mean		26.57
Median		27.00
Mode		27
Sum		24446

Figure 4.8 Measures of central tendency for total importance of brands

		Frequency	Percent	Valid Percent	Cumulative Percent
Valid	0	2	.2	.2	.2
	8	1	.1	.1	.3
	9	11	1.2	1.2	1.5
	10	4	.4	.4	2.0
	11	3	.3	.3	2.3
	12	5	.5	.5	2.8
	13	5	.5	.5	3.4
	14	7	.8	.8	4.1
	15	13	1.4	1.4	5.5
	16	13	1.4	1.4	7.0
	17	15	1.6	1.6	8.6
	18	32	3.5	3.5	12.1
	19	17	1.8	1.8	13.9
	20	22	2.4	2.4	16.3
	21	44	4.8	4.8	21.1
	22	37	4.0	4.0	25.1
	23	51	5.5	5.5	30.7
	24	53	5.8	5.8	36.4
	25	46	5.0	5.0	41.4
	26	46	5.0	5.0	46.4
	27	72	7.8	7.8	54.2
	28	54	5.9	5.9	60.1
	29	52	5.7	5.7	65.8
	30	67	7.3	7.3	73.0
	31	46	5.0	5.0	78.0
	32	38	4.1	4.1	82.2
	33	30	3.3	3.3	85.4
	34	32	3.5	3.5	88.9
	35	25	2.7	2.7	91.6
	36	18	2.0	2.0	93.6
	37	16	1.7	1.7	95.3
	38	12	1.3	1.3	96.6
	39	11	1.2	1.2	97.8
	40	8	.9	.9	98.7
	41	7	.8	.8	99.5
	42	2	.2	.2	99.7
	43	1	.1	.1	99.8
	44	1	.1	.1	99.9
	45	1	.1	.1	100.0
	Total	920	100.0	100.0	

Figure 4.9 Frequency distribution of scores for total importance of brands

For the variable total importance of brands the median is 27. You would have to imagine putting respondents in a row in order of the total score for brand importance and taking 921/2, so 50 per cent of cases are below a value that is midway between 460 and 461 and 50 per cent are above. One advantage of the median is that it is unaffected by extreme values.

The **mode,** as was explained above, is the most commonly occurring value. For metric variables, however, certainly if they are continuous, each value may occur only once. If the values are grouped, then there will be a modal class interval, but that may change with different intervals being used. For a discrete metric variable with many cases, the frequencies may be important. The modal score for total brand importance, for example, is 27, which has the highest frequency of 72, as can be seen from Figure 4.9.

Although SPSS offers the mean, the median and the mode as measures of central tendency, there are other measures that may be more appropriate on certain types of data. There are, for example, different kinds of mean apart from the arithmetic mean, including the geometric mean, the harmonic mean, the trimmed mean and the quadratic mean. There are also different kinds of median and different kinds of mode. However, the three measures offered by SPSS are the most commonly used and in practice the researcher is unlikely to need other measures of central tendency.

Dispersion

Dispersion refers to the amount of spread or variability in the data. The simplest measure is the range, which is the difference between the maximum and the minimum values. For total brand importance score this is 45 (Figure 4.10). The problem with the range, however, is that it does not tell us anything about the spread or shape of individual values either side of some central point.

A different approach is to measure the extent to which values depart either side of a central measure. The average of deviations about the arithmetic mean is a measure of dispersion called the mean deviation. This, statistic is, however, little used because it entails ignoring the sign, otherwise positive and negative deviations will, by definition, cancel one another out. It is more usual to square the deviations between each value and the arithmetic mean, sum the result, divide by the number of values and then take the square root. This is the **standard deviation**, which gives greater weight to extreme scores (since differences are squared) and will normally be greater than the mean deviation. The average score for total importance of brands is, as we have seen, 26.57. From Figure 4.9, you can see that

	N	Range	Minimum	Maximum	Std. Deviation
Total importance of brands	920	45	0	45	6.687
Valid N (listwise)	920				

Figure 4.10 Measures of dispersion for total importance of brands

one person has a total score of 45. This is 18.43 above the mean. The square of this is 339.66. This calculation is made for each case and the grand total is divided by 920 and then the square root taken, giving a standard deviation of 6.687 (see Figure 4.10). This is the average amount of dispersion about the mean and it tells us how well the mean represents the distribution of that variable. Provided the distribution of values has a central concentration, most of the values will be within two standard deviations of the mean, in this case between 13.2 and 39.9. The standard deviation is a statistic that is used a lot in calculations of statistical inference, as we will see later in this chapter and in later chapters.

Distribution shape

You can get a good idea of the actual distribution of a metric variable by getting SPSS to draw you a histogram, as in Figure 4.4, which shows the distribution of the scores for total importance of brands for a variety of products. Most of the values are close to the average score of 26.57. If a distribution is completely symmetrical, then the values of the mean, the median and the mode will be the same. However, if the distribution is skewed, they will not. If positively skewed, the mean will be higher than the median and the median will be higher than the mode. The reverse will be true if the distribution is skewed to the left. This gives a basis for a measure of the extent of **skewness**. This uses simply the difference between the mean and the median and is a measure of the asymmetry of the distribution. It will have a value of 0 if the median and the mean are the same.

A distribution that is of particular importance to statisticians is the **normal distribution**. This has a number of important features:

- it is symmetrical;
- it has a single hump;
- it forms a smooth curve;
- it has a zero or near-zero measure for skewness;
- it has two tails that in theory extend out to positive and negative infinity, never quite touching the baseline.

On account of the last feature, no distribution can be an exact normal curve since the actual highest and lowest scores will be finite. The normal curve is often described as 'bell-shaped', but may not be shaped like a bell – it may be very peaked or very flat. A measure of skewness which is less than one (or minus one) is generally taken as an indicator that the distribution is approximately normal in shape. The value calculated by SPSS for total importance of brands is minus 0.315, which shows that the distribution can be regarded as normal in shape.

Because of the shape of the normal distribution, over two-thirds of all observations will be within one standard deviation, over 95 per cent within two standard deviations and nearly all observations will lie within three standard deviations. The mean and the standard deviation together allow statisticians to describe the shape and position of any distribution. Furthermore, because the normal curve has a standard shape, it is possible to treat the area under the curve as representing total certainty that any observation will be encompassed by it. We can say, furthermore, that 50 per cent of the area is above the mean for that variable, and 50 per cent below. In other words, there is a 50 per cent chance that any observation will be above (or below) the mean. This argument can be taken further so that we can calculate exactly the area under a perfect normal curve between the mean and one standard deviation. The area is, in fact, 34.1 per cent (see Figure 4.11). Thus just over two-thirds or 68.2 per cent of the area is between plus one standard deviation and minus one standard deviation from the mean. Thus if the mean score of a set of cases is 20 with a standard deviation of 6, then just over two-thirds of the area (and, by implication, of the observations) would be within 20 ± 6 or between 14 and 26. Figure 4.11 also shows that all but 4.6 per cent of the area lies between plus and minus two standard deviations. There are, in fact, tables of areas under the normal curve, so that if we wished to know how many standard deviations encompassed exactly 95 per cent of the area, we could look it up and discover that 1.96 standard deviations either side of the mean do so. This characteristic becomes very important when we move on to consider statistical inference.

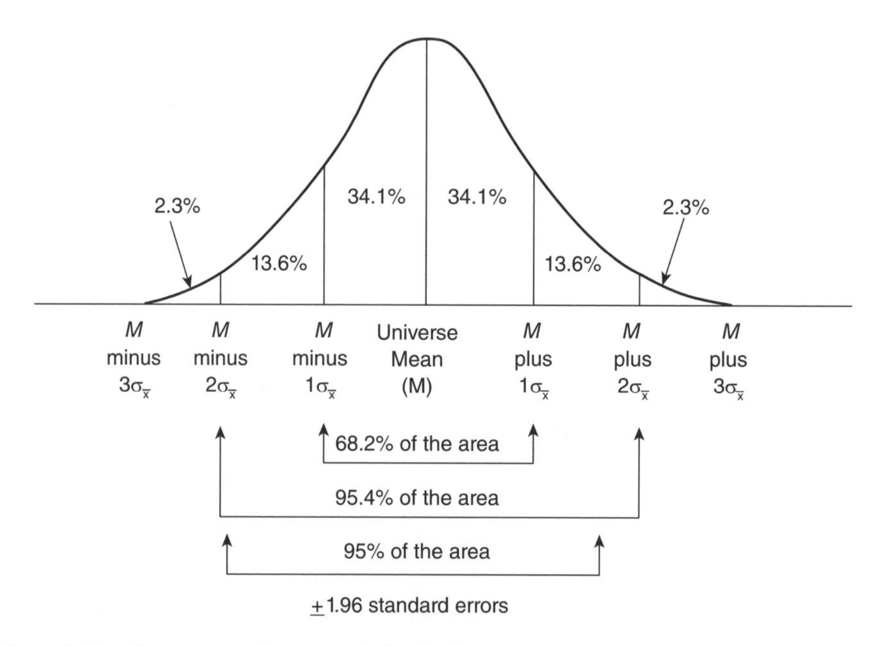

Figure 4.11 Areas under the normal distribution

Key points and wider issues

The statistical techniques used to produce data summaries for metric variables are fairly standard procedures that are covered in all texts on statistics or data analysis. The arithmetic mean and the standard deviation are perhaps the most commonly used of these. They can, however, be calculated only for metric variables (discrete or continuous) or for variables derived from summated ratings and assumed to be metric, that is the intervals between the categories can justifiably be taken as equal. They only make sense, furthermore, if the distribution can be accepted as normal or roughly normal. If the distribution is not normal in shape, then calculating the mean can hide more than it reveals. In an extreme case, an inverted bell-shape with the highest values at the extremes will, for example, give the same mean as a normal distribution. If the distribution is skewed, then the areas under the normal curve will be inaccurate.

For categorical variables, summary measures of central tendency, dispersion or distribution shape are of limited value. SPSS will happily calculate these, but beware when SPSS is in effect using the codes allocated to categories as if they are metric measures. Such procedures may or may not be justified, but in any event need to be carefully reviewed.

Box 4.2 Data summaries using SPSS

Data summaries for categorical variables can be obtained by using the `Frequencies` procedure to get percentages of each category and to obtain the modal category. However, note that for ordered category variables, SPSS will give you the median value, not the median category. This it does by treating the codes allocated to each category as metric values – which may be inappropriate.

There are two ways of obtaining univariate data summaries for metric variables in SPSS. One is to use the `Statistics` button in the `Frequencies` dialog box. Select `Analyze|Descriptive Statistics|Frequencies`. Put one or more metric variables in the `Variable(s)` box and then click on the `Statistics` button. Just put a tick in the box against the statistics you want by clicking with the left mouse button, then `Continue` and `OK`. The other procedure is found under `Analyze|Descriptive Statistics|Descriptives` and gives a quick summary of each variable that includes the minimum and maximum scores and the mean and standard deviation. This is a more useful layout if there are many variables since they are listed by column rather than across the page.

Statistical inference for univariate hypotheses

There are many situations faced by researchers where the number of cases to be investigated is very large and it is not feasible to study them all, so a sample is taken. The sample is then analysed in detail using the procedures that have been explained earlier in this chapter to produce tables, charts and summary measures. It would, of course, be possible to leave the analysis there and imply or assume that the results are likely to be typical of the total population of cases from which the sample was drawn. However, researchers are usually interested in making exact statements about the population based on the evidence from the sample – statements that are both precise and accurate and take account of the probability that sampling errors will be of a given magnitude. The calculations involved in making such statements are usually referred to as **statistical inference**.

Researchers may take their samples in many different ways, but statistical inference depends on samples that are selected at random. This means that cases are selected by methods that are independent of human judgement and each will have a calculable and non-zero probability of being selected. Thus a sample of 500 selected at random from a population of 5,000 has a 10 per cent chance or probability of 0.1 of being selected. The size of the population from which the sample is selected may, however, be unknown, so 'calculable' really means calculable in principle. Various kinds of error are likely to arise in this process; for example, because some cases selected cannot be contacted or, if the cases are individuals, they may refuse to answer. However, even if such errors are ignored or assumed to be unimportant, there will still be random fluctuations from one sample to another. For example, 100 samples taken at random from the same population will give slightly different results. Such error will be random and may result in the over- or under-representation of certain types of cases in a particular sample compared with the population. Over a large number of samples, the random errors will tend to cancel one another out. The result is that the average value of any given summary measure (e.g. the proportion, the mean, the standard deviation) from all possible samples of a given size that could be taken from a population of cases will be identical to the true population value.

A concept basic to all statistical inference is the **sampling distribution**. This is a theoretical distribution of statistical values for a given summary measure from all possible samples of a given size that could be taken from a population of cases. Imagine that we take lots of samples of a given size and calculate a particular statistic for each sample, say the proportion who purchased a chocolate bar in the last seven days. If the real proportion in the population is, for example, 46 per cent, we may take a sample and obtain a proportion of 47 per cent. Another may come out as 45 per cent. In fact, most sample results will cluster around 46 per cent with relatively few producing 'rogue' results of, say, 39 per cent. The tendency for this to happen will decline as the sample gets larger and increase as it gets smaller. If we plot the results as a distribution, we might obtain something like Figure 4.12. If we in fact took every conceivable sample of size n from a population and plotted the results we would obtain a **normal distribution** like the one imposed on Figure 4.12 by SPSS. In practice we would not, of course, actually do this, but the distribution can be derived statistically by calculating a theoretical sampling distribution.

The sampling distribution has its own standard deviation called the **standard error**. Since taking samples will result in less variation about the population parameter than picking cases one at a time, the standard error will always be less than the standard deviation for the population. To obtain the standard error for

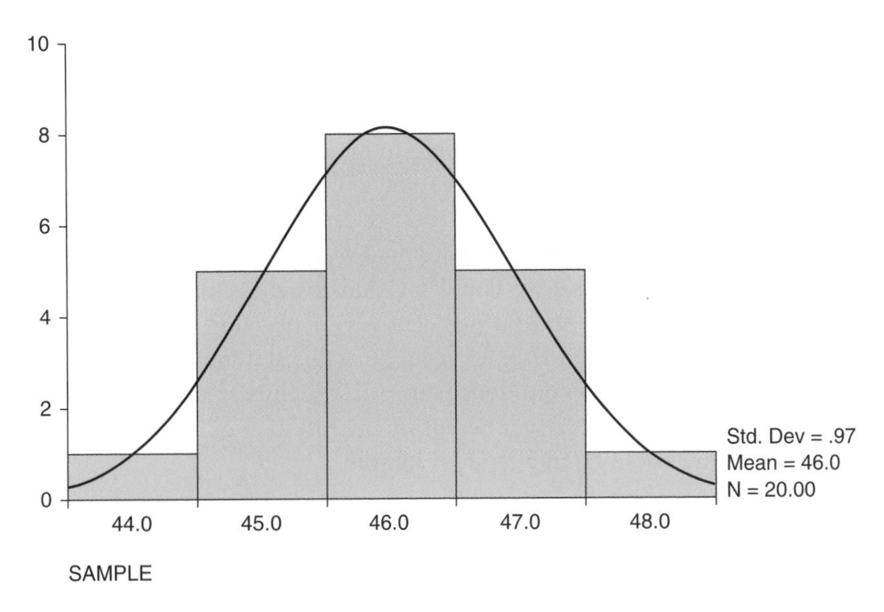

Figure 4.12 A sampling distribution of sample size *n*

metric variables, we divide the standard deviation in the population by the square root of the sample size, so that we halve the likelihood of random error in a sample of 4, it is one-third in a sample of 9, one-fifth in a sample of 25, and so on. As the size of the sample grows, so the standard error gets smaller, but by a declining amount.

Sampling distributions will always be normal in shape even if the distribution of the variable itself is skewed. This means it is possible to use the idea of areas under the normal curve to calculate the probability that, for any one *individual* sample, errors will be of a certain size. Such knowledge can be used in two rather different ways. In one situation the researcher may have little idea of the summary measure for the population (often called a population 'parameter'), so the sample is used to make an estimate of it. In the other situation, the researcher may feel that he or she knows the value of the population parameter and the sample statistics are used to test a hypothesis to this effect.

Estimation

Estimation is the process of using the value of a summary measure (like a proportion or a mean) derived from a sample to estimate the value of the corresponding population parameter. If the sample taken is a random sample and there are no other sources of error, this estimate should be reasonably close. However, since the sample statistic is unlikely to be *exactly* the same as the parent population, statistical inference is used to attach a degree of uncertainty to this process.

Any statement we make about the population of cases should have two properties: it should be precise and it should be correct. If we took a random sample of young adults aged between 16 and 19 and found that 46 per cent owned a smartphone, we could say that the percentage in the population of all young adults is just the same. This statement is very precise and is called a **point estimate**. Point estimates, however, are seldom likely to be correct. It would be very unusual for the value found in the sample to be exactly the same as the actual value in the population. As an alternative, we could say that the percentage who owned a smartphone is between 0 and 100 per cent. This statement is undoubtedly correct, but not very precise. A statement of this kind is called an **interval estimate**. To be more precise, we could say that the percentage who owned a smartphone is between 42 and 50 per cent (i.e. 4 per cent either side of the sample result). There is still a risk, however, that this statement is wrong. What statisticians do is calculate a **confidence interval**, which is an attempt to provide an illustration of the uncertainty inherent in any estimate of a population value based on the value obtained in a random sample.

Confidence intervals for metric variables

The standard error of the sampling distribution of the mean is calculated by taking

$$\frac{\sigma}{\sqrt{n}}$$

where σ (the Greek letter sigma) is the standard deviation of the population and n is the sample size. Thus the standard error increases with an increase in the standard deviation for the population, and decreases with an increase in sample size, but as the square root of the sample size. Where the standard deviation for the population is unknown – which is usually the situation – the standard deviation found for the sample is taken as an estimate of the population standard deviation. In this case we need what is called an **unbiased estimate**, which divides the sample standard deviation by the square root not of n, but $n - 1$:

$$\frac{s}{\sqrt{n-1}}$$

A more recent technique for estimating the standard deviation for the population from a sample is called **bootstrapping**. This involves taking lots of smaller samples from the main sample (putting the data back before a new sample is taken) and using these to estimate the population standard deviation. SPSS offers a procedure for bootstrapping on many of its procedures that require statistical inference.

In constructing a confidence interval, statisticians take the achieved sample result for a summary statistic like the mean and, for the 95 per cent level of confidence, subtract and add 1.96 standard errors to the achieved result. Thus if the mean score of a sample of 60 cases is 20 with a standard deviation of 6, then the confidence interval is

$$20 \pm 1.96\left(\frac{6}{\sqrt{60-1}}\right)$$

or between 18.5 and 21.5. Subtracting one from the sample size makes a meaningful difference only if the sample is quite small. Some statisticians will argue that if the standard deviation of the population is unknown we should use not the normal distribution but what is called the *t*-distribution. For samples over 30 or so, however, this makes very little difference.

SPSS will calculate confidence intervals for you for metric variables. How this is done is explained in Box 4.3.

Confidence intervals for categorical variables

If the population parameter being estimated is the proportion of the population that possesses a particular characteristic, then it is possible to plot a sampling distribution of the proportion. The standard error of this distribution is the **standard error of the proportion**. It can be shown (using binomial theory, which we will not go into here) that this is given by

$$\sqrt{\frac{p(1-p)}{n}}$$

where p is the proportion possessing the characteristic and $(1 - p)$ is the proportion not possessing that characteristic. The result of $p(1 - p)$ is an index of variability so that 0.5 multiplied by 0.5 gives maximum variability. Thus if a sample of 300 found that 40 per cent had purchased Brand A in the last week, then the confidence interval will be

$$0.4 \pm 1.96\sqrt{\frac{(0.4)(0.6)}{300}} = 0.4 \pm 1.96(0.028)$$
$$= 0.4 \pm 0.055$$

that is, between 0.345 and 0.455 (34.5 per cent and 45.5 per cent).

For a given sample size the closer the proportion is to 50 per cent, the greater will be the sampling error. For example, if only 1 per cent of a random sample of 300 said they had purchased a particular brand, the error will be roughly plus or minus 1.13 per cent at the 95 per cent level of confidence, that is

$$1.96\sqrt{\frac{(0.01)(0.99)}{300}} = 0.01126$$

If 50 per cent said they had made such a purchase, the error will be approximately plus or minus 5.7 per cent at the 95 per cent level, that is

$$1.96\sqrt{\frac{(0.5)(0.5)}{300}} = 0.0566$$

A larger sample would mean that the standard error of both of these results would be lower.

For categorical variables, SPSS does not, unfortunately, generate calculations based on the standard error of the proportion.

Key points and wider issues

Confidence intervals are widely calculated in the social sciences, but their interpretation is open to some dispute. Some researchers, some statisticians and some writers of textbooks on statistics or research methods will say that a confidence interval at the 95 per cent level of confidence means that we can be 95 per cent sure that the real population mean or the real population proportion lies between the limits calculated. Others will argue that this is incorrect – that the particular interval constructed is only one of many possible intervals that could be constructed based on different samples and that 95 per cent of all possible intervals constructed in this way will include the parameter concerned (and thus the particular interval involved has a 95 per cent chance of being one of them). Gorard (2013) takes this even further and argues that these calculations are correct only if we make the assumption that the real population mean is the same as the mean achieved in the sample, which makes the whole calculation tautological and circular. Furthermore, argues Gorard, the calculations make a number of assumptions that in real life are hardly ever met: that the sample is a random one; that there are no errors arising from non-response, non-contact, and so on; that the sample values are normally distributed (otherwise the 1.96 area under the normal curve is wrong); and that there is no bias or measurement error.

At best, confidence intervals are of limited value and, if calculated, the results need to be approached with caution. Some of the same kinds of issues arise when statistical inference is used to test hypotheses.

Testing hypotheses for statistical significance

Research in the social sciences is normally focused around generating and evaluating or testing some key idea or set of inter-related ideas. For research based on quantitative data, these ideas will usually be formulated by the researcher into explicit and testable statements in the form of hypotheses. A hypothesis is a carefully worded statement, as yet untested, about one or more properties of a set of cases. Hypotheses need to be carefully worded since the manner of their evaluation or testing will depend on how they are phrased. They are as yet untested since, once tested and found to be in some manner valid or at least partially substantiated, they become 'findings' or 'results'. Even hypotheses that have been rejected can still be considered as findings.

Hypotheses presuppose that researchers have some hunch or expectation about what they will find in the data, otherwise they remain as research questions. Hypotheses in variable-based analyses may refer to a single variable or to two or more. A single-variable or **univariate hypothesis** makes a statement about one variable; for example, a doctor hypothesizes that 'Seventy per cent of my patients are male' or a researcher that 'The average consumption of alcohol in the UK for men is 18.7 units per week'. Notice that the last statement is very precise and could easily be rejected if a different figure is found to be the case. If it had read '... over 18 units per week' then the conclusions might be quite different.

How researchers can or should go about evaluating their hypotheses against their data is an issue that is taken up in more detail in Chapter 10. The focus in this chapter is on situations where a researcher has taken a random sample and wishes to know how likely it is that any difference between the result in the sample and the hypothesized value is due to random sampling fluctuations, ignoring all other sources of error that were explained in Chapter 1. If the difference is small, given the variability of that variable and the size of the sample that was taken, then it is deemed likely that the difference could have been a random sampling fluctuation and the hypothesis stands – or, at least, is not rejected.

Testing hypotheses for metric variables

In the alcohol marketing dataset, respondents are asked how important or unimportant they think it is to choose popular well-known brands when deciding to buy each of nine different products, one of which is alcohol. Answers are coded from 5 for very important down to 1 for very unimportant, while 'Don't know' is coded as 0. If these codes are treated as metric values and are totalled, the maximum possible score then is 45 and the minimum is 0. Suppose on the basis of past research, other research findings, hunch or just taking all the middle values of 3, the researcher hypothesizes that the mean score for the population is 27. The result of an SPSS calculation is that the mean for the sample of 920 is 26.57 with a standard deviation of 6.69. The question that now arises is: could that result easily have come from a population in which the real mean is 27? The standard error of the mean is

$$\frac{s}{\sqrt{n-1}} = \frac{6.69}{\sqrt{920-1}} = 0.22$$

Assuming that the hypothesis is true, 95 per cent of samples will produce means in the range

$$27 \pm 1.96(0.22) = 27 \pm 0.43 \text{ or between } 26.57 \text{ and } 27.43$$

Since the achieved result of 26.57 is just in this range, there is insufficient evidence from the sample to *reject* the hypothesis that it is 27 at the 95 per cent level. This implies that the sample result may have come from a population of cases in which the mean score is 27. Note that, although a statistician will now say that, in this result, the hypothesis is 'accepted', this does not *prove* that it is true. The result does not, for example, prove that the population mean is 27, only that from the sample evidence we cannot say that it is not 27. Furthermore, there is still a 5 per cent probability that we might have accepted a hypothesis that is, in fact, false. So, remember that if we say we 'accept' the hypothesis – which appears to be the standard terminology – we actually mean that there is insufficient evidence to reject it. We have to behave as though the hypothesis still stands.

An alternative way of undertaking tests of significance for univariate metric hypotheses is to construct what is sometimes called a **z-test**. The result is exactly the same, but the procedure is a little different. Instead of constructing intervals under the assumption that the hypothesis is true, we calculate differences between the sample statistic and the hypothesized population parameter in terms of standardized units, that is we take, in this example, the difference between the sample mean and the hypothesized mean and divide by the standard error:

$$z = \frac{X - \mu}{s / \sqrt{n-1}} = \frac{27 - 26.57}{6.69 / \sqrt{920-1}} = 1.95$$

The critical value of z at $p<0.05$ is 1.96. Since 1.95 is just under this figure, we cannot reject the hypothesis that the population value is 27.

SPSS, however, follows *neither* of these procedures, but calculates the *actual* probability of getting a difference as big or bigger than this. It does this using what is usually called a **one-sample *t*-test**. It is 'one-sample' because just one variable is involved, and it calculates the probability of getting a difference between an achieved mean and some specified test value (the hypothesized value) as big or bigger than the one achieved resulting from random sampling fluctuations. If 27 is entered as the test value, then SPSS calculates that the probability is 0.052 or 5.2 per cent. How to get SPSS to do this is explained in Box 4.3. The result implies that there is a 5.2 per cent chance that in all possible samples of size 920, a difference as big or bigger than 0.43 (or 27 – 26.57) will be obtained. This probability is called a *p*-value.

The ***p*-value** is the probability of getting a difference between a sample result and some specified test value as big or bigger than the one we actually get. The *p*-value is normally used in the following way. If it is *less* than the significance level chosen for the test (e.g. 0.05) then the hypothesis is rejected; if it is *more* than the significance level then it is accepted. In the example above, the *p*-value of 0.052 is larger than 0.05 and implies that there is not enough evidence in the sample to reject the hypothesis. Note that the *p*-value is a conditional probability – it is calculated based on assuming that the stated hypothesis is true for the population being sampled. It is, furthermore, the probability of getting the sample result, given the stated hypothesis. It says nothing about the probability of the hypothesis being true, given the sample result. This issue arises again with greater force when we consider testing bivariate hypotheses in the next chapter.

When a hypothesis is rejected, the alternative may be non-directional so that it is unknown whether the real population value is above or below the rejected value. The area under the normal curve that lies outside the specified confidence level may be in either tail as in Figure 4.11, so it is called a 'two-tailed' test. SPSS usually assumes that a test is two-tailed. However, when there is more knowledge about the nature of the hypothesized phenomenon, the researcher may be able to specify that if the hypothesis is rejected, the alternative can only be in one direction, so the area under the normal curve that lies outside the specified confidence level is in one tail and lies within 1.645 rather than 1.96 standard deviations. A consequence of this is that a hypothesis is more likely to be rejected in a one-tailed test at the same level of confidence. A corollary is that if a hypothesis is rejected in a two-tailed test, it will also be rejected in a one-tailed test, but not necessarily the other way round.

The p-value provides a bit more information on how far down in the significance region a result lies. Articles in scientific journals often discuss their results in terms of p-values. They make statements like $p < 0.05$ or $p < 0.01$. However, it may be better to report the actual p-values (e.g. $p = 0.016$) rather than to use the conventional cut-off points, for example $p < 0.05$. This would allow readers to form their own judgements about the significance of the result and the strength of the evidence for or against the suggested hypothesis. Different people may, for example, feel that different levels of significance are appropriate.

Statisticians normally insist that the level of confidence (or **alpha value**, e.g. 0.05, 0.01 or 0.001) be established in advance of the test and that the hypothesis is then subsequently accepted or rejected. To decide on the confidence level after the results have been seen is generally regarded as not a legitimate practice since this amounts, in effect, to testing the hypothesis on the same data that suggested it in the first place. However, as we saw above, modern computer programs give not a decision, but the p-value. While reporting the p-value provides more information on how far down in the significance region a particular result lies than does simply accepting or rejecting a result, such a procedure may nevertheless compromise the idea of setting up the test criteria in advance. If a researcher produces a lot of results without prior hypotheses, then simply looking for significant results will probably involve just picking out all or most of the sampling accidents.

Testing hypotheses for categorical variables

A researcher studying the alcohol marketing dataset may have reason to believe – on the basis of a deduction from theory, on the basis of past research or from personal experience – that 60 per cent of young teenagers have never had a proper alcoholic drink. The dataset, however, produces a percentage of 65.4 per cent. Could this proportion of responses in the sample easily have been obtained from a population in which the real proportion is in fact 60 per cent? The standard error of the proportion is

$$\sigma_p = \sqrt{\frac{p(1-p)}{n}} = \sqrt{\frac{(0.6)(0.4)}{920}} = 0.016$$

If the stated hypothesis is true, 95 per cent of all samples of size 920 will produce results that vary between

$$0.6 \pm 1.96(0.016) \text{ or between 56.8 and 63.2 per cent}$$

Since the original sample result was 65.4 per cent, it lies outside this range and so is unlikely to have come from a population in which $p = 0.6$. The hypothesis is rejected – at the stated level of confidence. In other words, random error is an unlikely explanation of the difference between the sample result of 65.4 per cent and the predicted or hypothesized 60 per cent. Note that we cannot conclude from the sample result that the real proportion is 65.4 per cent, but only that it probably did not come from a population in which the hypothesis is true.

Why, it may be asked, did we not simply use the sample result for estimation purposes rather than going to the trouble of setting up a hypothesis prior to the analysis? The answer is that if we have a theory, a principle or a wealth of experience that we think allows us to make a prediction, it is this prediction (the hypothesis) that we want to test. If we wanted, for example, to compare the results of two or more pieces of research, then it is their support or questioning of the theory or principle or experience that we want to compare. Calculating confidence intervals does not test the hypothesis because the results could be consistent with a large number of hypotheses or predictions.

Be clear that it is the stated hypothesis that is being tested. Furthermore, it is being tested only for statistical significance, that is, for the probability that the achieved result came about as a result of random sampling fluctuation. If the hypothesis is rejected, it does not mean that it is disproven, only that it cannot be accepted, given the difference between the sample result and the hypothesized value, and the level of confidence chosen.

The standard error of the proportion only works if the variable is binary or nominal with two categories. If the variable is nominal with three or more categories, it is possible to use a statistic called **chi-square** (whose symbol is χ^2). This measures the overall extent of departure of a set of observations from some theoretical expectation. The difference between the observation and the expectation for each value is squared and taken as a proportion of the expectation by dividing by that expectation. Chi-square is then the sum of these calculations.

Suppose a company has four brands of a product and these are measured for preference in a survey of 100 respondents. The theoretical expectation might be that there is equal preference for the brands, so, in theory, we would expect 25 to prefer each brand. The results, however, suggest that 15 prefer Brand A, 23 Brand B, 27 Brand C and 35 Brand D. The differences between the observed frequencies (f_o) and expected frequencies (f_e) are 10, 2, 2 and 10, so chi-square is

$$\begin{aligned}\chi^2 &= \Sigma\frac{(f_o - f_e)^2}{f_e}\\ &= \frac{10^2}{25}+\frac{2^2}{25}+\frac{2^2}{25}+\frac{10^2}{25}\\ &= 4+0.16+0.16+4\\ &= 8.32\end{aligned}$$

Like all other statistics we could, in principle, calculate chi-square for a large number of samples and plot their distribution. This would be a **sampling distribution of chi-square**. This distribution, however, is not a single probability curve, but a family of curves. These vary according to the number of categories for which frequencies can be varied without changing marginal totals. Thus if there are four brands, then once the frequencies for three of the brands have been determined, the frequency in the fourth is fixed, so it is said to have three **degrees of freedom**. The probability of obtaining chi-square values of a given magnitude can be looked up in a table of critical values for chi-square. A simplified table is illustrated in Table 4.2.

This shows, for example, that for three degrees of freedom the critical value of 7.815 will not be exceeded more than 5 per cent of the time in a random sample. A value of 11.341 will not be exceeded more than 1 per cent of the time. We can conclude that the sample result of $\chi^2 = 8.32$ was unlikely to have occurred with a probability greater than $p = 0.05$ since $8.32 > 7.815$. So, at the 5 per cent level, we can reject the hypothesis that there is equal preference between the brands and conclude that the difference between the sample result and our expectation was unlikely to have been a result of random sampling fluctuations. It is the hypothesis of equal preferences that has been rejected. So preferences are, in all likelihood, not equal. The researcher, manager or client

Table 4.2 Critical values of chi-square

Probability	**0.05**	**0.01**	**0.001**
Degrees of freedom			
1	3.841	6.635	10.827
2	5.991	9.210	13.815
3	7.815	11.341	16.268

may, of course, sincerely believe that preferences are not equal; or perhaps more specifically that Brand D is the preferred brand. However, it still makes sense to test the statement that they are equal, because this statement has the virtue that it is precise. If it is rejected, the researcher can conclude that there is a 'real' difference (i.e. not one brought about by sampling fluctuation) and that by looking at the data, Brand D is indeed the preferred brand. Note, too, that in the above example, if the 1 per cent level of confidence had been selected, then the hypothesis would have been accepted since the critical value is 11.341. At this level, the differences between the brands are not such that random sampling error can be ruled out.

Key points and wider issues

Testing a hypothesis that relates to a summary statistic for a single variable in a situation where a random sample has been taken entails calculating intervals based not on the sample result (as with estimation), but on the assumption that the hypothesized value is true. The hypothesis is rejected if the actual sample result lies outside these limits. The calculation is based on the standard error of the mean for metric variables, on the standard error of the proportion for binary and nominal variables with two categories, and on the chi-square statistic for categorical variables with three or more categories. Beware that if *p*-values are used, a hypothesis is rejected if it is less than the chosen level of confidence; if critical values are used, as with chi-square, a hypothesis is rejected if the calculated chi-square value is greater than the critical value. However, if you use SPSS, the result, even for chi-square, is always in terms of *p*-values. Tables of critical values were created before the days of computers and researchers had to look up results in tables produced, for example, by Fisher in 1925.

Some of the issues raised in relation to the calculation of confidence intervals apply also to the calculations involved in testing for statistical significance. Statements are 'tested' only for the likelihood that any difference between a statement and the sample result cannot be explained by random sampling fluctuations. If the sample is not random, or if the cases used are a total population or an attempt at a total population, then such tests are not meaningful and should not be used. Even for random samples, such tests ignore any errors other than those potentially arising from random sampling fluctuations, for example from non-response, response error, interviewer errors, measurement error or errors arising from any biases in sample selection.

Box 4.3 Statistical inference using SPSS

SPSS provides confidence intervals for the estimation of metric variables under `Analyze|Descriptive Statistics|Explore`. This gives you the `Explore` dialog box. Put a metric variable (like `Total importance of` brands) in the Dependent List and click on OK. You should obtain the table illustrated in Figure 4.13. This provides many different statistical summaries, but the confidence interval for the mean at the 95 per cent level is 26.14–27.00. If you click on the Statistics box you can change the level of confidence, for example to the 99 per cent level. All these statistics can be split by a number of factors, for example gender of respondent, in which case you get separate tables for each. Just put the variable in the Factor List box.

Descriptives

			Statistic	Std. Error
Total importance of brands	Mean		26.57	.220
	95% Confidence Interval for Mean	Lower Bound	26.14	
		Upper Bound	27.00	
	5% Trimmed Mean		26.70	
	Median		27.00	
	Variance		44.720	
	Std. Deviation		6.687	
	Minimum		0	
	Maximum		45	
	Range		45	
	Interquartile Range		9	
	Skewness		-.315	.081
	Kurtosis		.367	.161

Figure 4.13 SPSS output for total importance of brands for Explore

Unfortunately, one thing that SPSS does not do is calculate the standard error of the proportion, which means that it cannot give you the corresponding confidence intervals for categorical variables. It can, however, test for differences between achieved sample proportions and hypothesized values. Under Analyze|Nonparametric Tests|One Sample, SPSS uses either a one-sample binomial test or a one-sample chi-square test as appropriate to assess the *p*-value for the difference between the sample result and the hypothesized values, which SPSS assumes are equal proportions in each category. To change the hypothesized proportions, select Legacy dialogs instead of One Sample and either Binomial, if the variable is binary, and change the Test Proportion or select Chi-square, if nominal with three or more categories, and change the Expected Values.

For metric variables, SPSS offers a one-sample *t*-test. This is found under Analyze|Compare Means|One-Sample T-Test. Put the metric variable (e.g. Total importance of brands) into the Test Variable(s) box. Enter the Test Value (e.g. 27) which is the hypothesized value and click on OK. The result is shown in Figure 4.14. This gives a *p*-value of 0.052 under Sig.(2-tailed).

(Continued)

(Continued)

One-Sample Test

	Test Value = 27					
					95% Confidence Interval of the Difference	
	t	df	Sig. (2-tailed)	Mean Difference	Lower	Upper
Total importance of brands	-1.942	919	.052	-.428	-.86	.00

Figure 4.14 SPSS output for total importance of brands for `One-Sample T-Test`

Other statistics and survey analysis packages

Computer programs relevant to the analysis of non-experimental datasets fall into two main groups: statistical packages and survey analysis packages. There is a bewildering variety of both. The former handle numerical datasets by applying variable-based and, less frequently, case-based analyses. Wikipedia, for example, lists over 50 different commercial statistical packages, many of which are general statistical packages, while some are more specialized, for example on data mining, visual analytics, statistical modelling, econometric analysis or time series. In addition there are over 30 open-source packages and 10 are either public domain or freeware. Some of the packages, like IBM SPSS Statistics, use a graphical user interface (GUI) that allows users to interact with the program through graphical icons, pull-down menus, dialog boxes, and so on. Others, such as SAS offered by the SAS Institute, have a command line interface (CLI) that requires commands to be typed in using the keyboard. These tend to be more flexible, but have a much steeper learning curve. However, some of these have an optional GUI front-end available. Others, like Stata, genuinely integrate both a GUI and a CLI. R, developed at the University of Auckland, New Zealand, is an open-source, free programming language that is constantly being developed by users themselves. It compares well with SPSS, Stata and SAS, but requires the user to have some facility with programming, although there are many add-on GUIs available. Both Stata and R have a downloadable or purchasable option to conduct fuzzy set analysis, which is explained in Chapter 7.

Survey analysis packages give you the tools you need to create a survey, design a questionnaire, organize the results and prepare a detailed report. However, the statistical techniques that are available tend to be more limited than for the statistical packages. The more popular ones include Snap Surveys, The Survey System, KeyPoint, Fluid Surveys, StatPac, SurveyGold, SurveyMonkey, Survey Crafter Professional, SurveyGizmo and SurveyPro. These packages are reviewed at http://survey-software-review.toptenreviews.com/. Some, like SurveyMonkey, are designed specifically for online use, while others are geared more to paper questionnaires, but there are usually optional extras to facilitate online use.

Implications of this chapter for the alcohol marketing dataset

Since this chapter focuses on variables one at a time, it does not directly address the stated hypotheses for the research, which are bivariate. Bivariate analysis is explained in the next chapter. However, a review of the key variables suggests that the two dependent variables, whether or not respondents have consumed alcohol and their perceived likelihood that they will do so in the next year, show that very nearly two-thirds claim that they have never had a proper alcoholic drink and over half (53.6 per cent) said they definitely or probably would not have an alcoholic drink in the next year; 29.2 per cent said they definitely or probably would. A pie chart shows the distribution of the answers rather well, but it is not reproduced here since it would not come out well in black and white.

In terms of awareness, nearly half of respondents have seen adverts for alcohol on up to five channels and nearly three in four of respondents up to seven channels (Figure 4.15). In terms of summary statistics, SPSS calculates the

Total number of channels seen

		Frequency	Percent	Valid Percent	Cumulative Percent
Valid	1	40	4.3	4.5	4.5
	2	56	6.1	6.3	10.8
	3	84	9.1	9.5	20.3
	4	104	11.3	11.7	32.0
	5	137	14.9	15.4	47.5
	6	121	13.2	13.6	61.1
	7	106	11.5	12.0	73.1
	8	83	9.0	9.4	82.4
	9	71	7.7	8.0	90.4
	10	43	4.7	4.8	95.3
	11	24	2.6	2.7	98.0
	12	3	.3	.3	98.3
	13	11	1.2	1.2	99.5
	14	3	.3	.3	99.9
	16	1	.1	.1	100.0
	Total	887	96.4	100.0	
Missing	0	33	3.6		
Total		920	100.0		

Figure 4.15 Number of channels on which adverts for alcohol have been seen

mean as 5.87, but remember that this is a discrete metric variable, so fractions have no meaning. If this is rounded up to six channels, it coincides with the median, which is also six. Note that the 33 who indicate 'None' have been treated as user-defined missing values. Whether or not these should be included can be argued either way, but in fact makes only a very small difference to the mean (which goes down to 5.66). For involvements in marketing, however, there were 396 or 43 per cent who indicate 'None'. Whether or not these are included in the calculation of the mean does make a considerable difference, but the median remains at a single form of involvement.

In short, a univariate analysis of the dataset for the key variables suggests that a majority of respondents say that have never had a proper alcoholic drink and have no intention of doing so in the next year. However, there is high awareness of the marketing of alcohol, but little actual involvement.

Chapter summary

Data analysis is an iterative process and the first move in this iteration is usually to undertake univariate analysis. This means inspecting the data one variable at a time and will include displaying categorical and metric variables in tables, charts and graphs, calculating summary measures that will pinpoint key characteristics of each distribution, and perhaps, where the data are derived from a random sample, evaluating the likely accuracy of estimates made from it or the statistical significance of hypotheses that have been put forward. At this point, the researcher may, in addition, reflect upon the implications of this univariate analysis for the client, for the research objectives and for the next steps to be undertaken in the analysis of the data.

The most commonly used summary measures for metric variables (discrete or continuous) or for variables derived from summated ratings and assumed to be metric are measures of central tendency, dispersion and distribution shape. Summary measures for categorical variables are somewhat limited to percentages, proportions and in some cases the modal category.

Once univariate analysis is complete, researchers will usually proceed to the next step, which is looking for patterns of relationship between variables, initially two at a time. This is the focus of the next chapter.

Exercises and questions for discussion

1. Is there a danger that the procedures used to analyse a dataset become largely a function of the procedures that happen to be available on a particular computer package like SPSS?
2. Do pie charts have any advantages over bar charts?
3. You can get SPSS to produce any kind of nonsense. The trick is to know what counts as 'nonsense'. Suggest some of the main ways in which the researcher might produce nonsensical tables and charts.

4. Get SPSS to produce a one-way table for each of the variables either in the alcohol marketing dataset (available at https://study.sagepub.com/kent) or in the `Trinians` dataset (see Chapter 2, Exercise 6 for instructions). Look at the distribution of each and think about which ones might require some data transformations.
5. Try out the `Explore` and `Descriptives` functions in SPSS on some of the variables.

Further reading

Argyrous, G. (2011) *Statistics for Research with a Guide to SPSS*, 3rd edn. London: Sage.

Chapters 3 and 4 cover the graphical and tabular description of data, all explained very clearly with lots of examples and screenshots using SPSS. Chapters 9, 10 and 11 explain measures of central tendency, measures of dispersion and the normal curve. For estimation look at Chapter 17 and for an introduction to hypothesis testing try Chapter 15.

Diamantopoulos, A. and Schlegelmilch, B. (1997) *Taking the Fear Out of Data Analysis*. London: Dryden Press. Republished by Cengage Learning, 2000.

Chapters 7–11 cover most of what is in this chapter, clearly in more detail, but do not use SPSS. The authors do, however, include the treatment of proportions, which Argyrous does not.

De Vaus, D. (2002) *Analyzing Social Science Data: 50 Key Problems and Data Analysis*. London: Sage.

A rather different approach. It is a bit like a series of answers to frequently asked questions. Part Five considers how to analyse a single variable. It illustrates using SPSS but does not cover proportions.

Field, A. (2013) *Discovering Statistics Using IBM SPSS Statistics*, 4th edn. London: Sage.

At 952 pages, this book is comprehensive and uses SPSS throughout. Very good at explaining SPSS outputs. Probably a bit too advanced for readers of this book and it is geared more to analysing experimental data. However, Chapter 2, 'Everything you ever wanted to know about statistics', is well worth a read and also Chapter 4 on exploring data with graphs. There are companion volumes on using SAS and using R. If you are interested in Stata try Longest, K.C. (2012) *Using Stata for Quantitative Analysis*. Thousand Oaks, CA: Sage.

Yang, K. (2010) *Making Sense of Statistical Methods in Social Research*. London: Sage.

This book focuses more on the conceptual foundations of statistical thinking than on how to calculate statistics. A lot more words and very few equations. Chapter 5 on estimating and measuring one important thing is well worth a read.

Suggested answers to the exercises and questions for discussion can be found at the end of this text, pp. 293–321, and on the companion website, (https://study.sagepub.com/kent), which also give links to relevant free online Sage journal articles, PowerPoint slides, an overview of data analysis packages, an introduction to SPSS and weblinks to alternative datasets.

5

BIVARIATE ANALYSIS

Learning objectives

In this chapter you will learn about:

- the different ways in which two variables can be related;
- bivariate data displays for both categorical and metric variables;
- bivariate data summaries for both categorical and metric variables;
- testing the statistical significance of null hypotheses for variables two at a time;
- using SPSS for bivariate analysis;
- how a bivariate analysis of the alcohol marketing dataset suggests that there is a small tendency for those who had seen alcohol advertising on six or more channels to be more likely to have had an alcoholic drink than those who had seen it on fewer channels, and the more likely they are to say that they will probably drink alcohol in the next year.

Introduction

Chapter 4 focused on displaying, summarizing and drawing inferences from the distribution of values of variables one variable at a time. This chapter turns to ways in which researchers can explore or demonstrate how the distributions of values of variables two at a time may be patterned or related in some way. Such analyses may be used in one or both of two main contexts. First, the researchers may want to explore the extent to which there are any patterns; in the second context, the researchers may have ideas, hunches or even very specific hypotheses about what variables are in fact related and in what ways, and may use bivariate statistics to establish how far these hunches or hypotheses are supported by the data.

This chapter begins by examining the various ways in which two variables may be related. It then turns to the display of different patterns or relationships between two categorical or two metric variables and then considers how these patterns can be summarized statistically. Finally, it explains how, in a context where the data constitute a random sample from a population of cases, hypotheses relating to two variables can be tested for statistical significance.

The variety of relationships between two variables

The most basic pattern between two variables is simply a difference between two categories in terms either of two other categories or of a summary measure for a metric variable. Differences between categories may, for example, be expressed in raw frequencies (65 women said 'yes' compared with 55 men), in percentages (60 per cent of members said they were 'satisfied' compared with 40 per cent of non-members) or in terms of the size of the difference (there is a 10 per cent difference between men and women). Differences in summary measures might, for example, be in terms of mean, median or modal scores in two groups or the size of the difference in scores between them.

Differences need to be between just two categories of the variable acting as the basis for the comparison, so, ideally, it should be binary, otherwise there will be several differences involved, for example, between three categories, the differences between A and B, A and C, and B and C. If both variables are nominal with three or more categories, then talking about differences is not possible. However, there may be tendencies for particular combinations or clusters of categories to predominate, as in Table 5.1. This shows the maximum degree of category clustering – for each row and for each column, all the cases are in just one cell. Thus all those who are single prefer Product A, those who are married prefer Product C and those who are married with children prefer Product B. The degree to which there is a tendency to such a pattern can be measured using a statistic you have already met in the previous chapter, namely **chi-square**. How this is done is explained later in this chapter.

The pattern of relationships between two variables that is most often sought by standard variable-based analysis, however, is that of **covariation**. This means that there is a tendency for changes either of category or of metric value on one variable to be exactly mirrored by changes in category or metric value on another variable. If there is, for example, a tendency of male

Table 5.1 Category clustering

	Single	Married	Married with children
Product A	60		
Product B			65
Product C		55	

respondents in a survey to say 'yes' to a particular question, then there is a corresponding tendency for the females to say 'no'. This situation is illustrated in Table 5.2. There is a predominance of cases that lie 'on the diagonal'. The implication of covariation is that relationships are symmetrical. Even the simple 'If X then Y' proposition (e.g. 'If male then the answer is yes') implies two other things: the relationship is reversible, for example 'If Y then X' ('If yes then male'); and it is diagonal, for example 'If not X then not Y' ('If female then the answer is no'). In short, if all men say 'yes' then by implication all 'yeses' are men, all females say 'no' and all those who say 'no' are female. Note that this logic can be applied only if both variables are binary. If either or both variables are nominal (having three or more categories) then the notion of covariation does not apply. The relationship can only be in terms of **category clustering**.

Table 5.2 Covariation for two binary variables

	Gender	
	Male	**Female**
Yes	48	2
No	2	48

For ordered category, ranked or metric variables, covariation means that if high values on one variable tend to be found with high values on the other, then there is a corresponding tendency for low values on both to be found together, as in Table 5.3 and Figure 5.1. In Table 5.3, high product usage and high product satisfaction tend to be found together, and, at the same time, low product usage and low satisfaction are found together. In Figure 5.1, if respondents in a survey report higher than average television viewing per week, they will have higher than average spend on convenience foods. By the same token, those who have lower than average television viewing will have lower than average spend. Again, it is symmetrical – it is reversible and 'on the diagonal' – but for metric variables it is also linear in the sense of approximating a straight line. Linearity implies that as one variable changes by one unit, so the other increases (or decreases) by a fixed amount. As we will see later in this chapter, measures of the degree of covariation are based on assessing the extent of diagonality.

Another form of relationship between two variables follows an asymmetrical pattern. It may be the case that, for example, all university lecturers may have higher degrees, but not all those with higher degrees are university lecturers. All the men may say 'yes' to a question in a survey, but the women may say 'yes' or 'no'. For two binary variables this may be seen as a subset relationship. If all Xs are Ys, but not all Ys are Xs, then X is a subset of Y, as in Figure 5.2. Being a university lecturer is a subset of all those possessing higher degrees.

Table 5.3 Covariation for two ordered category variables

	Product usage		
Satisfaction	**High**	**Medium**	**Low**
High	10	8	4
Low	4	8	10

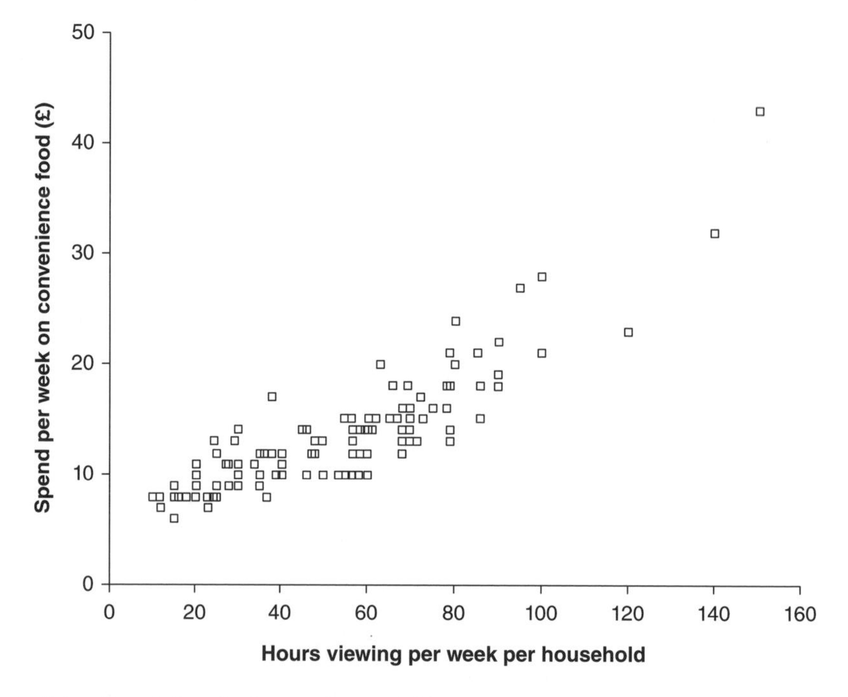

Figure 5.1 Covariation for two metric variables

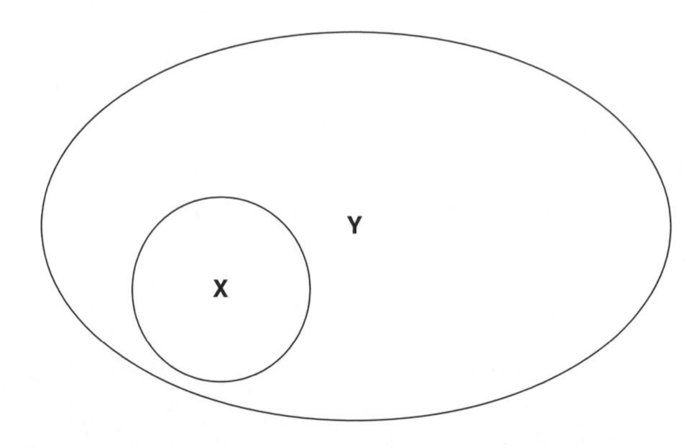

Figure 5.2 A subset relationship

In a subset relationship, the proposition 'If X then Y' does *not* imply 'If Y then X' and it says nothing about cases that are 'not X'. If, however, we include cases that are 'not X', then a different form of **asymmetrical relationship** may arise, for example the proposition 'If X then Y' does *not* imply 'If not X then not Y'. Table 5.4 shows what might be called a triangular (rather than diagonal) relationship. All of those whose television viewing has been classified as 'Heavy' have what has been classified as a 'Large' spend on convenience food. However, if television viewing is 'Light', then spend on convenience food may be either 'Large' or 'Small'. We can conclude (if it makes sense to do so for theoretical reasons) that while the presence of the condition (being a heavy television viewer) is sufficient to ensure the outcome (there are no cases of heavy television viewing with a small spend on convenience food), it is not a **necessary condition** since there are many cases of light television viewing who also have a large spend on convenience food. In short, the outcome may also be a result of other conditions.

Table 5.4 An asymmetrical triangular relationship

	Television viewing	
Spend on convenience food	**Heavy**	**Light**
Large	40	40
Small	0	30

Note that spend on convenience foods has been treated as the 'outcome' – in other words, the researcher is interested in what kinds of people spend a lot on convenience foods. One factor may be the amount of television they watch. Alternatively, the researcher may wish to study what kinds of people watch a lot of television. In this situation, television viewing is the outcome. People who spend a lot on convenience food may have more time to watch television, so this may be one potentially causal condition. In this situation, the conclusions about necessity and sufficiency are reversed.

For metric variables, subset analysis is not possible unless these variables are converted into binary form (e.g. income is put into the categories of 'high' and 'not high') or the researcher makes use of fuzzy sets in which it is possible to have degrees of membership of a category. Fuzzy set analysis is explained in Chapter 7. Triangular patterns for two metric variables are, however, quite possible – even quite common, as illustrated in Figure 5.3.

Large amounts of viewing hours per week per household are associated with high spend on convenience food, but limited hours viewing may be associated with either high or low spend on convenience food. The researcher might conclude from this that high viewing is sufficient to bring about a high expenditure on convenience food, but it is not a necessary condition since high expenditure can come about as a result of other factors.

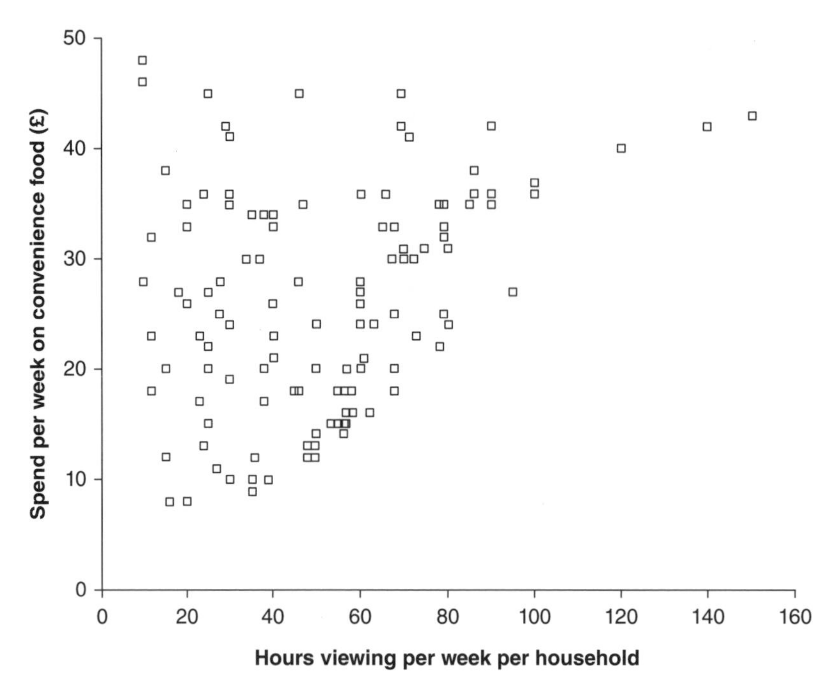

Figure 5.3 A triangular pattern for two metric variables

> ### Key points and wider issues
>
> There are many different kinds of relationship between two variables. Variable-based statistical techniques tend to address only differences, category clustering and covariation. They cannot handle asymmetrical patterns either as subsets or as triangular patterns. One approach that can handle these is to use configurational and fuzzy set analysis, which is explained in Chapter 7. We will also see in the next chapter that when we move from bivariate to multivariate analysis, the variety of relationships between variables expands considerably. We will also see in Chapter 10 that while data analysis procedures can establish the extent to which patterns exist between two or more variables, they cannot explain how or why those relationships hold.

Bivariate data display

Differences, category clustering and covariation between two categorical variables can be displayed as **bivariate crosstabulations**. These are sometimes referred to as 'contingency' or 'two-way' tables. A **crosstabulation** is a particular kind of table in which the frequencies or proportions of cases that combine a value on one categorical variable with a value on another are laid out in combinations of rows and columns. In other words, a crosstabulation presents the frequencies of two variables, the values of which are interlaced. Tables 5.1 to 5.4 earlier in this chapter are examples of such tables. The smallest table crosstabulates two binary variables. Figure 5.4 is taken from the alcohol marketing dataset and is produced by SPSS. How to do so is explained in Box 5.1 below. There are four cells created by the two variables, each having two categories, so the table would normally be called a 'two-by-two' table. The totals at the end of each row (602 and 318 in this table) are called **row marginals** and the totals at the bottom (213 and 707) are **column marginals**.

There are 262 who answered 'Yes' to the question 'Have you ever had a proper alcoholic drink?' and who have seen ads for alcohol on TV or in the cinema. That represents 37.1 per cent of those who have seen ads for alcohol on television or in the cinema compared with 26.3 per cent of those who had not seen such ads. So, those who have seen such ads are a little more likely to be those who have had a proper alcoholic drink. Measures of the extent of this tendency are considered later in this chapter.

Count

		Seen ads for alcohol on tv/cinema		
		Not seen/DK	Seen	Total
Have you ever had a proper alcoholic drink?	No	157	445	602
	Yes	56	262	318
Total		213	707	920

		Seen ads for alcohol on tv/cinema		
		Not seen/DK	Seen	Total
Have you ever had a proper alcoholic drink?	No	73.7%	62.9%	65.4%
	Yes	26.3%	37.1%	34.6%
Total		100.0%	100.0%	100.0%

Figure 5.4 Crosstabulation of drink status by seen ads for alcohol on TV or cinema

Crosstabulations may be of any size in terms of rows and columns. As tables get larger, however, they become increasingly difficult to interpret and they require many more cases to avoid having very small numbers in most of the cells. The only solutions are either to increase the number of cases or to reduce the size of the table by, for example, using the SPSS `Recode` procedure explained in Chapter 2. Figure 5.5(a) shows a 'five-by-three' crosstabulation with 15 cells. Even with 774 cases (there are 146 missing; these were those who answered 'Don't know'), some of the frequencies are very small. In Figure 5.5(b) this has been reduced to a two-by-three. It looks as though there may be a small tendency for those who think branding is important for alcohol to be more likely to drink alcohol in the future (37.9 per cent compared with 28.2 per cent).

(a)

Count

		Brand importance for alcohol					Total
		Very unimportant	Quite unimportant	Neither important nor unimportant	Quite important	Very important	
Intention to drink alcohol	Definitely or probably yes	76	35	64	42	16	233
	Not sure, DK or no answer	53	23	30	24	7	137
	Definitely or probably not	207	67	66	51	13	404
Total		336	125	160	117	36	774

(b)

			Brand importance for alcohol		Total
			Important	Not important	
Intention to drink alcohol	Definitely or probably yes	Count	58	175	233
		% within Brand importance for alcohol	37.9%	28.2%	30.1%
	Not sure, DK or no answer	Count	31	106	137
		% within Brand importance for alcohol	20.3%	17.1%	17.7%
	Definitely or probably not	Count	64	340	404
		% within Brand importance for alcohol	41.8%	54.8%	52.2%
Total		Count	153	621	774
		% within Brand importance for alcohol	100.0%	100.0%	100.0%

Figure 5.5 Crosstabulation of intention to drink alcohol by brand importance

In terms of table presentation, note that the independent variable (or the variable being treated as 'independent' for the purpose of the analysis) is usually put in the columns and the dependent variable in the rows. The calculation of percentages should always be in the direction of the independent variable, so percentage downwards in the columns, but compare the percentages across. In the title of the table, it makes sense to mention the dependent variable first (this, after all, is the variable being studied) 'by' whatever independent variables are being used to analyse it.

In the research on alcohol marketing, the researchers have assumed that intention to drink alcohol in the next year is the 'dependent' variable that may be affected by a range of other factors including importance of branding and awareness of alcohol advertising. However, it could be argued that the relationship is the other way round and that those who intend to drink alcohol in the next year are more likely to be aware of alcohol advertising and see branding as more important. On this argument, intention (and maybe drink status) become the independent variables and would go at the top of the crosstabulation. The status of variables as dependent or independent is determined by the researcher according to the theories being applied or understood.

The relationships between two metric variables can be illustrated using a **scattergram**. The horizontal axis is used to represent the values of one of the metric variables and the vertical axis to represent the other. The combination of two values is then plotted for each case, as in Figure 5.1. This shows that the relationship is approximately symmetrical and linear. We will see later in this chapter how this property can be used to calculate a measure of correlation. Note that if one of the variables is considered to be independent (like hours viewing per week per household) then it is plotted along the horizontal or X-axis. The dependent variable (spend per week on convenience food) is put on the vertical Y-axis. For the alcohol marketing dataset, if total importance of brands and total number of channels are plotted as a scattergram, then Figure 5.6 shows the result. Any relationship between the two is certainly not symmetrical or linear, but possibly triangular. The interpretation of triangular patterns must wait until Chapter 7.

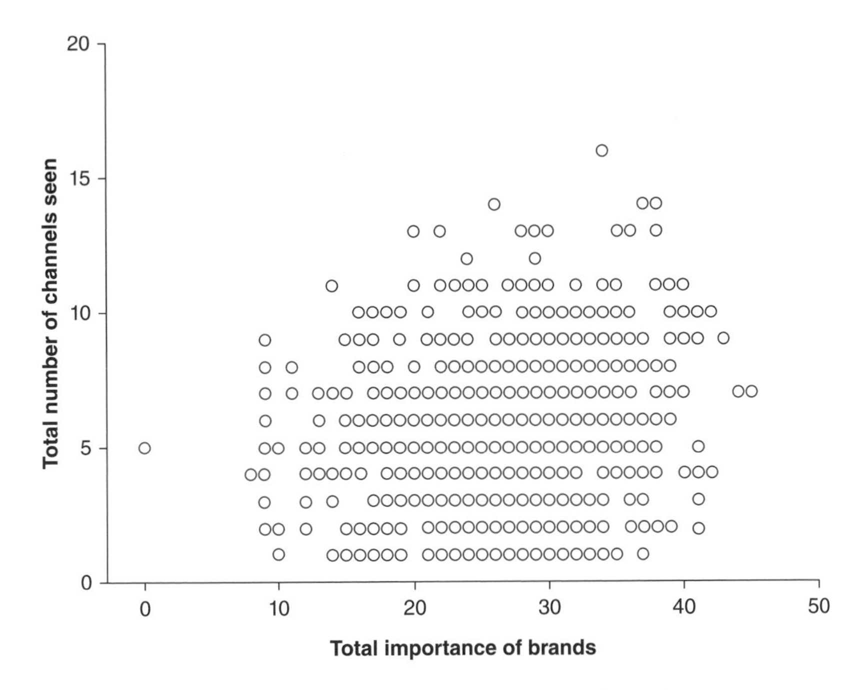

Figure 5.6 Total importance of brands by total number of channels seen

Box 5.1 Using SPSS to produce crosstabulations

Select `Analyze|Descriptive Statistics|Crosstabs` to obtain the `Crosstabs` dialog box. Enter your dependent variable (e.g. intention to drink alcohol in the next year) in the `Rows` box so it will appear at the side, and enter the independent variable (e.g. brand importance for alcohol) in the `Columns` box. This should produce the crosstabulation shown in Figure 5.5. If you put several variables in each box, then you will obtain a crosstabulation of each combination. To obtain column percentages, click on the `Cells` button in the `Crosstabs` dialog box to obtain `Crosstabs: Cell Display`. Click on `Column` in the `Percentages` check box, uncheck the `Observed` box under `Counts`, then on `Continue`, then on `OK`. If you wish to change the order in which the categories of the dependent variable are listed (so that, for example, 'Definitely or probably yes' appears at the top), click on the `Format` button in the `Crosstabs` dialog box and change `Row Order` to `Descending`. Click on `Continue`, then on `OK`.

Key points and wider issues

Differences, category clustering or covariation may be displayed in crosstabulations for two categorical variables, and in scattergrams for two metric variables. Metric variables may, however, also be displayed in crosstabulations provided they have been grouped into **class intervals** (see Chapter 4). Bivariate data display may be used by researchers in two main contexts. First, it may be used in a preliminary diagnostic context in which the researcher runs a number, perhaps a large number, of key crosstabulations to get a feel for the nature of relationships between variables two at a time, or plots scattergrams to check whether relationships between metric variables are roughly linear and whether there are any **outliers** – values that are substantially different from the general body of values. Second, in a presentational context, the researcher may use tables and graphs selectively to illustrate key findings in a report.

Bivariate data summaries: categorical variables

Data displays enable researchers to eyeball patterns in the relationships between variables two at a time. However, they may also want to be able to condense a bivariate relationship to a single summary figure, just as an average condenses a single univariate distribution to a single figure. For categorical variables, these summary measures may utilize the size of differences, the extent of category clustering or covariation.

Differences

Differences may be summarized in terms of the size of the difference between two metric values, two proportions or two percentages. Thus from Figure 5.4 (p. 119) we can say that the percentage difference between those who have and have not seen ads for alcohol on television or in the cinema in terms of drink status is 37.1 minus 26.3 per cent or 10.8 per cent. The percentage difference will range from 0 per cent, when the proportions are the same, to 100 per cent when there is the maximum difference possible. The percentage difference, however, only works in this way when both variables are binary, otherwise there are lots of differences. It is therefore of limited value; fortunately, there are alternatives that work much better.

Category clustering

Statisticians have developed an array of measures of association that measure either category clustering or covariation for crosstabulated variables. Measures of association are usually designed so that they vary either between zero (no association) and plus one (a perfect association), or between minus one (a perfect negative association) and plus one (a perfect positive association). The results of calculating such measures are usually described as **coefficients**. If both variables are nominal, the statistic should vary between zero and plus one – the notion of 'negative' association is nonsensical unless there is some kind of order between categories. If both variables are ordered category, then the statistic should be negative if higher order values on one variable are associated with lower order values on the other. It is not that applying a statistic that cannot take negative values to two ordered category variables is necessarily 'wrong' – it is just that the statistic will not reveal whether the association is positive or negative. This may be no big deal, however, since this is usually obvious just by looking at the table.

Category clustering can be measured in two main ways: as departure from independence or as proportional reduction in error. **Departure from independence** involves imagining what the data would look like if there were no clustering, and then saying that clustering is present to the extent that the observed data depart from this. Thus if 120 respondents consisted of 60 males and 60 females responding 'yes' or 'no' to a question, and equal numbers of each said 'yes' and 'no', the marginals would look like Table 5.5; 60 out of 120 are male, that is half. Therefore, if there were no association, we would expect half of the 60 'yeses' to be male. The expected frequency for any cell can always be found by multiplying the row and column marginal for that cell, and dividing by n, the number of cases. In this example

$$\frac{(60)(60)}{120} = 30$$

The expected frequencies for each cell are shown in Table 5.6. This shows that just as many men as women are likely to say 'yes' (or 'no'). Suppose, however, that all the men said 'yes' and all the women said 'no'. The result is shown in Table 5.7. The difference between the observed and expected frequencies for each cell is now either +30 or –30 (the maximum difference that is possible in this table). Simply

Table 5.5 Responses by gender

Answer	Male	Female	Total
Yes	–	–	60
No	–	–	60
Total	60	60	120

Table 5.6 Responses by gender: expected frequencies

Answer	Male	Female	Total
Yes	30	30	60
No	30	30	60
Total	60	60	120

adding these up will, of course, produce zero. If, however, we square the difference and take that as a proportion of our expectations for each cell, then we take account of whether the absolute difference is based on a large or small expectation. We can now add these up for each cell.

Table 5.7 Responses by gender: actual frequencies

Answer	Male	Female	Total
Yes	60	–	60
No	–	60	60
Total	60	60	120

Table 5.8 Calculating chi-square

f_o	f_e	$(f_o - f_e)$	$\frac{(f_o - f_e)^2}{f_e}$
60	30	+30	30
0	30	–30	30
0	30	–30	30
60	30	+30	30
			120

If f_o is observed frequency and f_e expected frequency, then the calculations are as shown in Table 5.8. In this table, 120 is the maximum value that the sum of the squared differences as a proportion of expected values can take. This is a statistic called *chi-square*, usually symbolized as χ^2, which has already been introduced in Chapter 4. Chi-square will be zero if there are no differences between observed and expected frequencies. The maximum value it can take depends on the number of cases and the number of cells. For a two-by-two table the maximum value for chi-square is always the number of cases, *n*. Chi-square can, however, be calculated for any size of table. For example, for Table 5.1, which is a three-by-three table with nine cells, chi-square again takes its maximum value, which in this example is 360.

While it is difficult to use chi-square itself as a measure of association, various adjustments to it have been proposed that result in more acceptable measures. These measures make no assumptions about the order in which the categories are placed – in fact, their calculation is unaffected if we list them in different ways. Accordingly, they are particularly appropriate for tables where both variables are binary or nominal. One feature of this basis for measuring association is that it does not require us to select one variable as the outcome or dependent variable; hence all measures based on chi-square are non-directional.

The related coefficients **phi** and **Cramer's V** are both based on chi-square and its measurement of departure from independence. As noted above, for a two-by-two table the maximum value for chi-square is always the number of cases, *n*.

If we simply divide chi-square by N and take the square root, we obtain a measure called phi. This has a minimum value of 0 when the two variables are independent. For two-by-two tables and for tables where either rows or columns have only two categories, it also has an upper maximum of 1 when the relationship between the two variables is perfect. For tables of this size, then, phi is a good measure. For the example in Tables 5.7 and 5.8, chi-square is 120 and the number of cases is also 120, so phi = 1 (the square root of 1 is still 1).

However, for larger tables, phi can attain a value considerably larger than unity. An adjustment to the formula was suggested by H.Cramer, and is known as Cramer's V. This divides chi-square by N multiplied by a value which is either rows minus one or columns minus one, whichever is the minimum. The formula is

$$V = \sqrt{\frac{\chi^2}{N \min(r-1)(c-1)}}$$

For Table 5.1, the minimum is 3 – 1 (for either rows or columns) so χ^2 is divided by 180 × 2 and $V = 1$. Notice that the value of $\min(r - 1)(c - 1)$ for a two-by-two table or any table where either variable is binary is 1, so V is identical to phi. Whether we choose to call the resulting measure phi or V is perhaps not important. For Figure 5.5(a), SPSS calculates chi-square to be 29.7 and Cramer's V to be 0.14, which is a low degree of association between perceived importance of brands for purchasing alcohol and intention to drink alcohol in the next year.

Cramer's V is, then, a good statistic. It can be used on any size of table. It can be used for binary, nominal or ordinal variables, or any combination like one binary and one ordinal. It is a fairly robust measure in that it can be applied in a wide range of circumstances and does not produce nonsensical results under certain conditions. For example, unlike some other statistics, it does not revert to zero simply because one or more cells are empty.

However, Cramer's V is not perfect and it has its limitations. First, although it is always zero when there is a complete lack of association between the two variables, when it is at the maximum value of 1.00, there may not be a 'perfect' association. If the table is square (e.g. a two-by-two, a three-by-three, a four-by-four, and so on) then $V = 1$ does indeed indicate that the variables are perfectly associated. However, if the table is non-square (e.g. a three-by-five) then V can reach unity even when there is clearly not a perfect association.

A second limitation is that since V depends on calculating chi-square, it must be amenable to the use of that statistic. This statistic assumes that the expected values are large. A rule of thumb that is commonly applied is that the statistic should be calculated only if fewer than about 20 per cent of all the cells have expected frequencies of less than five and that no cell has an expected frequency of less than one. (SPSS will warn you if this is the case.)

A third limitation is that the magnitude of achieved values of V has no direct interpretation, and, furthermore, it is not directly comparable with any other measure of association, except in the case of the two-by-two table, when it is identical to Pearson's r statistic, which will be explained in the next section.

Finally, the statistic is designed for use with binary and nominal variables. When it is being used for ordinal variables it does not violate any statistical assumptions and its use is legitimate, but it may be less informative than measures that are specifically for ordinal data. Thus *V* does not indicate a direction of association. This is fine for nominal variables where the notion of a 'negative' association does not apply. However, for ordinal variables it could happen that high values on one variable are associated with low values on the other. This would indicate a negative association, but *V* cannot be negative.

Despite these limitations, Cramer's V is an extremely useful coefficient because of its wide applicability. It does not make any distinction between dependent and independent variables. This may be seen as an advantage if the researcher does not wish to make such a distinction, but where he or she does, then other measures of association may be better.

Proportional reduction in error involves arguing that if two variables are associated then it should be possible to use knowledge of the values of one variable to predict the values of the other for each case. Thus if, in a survey, all the men say 'yes' they would purchase Brand X and all the women say 'no', then, for each person, if we know their gender, we can perfectly predict their answer to the question. In practice, of course, prediction is seldom perfect, but what we can do is to measure the extent to which knowledge of the value on one variable reduces the number of errors in predicting the other.

To generate a measure of association, we imagine that we are called upon to predict, for each case, which category of a dependent variable each is likely to exhibit. We do this first without any knowledge of the category of an independent variable for each case, and then see whether such knowledge enables us to improve our predictions. The proportion of errors that we can eliminate in this process is called the proportional reduction in error. There are many statistics that use this notion, and they may be referred to as 'PRE' statistics. Like those measures based on departure from independence, these measures are unaffected by the order of the categories and hence are also appropriate to two binary or nominal variables. Unlike departure from independence, however, all PRE measures require that we select one of the variables as the outcome to be predicted; choosing the other variable as the dependent one will produce a different result. In other words, all PRE measures are directional.

Table 5.9 Responses by gender

	Gender		
Answer	**Male**	**Female**	**Total**
Yes	90	20	110
No	10	80	90
Total	100	100	200

The appropriate PRE coefficient for binary or nominal variables is **lambda**. In Table 5.9, nearly all men say 'yes' and most of the women say 'no'. To calculate lambda, we need to calculate how many errors predicting the answer to the question we would make not knowing a person's gender. If we know the marginal totals of cases saying 'yes' and 'no', but not each individual case, then the 'best' guess we could make for each person is to predict that all answers are 'yes' (since there are more of these), and be wrong for 90 out of the 200 cases.

Suppose, now, we know a person's gender; then, knowing that a person is male and that 90 out of 100 males say 'yes', clearly it is sensible to predict 'yes' for all males and make only 10 errors. If, on the other hand, we predict 'no' for all 100 women, we make only 20 errors. That totals 10 + 20 = 30 errors in all. This may be compared with the original 90 errors. This is a 66 per cent reduction, or 0.66. Clearly, if we eliminate all errors, we will end up with a coefficient of 1.0. If we eliminate none, we will end up with 0.0.

It is difficult to generate a formula for lambda that you can use just to plug in selected values. Most formulae that have been generated are rather complex. It is probably best to be clear that lambda is based on the following calculation:

$$\frac{\text{number of errors eliminated}}{\text{number of original errors}}$$

In the above example, this is

$$\frac{90-30}{90} = 0.66$$

The SPSS output for lambda for Figure 5.5 (a), taking intention to drink alcohol as the dependent variable, shows a coefficient of 0.008. There is virtually no association according to this measure.

Lambda has an advantage over chi-square-based measures in that its interpretation is much clearer. Thus a result of 0.66 means that 66 per cent of errors can be eliminated by using knowledge of categories of the independent variable to predict categories on the dependent variable. The main drawback of lambda is that it will take on a numerical value of 0 in instances where all the other measures will not be zero and where we would not wish to refer to the variables as being statistically independent. This may occur where one of the row marginals is much larger than the rest, so no matter what the category of the independent variable, the prediction of the dependent variable category will be the same. In large tables, even if a single marginal total does not dominate, it is likely that some of the less numerous categories will not enter into the computation of lambda at all.

Covariation

Both departure from independence and proportional reduction in error reflect a tendency for binary or nominal categories to cluster together. Measures based on either of these approaches do not measure covariation, nor do they assess subset relationships or triangular patterns. However, making **pair-by-pair comparisons** assesses covariation and requires that both variables are ordered categories. This relies on measuring the tendency for all possible combinations of pairs of cases to show similar orderings on both variables. If there is a general tendency for this to happen, then there is a positive association. Relatively high values on one variable are associated with relatively high values on the other. Negative association means a tendency for relatively high values on one variable to be associated with relatively low values on the other.

SPSS lists four coefficients as appropriate for ordered category variables. Fortunately, they all are based on the principle of pair-by-pair comparisons. The key difference between them is in the way they treat 'tied' pairs. The simplest is **gamma**, which was developed by Goodman and Kruskal, and it calculates the number of pairs in a crosstabulation having the same rank order of inequality on both variables (concordant pairs) and compares this with the number of pairs having the reverse rank order of inequality on both variables (discordant pairs). Gamma, G, takes the difference between concordance and discordance and divides this by the total number of both concordant and discordant pairs. Its formula is

$$G = \frac{C - D}{C + D}$$

To obtain the number of concordant pairs, take each cell and multiply its frequency by the sum of the frequencies of all the cells below and to the right. Add together the results for each cell. Begin in the top left hand cell. In Table 5.3, only two cells are below and to the right. The next cell along has only one cell below and to the right. No other cells have others both below and to the right, so the calculation is

$$\begin{aligned} C &= 10(8+10) + 8(10) \\ &= 180 + 80 \\ &= 260 \end{aligned}$$

To calculate the number of discordant pairs the procedure is the same, except that we take cells below and to the left. So, begin in the top right hand cell. The calculation is

$$\begin{aligned} D &= 4(4+8) + 8(4) \\ &= 48 + 32 \\ &= 80 \end{aligned}$$

$$G = \frac{C - D}{C + D} = \frac{260 - 80}{260 + 80} = \frac{180}{340} = 0.52$$

Gamma varies between minus one and plus one. Zero indicates no association – the number of concordant pairs equals the number of discordant pairs. A result of 0.52 indicates a 52 per cent predominance of concordance over discordance, so there is a moderate degree of association between the two variables. Remember that G is a non-directional statistic – we do not need to distinguish between dependent and independent variables. Notice also that it is 'margin-free' – its value does not depend on the row or column marginals. Gamma can be applied to any size of table, but the variables must be in categories that can be ordered. It has the drawback, however, that it is overly sensitive to empty cells. This is because no matter how many cases in a cell, if multiplied by zero in an empty cell, the result will be zero. Gamma is simple to calculate and easy to understand. However, it is insensitive to types of pairing other than concordance and discordance.

The coefficient **Somers' d** is similar to gamma except that it takes account of those pairs that are tied on one variable but not on the other. Accordingly, it is a directional measure since there will be two results depending on which variable is taken as the independent variable. **Kendall's tau-b** is a still more stringent measure of association and includes ties in both directions, so it is a non-directional measure. However, it varies between minus one and plus one only when the table is square, that is the number of rows equals the number of columns. **Kendall's tau-c** makes an adjustment for the number of rows and columns so that it varies between minus one and plus one for any size of table.

Where both variables are ranked, SPSS offers only one coefficient: **Spearman's rho**. Recall that in a ranked variable each case is assigned a number from 1 to n to reflect its standing relative to other cases. There are as many ranks as cases to be ranked. Consider five types of ground coffee ranked 1–5 in terms of taste and aroma by five people as in Table 5.10. Spearman's coefficient is calculated from the formula

$$\text{rho} = 1 - \frac{6\sum d^2}{n\left(n^2 - 1\right)}$$

where n is the number of paired observations and d is the difference between ranks for each pair of observations. Where the ranks are identical then $d = 0$

Table 5.10 Coffee ranked by taste and aroma

Respondent	Taste	Aroma
A	5	3
B	3	4
C	1	1
D	2	2
E	4	5

Table 5.11 Calculation of rho

Respondent	Taste	Aroma	*d*	d^2
A	5	3	2	4
B	3	4	1	1
C	1	1	0	0
D	2	2	0	0
E	4	5	1	1
			Total	6

$$\text{rho} = 1 - \frac{6(6)}{5(25-1)} = 0.7$$

and rho = 1. The calculation for Table 5.10 is set out in Table 5.11. Rho varies between +1 and –1. A value of 0 indicates no association. To use rho on metric data it is necessary to convert the metric values into ranks. It has the advantage that, by doing so, it is unaffected by extreme values.

SPSS suggests one coefficient for situations where the independent variable is binary or nominal and the dependent variable is metric (or 'interval' or 'scale' in SPSS terminology), namely **eta**. Eta-squared is sometimes known as the **correlation ratio** and can be interpreted as the proportion of the total variability in the dependent variable that can be accounted for by knowing the categories of the independent variable. For other combinations of measures like one ordered category and one metric, there are other coefficients, but these are not available in SPSS. If you do have these combinations of measures, it would be more usual to treat the higher order measure at a lower level of complexity. So, for example, for an ordered category by nominal combination, treat them as two nominal measures.

Box 5.2 Using SPSS to obtain measures of association for categorical variables

For bivariate analysis of two categorical variables, click on the `Statistics` button in the `Crosstabs` dialog box and you will obtain the `Crosstabs:Statistics` dialog box.

SPSS features no fewer than nine coefficients of association that may be used to create bivariate summaries for categorical data. However, some of these are for situations where both variables are nominal, some where both are ordered category, and one for a mixture of nominal and metric. Select any of the statistics you require, click on `Continue` and then `OK`.

Spearman's rho is found under the `Correlation Coefficients` procedure which is explained in Box 5.3.

Key points and wider issues

For summarizing the relationships between two categorical variables, the researcher can think in terms of the size of the difference between two categories or groups, in terms of category clustering or in terms of covariation. Where either or both measures are nominal with three or more categories the researcher is restricted to category clustering, which can be assessed based either on departure from independence using a coefficient like Cramer's V or on proportional reduction in error using the coefficient lambda. Assessing covariation requires either two ordered category variables for which pair-by-pair comparisons can be made and incorporated into a coefficient like gamma, or two ranked variables for which the coefficient Spearman's rho would be used. Where both variables are binary (or nominal with two categories), then in addition to those coefficients appropriate for two nominal variables, ordered category coefficients can be used since two categories cannot be put 'out of order'. Where the two variables are based on different levels of measurement complexity, the situation then becomes rather more complex.

Where there is a choice of appropriate coefficient, it is likely that each will give a different answer on the same data. Accordingly, the same coefficient should be used if comparisons are being made between degrees of clustering or covariation for different pairs of variables. Ideally, a coefficient should be chosen that is appropriate for all the crosstabulations to be compared whatever their size and whatever the type of categorical measure. This means that the variable with the lowest level of measurement complexity will determine which coefficient to use in all the analyses. Any coefficient used should also vary between zero and one (or minus one) and it should be robust, that is, it should not be affected by the marginal distributions, by empty or low frequency cells, or by the table not being square. The coefficient Cramer's V is probably the best for meeting these criteria. However, it is more difficult to interpret since the result is only an index of a general tendency for the table to depart from what would be expected if there were no association. Lambda is easier to interpret, but it is directional. SPSS will produce three statistics, one for each variable taken as 'dependent', and one that combines the two, so you will need to pick the appropriate coefficient carefully – SPSS will not do it for you. If your research, for example, is looking at a range of factors that may be associated with a particular dependent variable, then it is better to use a directional measure, selecting this as the dependent variable for the appropriate statistic. If you are looking largely for patterns of relationships or interconnections between variables then choose a non-directional measure.

If a researcher constructs large crosstabulations with many cells, particularly when the number of cases is small, then there will be many empty cells or very small numbers in some cells. In these circumstances all coefficients lose their cool! It is probably wiser to collapse the tables to two-by-two or two-by-three before calculating measures of association. Also, bear in mind that few measures meet all the criteria for a good statistic, so what counts as the 'best' statistic to use in any particular situation is not always evident.

Bivariate data summaries: metric variables

Data summary for two metric variables involves calculating a coefficient similar to those for crosstabulations. The good news is that there is only one coefficient that is universally used – Pearson's r, otherwise known as the correlation coefficient or the product-moment correlation coefficient. The bad news is that it is a bit more difficult to explain than the measures we have looked at so far. If you have access to SPSS then it can be calculated for you at the click of a mouse button. However, the point is to understand how the statistic is calculated and what it is telling you.

If two metric variables are correlated, then high values on one variable are associated with high values on the other for a positive correlation and with low values for a negative one. 'High' or 'low' in this context is measured by the distance of any given value above or below the arithmetic mean for the variable concerned. This is achieved by subtracting the mean from each value. Each case will then have a corresponding distance from the mean for both variables. If a case is above the mean for both variables and we multiply the two distances, the result will be positive; similarly if a case is below the mean for both variables. If we add all these products for all cases together, we will get a large sum. However, if a case is below the mean on one variable and above on the other, the result will be negative. If this is true for many cases, we will get a large negative sum. If there is little correlation, then positive and negative values will tend to offset one another. So, the sum we are calculating – the **covariation** in technical parlance – is an indication of the extent to which the two variables are correlated. The formula is

$$\Sigma\left(X-\overline{X}\right)\left(Y-\overline{Y}\right)$$

where X and Y are individual values, and $\overline{X}$ and $\overline{Y}$ are the means of the X-values and Y-values respectively. If we divide this covariation by the number of cases, n, we get an average covariation called the **covariance**, which takes account of the fact that the mean of X and the mean of Y may be different. The two variables, however, may be in different units, for example age in years and income in euros. To take this into account, the covariance is divided by the **standard deviation** of X multiplied by the standard deviation of Y. This standardizes the covariance. If we divide both the numerator and the denominator by n, the formula for r becomes

$$r=\frac{\Sigma\left(X-\overline{X}\right)\left(Y-\overline{Y}\right)}{\sqrt{\left[\Sigma\left(X-\overline{X}\right)^2\right]\left[\Sigma\left(Y-\overline{Y}\right)^2\right]}}$$

Suppose four students (cases A, B, C and D) take two tests that are scored out of 10. The results are given in Table 5.12. You can see that there is a tendency for those who perform well on Test X also to perform well on Test Y. The calculation of Pearson's r for Table 5.12 is shown in Table 5.13.

Table 5.12 Scores of A–D on two tests

Individual	Test X	Test Y
A	0	1
B	6	2
C	6	4
D	8	5
Total	20	12

Table 5.13 The calculation of *r* from Table 5.12

	X	Y	$(X-\bar{X})$	$(Y-\bar{Y})$	$(X-\bar{X})(Y-\bar{Y})$	$(X-\bar{X})^2$	$(Y-\bar{Y})^2$
A	0	1	−5	−2	+10	25	4
B	6	2	+1	−1	−1	1	1
C	6	4	+1	+1	+1	1	1
D	8	5	+3	+2	+6	9	4
Total	20	12			16	36	10

$\bar{X} = 20/4 = 5$

$\bar{Y} = 12/4 = 3$

$$r = \frac{16}{\sqrt{36(10)}} = 0.843$$

Pearson's r itself is a little difficult to interpret (it is the coefficient of the slope for a standardized variable), but if we square the result to produce r^2 then we have what is often called the **coefficient of determination**, which gives the proportion of the variance on the Y observations that is accounted for or 'explained' by variations in X. Thus if $r = 0.84$ then r^2 is 0.71. This means that 71 per cent of the variance in Y is accounted for by the variance in X. If the value calculated for r^2 is very low or 0, do not assume, however, that there is no relationship between X and Y – it might be that the relationship is not linear but curvilinear. The best way to check is to draw or get SPSS to draw the scattergram.

Regression takes correlation a little further so that the researcher can actually predict the values of one variable from values of the other. It constructs an imaginary line called a regression line that goes through all the dots on a scattergram in such a way that the distances between all the points and the line are minimized. This is the 'best-fitting' line, which for the four students A–D in Table 5.12 is shown in Figure 5.7. The regression line may be described by a formula whose general form is

$$Y = a + bX$$

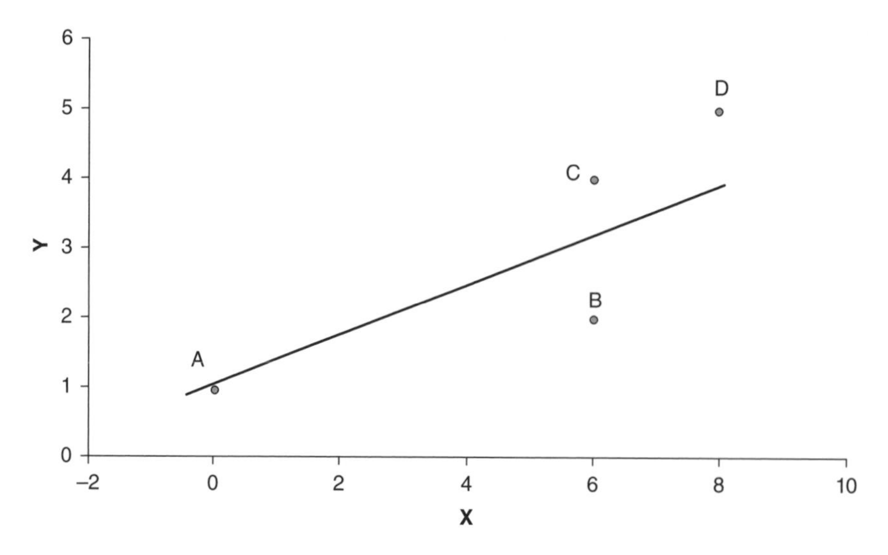

Figure 5.7 A scattergram of *X* on *Y* and regression line

where *Y* and *X* are values from the two metric variables and *a* and *b* are constants. The constant *a* indicates at what point the regression line cuts the vertical axis (it is the value of *Y* when *X* is zero), and *b* is the steepness of the slope of the line. In order actually to make a prediction of the value of *Y* from a particular value of *X* you would need to know what these constants are. The regression output from SPSS is shown in Figure 5.8. The constant *a* is 0.778 and *b* is 0.444, so the full equation is

$$Y = 0.778 + (0.444)X$$

This means that for any given value of *X* we can calculate what *Y* would be from the equation – and vice versa. Thus if $X = 6$, then *Y* is estimated as $Y = 0.778 + (0.444)6$ or 3.44. This is a fair estimate, since from Table 5.12, when $X = 6$, $Y = 2$ for one case and 4 for the other.

Correlation and regression are clearly related statistics, but they serve different purposes. Correlation is concerned with measuring the strength of the relationship between variables. It determines the extent to which the values of two or more metric variables covary. The square of the correlation coefficient

Coefficients[a]

Model		Unstandardized Coefficients	
		B	Std. Error
1	(Constant)	.778	1.168
	VAR00001	.444	.200

a. Dependent Variable: VAR00002

Figure 5.8 SPSS regression coefficients

measures how near on average each data point is to a hypothetical line that runs through the points in a way that minimizes the distances between each data point and the line. The procedure makes no distinction between dependent and independent variables – both variables have equal status.

Regression, by contrast, helps the researcher to determine the form of the relationship. The objective is to be able to predict or estimate the value of one variable corresponding to a given value of the other. It is concerned with identifying the parameters of the regression so that it can be used to make such predictions. However, which variable is chosen as the dependent variable makes a difference to the regression line that is derived. Regression is thus directional, but it is still based on the concept of covariation and in this sense is symmetrical.

In the alcohol marketing dataset a correlation between total importance of brands and total number of channels seen was carried out using SPSS. The results are shown in Figure 5.9. Notice that the lower box is a mirror image of the top box. Clearly each variable correlates with itself perfectly with a value of 1.000. The correlation coefficient between the two variables is 0.143. The correlation is based on 887 cases since some respondents did not answer the number of channels seen question. If you were to enter more variables in the `Variables` box you would get a series of bivariate correlations for each combination of variables.

The results of a regression analysis, taking total importance of brands as the 'independent' variable, are shown in Figure 5.10. The first table tells you what variables have been entered as independent variables and what the dependent variable is. The next table is a model summary, which gives the values for Pearson's r (0.143) and r^2, which is 0.021. This is the amount of variation in number of channels seen that can be explained by total importance of brands. If the data relate to a small sample, then r^2 gives a biased estimate of r^2 for the population from which the sample was drawn. The adjusted *r* is an unbiased estimate provided by adjusting the equation for r^2 for the number of values that were used in the original calculation. For large samples, the difference is very small. The next table is an analysis of variance (which is considered later in this chapter), while in the last table the unstandardized coefficients under B give the

		Total importance of brands	Total number of channels seen
Total importance of brands	Pearson Correlation	1	.143**
	Sig. (2-tailed)		.000
	N	920	887
Total number of channels seen	Pearson Correlation	.143**	1
	Sig. (2-tailed)	.000	
	N	887	887

**. Correlation is significant at the 0.01 level (2-tailed).

Figure 5.9 SPSS output for Pearson correlation

constants a and b in the equation $Y = a + bX$. So, the regression of total number of channels seen on total importance of brands provides

$$Y = 4.312 + (0.058)X$$

How to do correlation and regression on SPSS is explained in Box 5.3.

Variables Entered/Removed[b]

Model	Variables Entered	Variables Removed	Method
1	Total importance of brands	.	Enter

a. All requested variables entered.
b. Dependent Variable: Total number of channels seen

Model Summary

Model	R	R Square	Adjusted R Square	Std. Error of the Estimate
1	.143[a]	.021	.019	2.681

a. Predictors: (Constant), Total importance of brands

ANOVA[b]

Model		Sum of Squares	df	Mean Square	F	Sig.
1	Regression	133.498	1	133.498	18.575	.000[a]
	Residual	6360.592	885	7.187		
	Total	6494.090	886			

a. Predictors: (Constant), Total importance of brands
b. Dependent Variable: Total number of channels seen

Coefficients[a]

Model		Unstandardized Coefficients		Standardized Coefficients	t	Sig.
		B	Std. Error	Beta		
1	(Constant)	4.312	.373		11.567	.000
	Total importance of brands	.058	.014	.143	4.310	.000

a. Dependent Variable: Total number of channels seen

Figure 5.10 SPSS regression output

Box 5.3 Correlation and regression on SPSS

To correlate two metric variables select `Statistics|Correlate|Bivariate`. To obtain Figure 5.9, put total importance of brands and total number of channels seen into the `Variables` box. Under `Correlation`

Coefficients you will find the Pearson box already ticked. Spearman's rho can also be obtained from this dialog box. Click on OK.

To obtain a regression analysis, select Analyze|Regression| Linear. Notice that you now have to make a selection of dependent and independent variables. The dependent variable goes along the Y-axis; it is the one the researcher is trying to predict. You can have one or more than one independent variable if you wish to undertake multiple regression. This is explained in the next chapter. In discussing the bivariate linear relationship between *X* and *Y* statisticians speak of the 'regression of *Y* on *X*'. In Figure 5.6, total importance of brands has been put along the X-axis, so is being treated as the 'independent' variable, while number of channels seen is the variable being predicted. Put these into the appropriate boxes and click on OK.

Key points and wider issues

To measure the degree of correlation between two metric variables, Pearson's r, or its derivative, the coefficient of determination r^2, is the standard statistic. Regression allows the researcher actually to make predictions from one metric variable to another; the accuracy of these predictions, however, depends on how close the data points lie to the regression line. The researcher needs to be aware, furthermore, that this procedure is only evaluating the degree of **linear regression**, that is the pattern approximates a straight line. There are, however, procedures for applying **curvilinear regression**, and for **multiple regression** when there are two or more independent variables. Both correlation and regression assume that the variables involved are distributed normally, otherwise the mean and the standard deviation are not good descriptors of the data.

Testing bivariate hypotheses

We saw in Chapter 4 that univariate hypotheses could be tested for statistical significance in circumstances where the data being analysed represent a random sample from a wider population of cases. This was achieved by using the concept of a sampling distribution to calculate the probability that the sample data could have come from a population of cases in which the stated hypothesis is true. For bivariate hypotheses linking two variables, statisticians test what they call the **null hypothesis** that there is no difference, no association or no correlation in the population from which the sample was drawn. If the null hypothesis can be rejected, then researchers will conclude that the sample result was unlikely to have come from a population in which the null hypothesis is true, that random sampling fluctuation is an unlikely explanation of the discovered pattern and that there is, indeed, some difference, association or correlation.

While researchers would like to accept a univariate hypothesis, they are hoping to reject a bivariate null hypothesis. The benefits of doing it this way are: first, it means that the null hypothesis is clear; second, it is not necessary to specify in advance a degree of difference, association or correlation that is to be accepted or rejected; and third, that, logically, it is better to reject a proposition (or fail to reject it) than to accept it.

Testing bivariate hypotheses for categorical variables

If a null hypothesis of no difference or no association between two categorical variables is to be used, then the obvious basis on which to detect departures from this condition in the sample data is departure from independence. The statistic chi-square, you will recall, is based on the difference between observed frequencies and the frequencies from some theoretical expectation. The theoretical expectation for **bivariate hypothesis** is that there is no difference or association between the two variables. The calculation of chi-square was explained earlier in this chapter. Using chi-square as a test of significance involves taking the calculated value of the statistic and looking up in a table of the distribution of chi-square the probability of obtaining from a random sample a value as large or larger than the one generated.

For the crosstabulation we saw earlier in Figure 5.5(a), SPSS calculated chi-square to be 29.7 and Cramer's V to be 0.14, which is a low degree of association between perceived importance of brands for purchasing alcohol and intention to drink alcohol in the next year. The SPSS output for the lower two-by-three table (Figure 5.5(b)) is shown in Figure 5.11. This shows a Cramer's V for 774 cases of 0.105, which is even lower than for the original five-by-three table. The number under `Approx.Sig.` (0.014) is the *p*-value, which was introduced in the previous chapter. The ***p*-value** is the probability of getting a difference between a sample result and the null hypothesis (which is $V = 0.0$) as big or bigger than the one we actually get, given the null hypothesis. If it is less than the significance level chosen for the test (e.g. 0.05) then the hypothesis is rejected; if it is more than the significance level then it is accepted. Here, the *p*-value is less than 0.05 so the null hypothesis is rejected. All this means is that, even though the level of association is very low, given the large sample size, the achieved result of 0.105 cannot be explained as being the likely result of random sampling fluctuation. The value of chi-square for this table is 8.492, which is a lot less than 29.7 for the larger table. Larger tables will produce higher values of

		Value	Approx. Sig.
Nominal by Nominal	Phi	.105	.014
	Cramer's V	.105	.014
N of Valid Cases		774	

Figure 5.11 SPSS output for phi and Cramer's V for Figure 5.5(b)

chi-square simply because there are more differences between observed and expected frequencies.

Testing bivariate hypotheses for metric variables

Figure 5.6 (p. 121) shows the scattergram of the relationship between the two metric variables in the alcohol marketing dataset, and, clearly, there is not much of a pattern. However, the output from SPSS showing the Pearson's r correlation between the total importance of brands and the total number of channels seen is illustrated in Figure 5.9 (p. 135). It shows that there is a degree of linear correlation of $r = 0.143$. Even though r^2 is only 0.02 (2 per cent of the variance in number of channels seen is explained by the total importance of brands), it is still statistically significant with a *p*-value of less than 0.000. The researcher can reject the null hypothesis that there is no association between the two variables (or, more correctly, reject the notion that the result came from a population in which the null hypothesis is true). The test is usually a two-tailed test: this assumes that the test is non-directional and that there was no expectation as to the sign of the population value, positive or negative. Like all significance tests, the result is always sensitive to the size of the sample. Thus, for a sample of 30, the Pearson correlation needs to be greater than 0.36 to be significant at the 0.05 level. For a sample of 100 it needs to be larger than 0.2, and for a sample of 1,000 larger than 0.062. In short, statistically significant results for large samples do not necessarily imply that the degree of correlation is large or even of any interest to the researcher.

Testing metric differences for categories

There are many situations in social research where researchers have, or assume that they have, a metric dependent variable, like the results of deploying **summated rating scales**, and want to check whether there are any differences in mean scores between categories of respondents in a survey that are probably not due to random sampling error. Do the men, for example, have different mean scores from the women? Do those who grew up in a single-parent family give different average responses to those who grew up in families where both parents were together? The researcher has, in short, a metric dependent variable and one or more categorical independent variables. As was explained above, the categories may be described as 'groups' or 'samples' and the researcher will ask whether or not differences discovered among survey respondents between various groups (or samples) could have arisen from a population in which there is, in fact, no difference (or whether, in short, the result is statistically significant). In this situation, researchers will use a statistical procedure called **analysis of variance (ANOVA)**.

ANOVA recognizes that there will be variations in individual scores within groups as well as differences between groups. The logic of ANOVA is that it compares the variance of scores between the groups with the variance within the groups. If the variance between groups is larger, we conclude that the groups differ; if not, we conclude that the results are not statistically significant. The steps involved are as follows:

1. Calculate the mean score of each group.
2. Calculate to what extent individual scores within the groups vary around the group mean (this is done by calculating the variance – the standard deviation squared).
3. Calculate the overall mean for all the groups, that is the whole sample.
4. Calculate how much group means vary about the overall mean.

These steps enable us to derive two quantities: the between-group variance and the within-group variance. The ratio between the two is called the *F*-ratio, calculated by dividing the between-group variance by the within-group variance. When *F* is large, between-group variance is significantly greater than within-group and there are 'significant' differences between the groups. To determine whether the ratio is sufficiently large we need to compare it with a critical value derived from a table of the distribution of *F*.

When there is only one basis for classifying groups, then it is called **one-way ANOVA**. Two-way ANOVA is used when there are two independent variables; that is, groups are classified in two ways, for example by gender and employment status. This means that it can be considered as a multivariate technique since a total of three variables is involved, although what is normally referred to as **multivariate analysis of variance (MANOVA)** refers to situations when there is more than one metric dependent variable. Two-way, *n*-way and MANOVA are explained in the next chapter.

Figure 5.12 shows the SPSS output for each one-way ANOVA for the three metric variables in the alcohol marketing dataset taking gender as the two groups being compared. Only total number of channels seen is statistically significant; that is, given the null hypothesis that there is no difference between males and females, the likelihood of obtaining the difference between number of channels seen is less than 0.05 (in fact it is 0.011).

ANOVA

		Sum of Squares	df	Mean Square	F	Sig.
Total importance of brands	Between Groups	22.343	1	22.343	.499	.480
	Within Groups	41074.922	918	44.744		
	Total	41097.265	919			
Total number of channels seen	Between Groups	47.204	1	47.204	6.480	.011
	Within Groups	6446.886	885	7.285		
	Total	6494.090	886			
Total involvement	Between Groups	3.232	1	3.232	2.675	.103
	Within Groups	630.684	522	1.208		
	Total	633.916	523			

Figure 5.12 SPSS output for one-way ANOVA

ANOVA was developed for use mainly in experimental set-ups where the differences being tested are between experimental groups and control groups. To use ANOVA properly, four key conditions need to be met:

- the respondents whose results are being compared must be a random sample;
- the values of the dependent variables must be continuous metric and normally distributed within each group;
- the variances within the groups must be approximately equal;
- the groups or categories must be 'independent' samples.

The last term requires some explanation. Recall that the categories of binary or nominal variables are sometimes called 'groups' or 'samples'. So, in comparing the results, for example, of male and female respondents, we have two 'samples', one sample of males and another of females. Strictly speaking, these samples should be 'treatments'. Thus a researcher may decide to compare the crop yields of fields that have been randomly allocated with the use and non-use of genetically modified seeds. We cannot, of course, randomly allocate individuals to their gender, but provided the selection of respondents has been made randomly, we can consider this to be the equivalent of two separate samples of males and females even though in reality they are from the same sample of respondents. The samples are 'independent' in the sense that the scores of the males and the scores of the females are independent of one another. This might not be the case if, for example, the males and females involved were married couples and so their attitudes to a whole range of things may well be related.

Many researchers use ANOVA on the results of surveys even though it is very difficult to meet all the conditions necessary for its use. Samples are often not random or not entirely random, or random with very poor response rates, the distribution of values within groups may not be normal and the variances may well not be even approximately equal. The use of ANOVA on rating scales like Likert scales may be dubious if the variables cannot be considered to be approximately scaled at equal intervals. Remember that ANOVA depends on statistical inference. The end result of any analysis is always a *p*-value. Like all techniques that use probability theory, it takes account only of random sampling fluctuations, not any other source of error.

Box 5.4 Using SPSS to test for statistical significance

To test the statistical significance of a bivariate hypothesis linking two categorical variables, SPSS provides the statistical significance of chi-square. Check the tick-box against chi-square on the `Crosstabs:Statistics` dialog box (select `Analyze|Descriptive Statistics| Crosstabs|Statistics`).

(Continued)

(Continued)

To test the statistical significance of a bivariate hypothesis linking two metric variables, SPSS provides a *p*-value for Pearson's r which appears automatically on the `Pearson Correlation` procedure as illustrated in Figure 5.9 and is shown as 0.000. SPSS only goes to three decimal places so this result might be reported as $p < 0.001$.

One-way analysis of variance is to be found under `Analyze|Compare Means|One-Way ANOVA`. To obtain Figure 5.12, put the three metric variables, namely total importance of brands, total number of channels seen and total involvement, into the `Dependent List` box and sex of respondent under `Factor`. Click on `OK`.

Key points and wider issues

When testing bivariate hypotheses linking two variables, the null hypothesis is always that there is no association or correlation. Where both variables are categorical, chi-square may be used as a test of significance. Where both are metric, the statistical significance of Pearson's r will be generated by SPSS. Where measures are mixed, provided the independent variable is categorical and the dependent variable metric, then analysis of variance may be used. This is based on comparing the variance of the metric variable within the categories with the variance between the categories. Although analysis of variance is commonly used by researchers, the conditions for its proper use are seldom met, particularly when the data were derived from surveys rather than experimental designs.

Testing a null hypothesis for statistical significance only means establishing the probability that the survey result came from a population of cases in which the null hypothesis is true. A statistically 'significant' result does not mean that there is a high level of association or correlation, or even that the difference or the degree of covariation is worth looking at. For large samples of over 300 or so, even very small levels of covariation become statistically significant.

Statistical inference and bivariate data summaries

These two procedures are closely related together yet they perform very different functions. Statistical inference is concerned with the connection between a sample result and the corresponding value in the population from which the sample was drawn. It is trying to infer things about the population from evidence in the sample. Bivariate data summary, by contrast, utilizes descriptive statistics to pick out features in a dataset, whether these are a sample from a wider population, whether they constitute the entire population or whether they are the result of an attempt to reach the entire population. Thus measures of association like Cramer's V evaluate the strength of the relationship between two variables in the data we have before us.

It may help if you are clear about how the two procedures are related. If a researcher takes a random sample, it will be true that the stronger the relationship between the two variables – as measured by a particular coefficient – the more likely it is to be statistically significant. In other words, high values of Cramer's V, for example, are unlikely to have come from a population in which there is no association between the two variables. We saw in Figure 5.11 (p.138), for example, that a Cramer's V of 0.105 was significant at the 0.05 level, but it would not be at the 0.01 level (having a *p*-value of 0.014). This gives some idea of the value of *V* that separates significant and non-significant results. However, this cannot be applied throughout the research. This is because the significance of any one particular result will depend on the number of cases used in the calculation and on the size of the crosstabulation. We saw in Chapter 1 that the number of cases used in a given calculation does not necessarily equal the size of the sample and may well vary from table to table. This will affect the value of chi-square. Basically, if you double the number of cases, you will double the size of chi-square for a table that shows essentially the same pattern of association. The size of the table, furthermore, affects the number of degrees of freedom, which determines the critical value of chi-square. In short, simply because a Cramer's V of 0.105 is statistically significant on one table in a piece of research does not mean that it is on another.

Furthermore, a 'statistically significant' result does not mean that there is a high degree of association. It simply means that the result probably cannot be explained away as a result of random sampling fluctuations. As samples get larger, so smaller degrees of association become statistically significant. For a two-by-two table, for example, with one degree of freedom, the critical value of chi-square is 3.841 at the 0.05 level of probability, so any chi-square value greater than this will lead us to reject the null hypothesis. Cramer's V (or phi) for a sample of 100 cases when chi-square is 3.841 will be $\sqrt{(3.841/100)}$ = 0.196. This is a very small degree of association. To the researcher, this is more likely to be of value as a negative finding (the degree of association is very small) rather than as a positive finding (gee whiz, the association is statistically significant!). With larger samples, even smaller values of Cramer's V can be significant, so for a sample of 1,000, a *V* of 0.06 will be statistically significant for a two-by-two table at the 0.05 level. It can be argued that, for larger samples, the critical value for p should not be $p < 0.5$, but $p < 0.1$ or even $p < 0.001$.

The calculations for statistical inference explained in this chapter and in the previous chapter for univariate analysis make a number of key assumptions:

- the result being tested comes from a simple random sample;
- there is no bias in the selection procedure;
- non-sampling errors can be ignored.

If the sample is not a simple random one, for example it is a stratified, clustered or multi-stage sample, then researchers may do a number of things:

- adjust the standard error according to the particular sample design;
- apply a design factor;
- ignore the problem and perform the calculations anyway.

Stratification is the random selection of cases, but within strata, for example selecting a certain number of males at random and a certain number of females at random. This would be stratification by gender. The effect of stratification is to decrease the sampling error because it removes some of the variation that might otherwise arise from under- or over-representation of cases reflecting the stratification factor, for example having too many females or too many males. By contrast, clustering is the random selection of cases within defined geographical areas and has the effect of increasing sampling error.

Statisticians have suggested various ways in which the standard error can, indeed should, be adjusted when these techniques are used. However, the calculations are complex, while stratification and clustering are often in practice combined so the overall effect may be difficult to determine.

An alternative that is not commonly considered is to apply an overall adjustment to the standard error based on approximate rules of thumb. This overall adjustment is called a 'design factor'. Koerner (1980) suggests, for example, that simple stratified random samples should have a design factor of 0.8–0.9 (so the standard error is, in fact, decreased lightly), but if used in a multi-stage design should be between 1.05 and 1.9. Clustering increases error depending on the number of cases or sampling units in the cluster and should be between 1.02 and 1.26. Unsophisticated quota samples should have a design factor of about 2.0, but complex interlocking quotas are less prone to error and should be between 1.22 and 1.4 according to Koerner (1980).

In practice, many researchers will make estimates of population values or test null hypotheses irrespective of the sample design. However, ignoring the effects of stratification, clustering, imposing quotas or multi-staging should mean that the conclusions drawn from statistical inference should be treated with extreme caution. If the sample taken by the researcher was not selected by random methods (e.g. it was a convenience sample or a quota sample) then it would seem rather pointless to be fine-tuning our application of probability theory which is based on having a **probability sample**. Quota sampling is a non-random form of sampling because it is the interviewer who actually makes the selection of who to approach. Different approaches, however, are taken about the applicability of statistical inference to quota samples. One approach argues that quota sampling, if done carefully, is a good approximation of a random sample and the application of statistical tests tells us something that we would not otherwise know. However, a design factor should be applied. A more purist approach is to argue that, since quota sampling is non-probability sampling and statistical inference relies on probability theory, then statistical inference should not be applied at all.

Key points and wider issues

The procedures for testing hypotheses for statistical significance were developed largely for use on experimental datasets. Books on statistics are apt to devote the majority of their pages to such tests, often picking out particular tests for given experimental designs, for example tests for differences between means for one sample, two samples or three or more samples, tests for independence, binomial tests, tests for rank order or for correlation, tests for repeat measures, or non-parametric tests. Diamantopoulos and Schlegelmilch (1997), for example, consider six tests for univariate hypotheses and 12 for bivariate; they do try to explain the circumstances in which each should be used. Most tests are available on SPSS and they all end up with a *p*-value as the final result. They are all based on the same logic, namely calculating the probably of obtaining a difference, a clustering or a covariation as large or larger than the one between the sample result and some hypothesized value, as a result of random sampling fluctuations, assuming a perfectly random sample with no other sources of error.

Significance tests cannot be used with population data, or with incomplete or non-random samples. They do not address missing data, measurement error, bias in sample selection, or any of the many sources of non-sampling error. In short, they are seldom relevant in real-life social research in the absence of experimental controls. Even in their own terms and thinking only about random sampling error, they are heavily circumscribed in terms of what they mean. They do not provide a probability that any hypothesis is true or false, given the data; rather they calculate the probability of obtaining the data in the sample, given the hypothesis that is being assumed to be true. Unfortunately, they are often treated as 'tests' of a finding, rather than the probability of making a mistake if the null hypothesis is assumed to be true. Rejecting a null hypothesis does not 'prove' the alternative; indeed, it says nothing about the alternative – it only suggests that the alternative may be worth investigating further. This may be by way, for example, of evaluating hypotheses against other criteria (which are considered in Chapter 10) or elaborating a potential bivariate relationship by controlling for other factors using multivariate analysis, which is considered in the next chapter. In short, a 'statistically significant' result may not be an important or an interesting result, particularly since, for large samples, even very small differences may lead to the rejection of a null hypothesis.

Implications of this chapter for the alcohol marketing dataset

The key research hypotheses as stated by the researchers (Gordon et al., 2010a) are that the more aware of and involved in alcohol marketing that young people are, the more likely they are to have consumed alcohol, and the more likely they are to think that they will drink alcohol in the next year. The implications of this chapter are that we need to think about what sorts of relationships are being proposed here. The hypotheses as stated can be seen as a series of bivariate hypotheses, for example:

H_1 The more aware young people are of alcohol marketing, the more likely they are to have consumed alcohol.

H_2 The more young people are involved in alcohol marketing, the more likely they are to have consumed alcohol.

H_3 The more aware young people are of alcohol marketing, the more likely they are to drink alcohol in the next year.

H_4 The more young people are involved in alcohol marketing, the more likely they are to drink alcohol in the next year.

In H_1 and H_2 the variable drink status, as measured by the answer to the question 'Have you ever had a proper alcoholic drink – a whole drink, not just a sip?', is binary, so the relationship could be seen as a difference in awareness (or involvement) between those who said 'Yes' and those who said 'No'. Awareness is measured by totalling the number of channels on which young people have said they have seen alcohol advertising, so it is discrete metric. The average total for those who said 'Yes' is 6.35 and 5.61 for those who said 'No'. So, drinkers are a bit more likely to be aware of alcohol advertising. For H_2, similarly, drinkers are a bit more likely to be involved in alcohol marketing.

The way the researchers have phrased the hypotheses implies that drink status is the dependent variable and that awareness and involvement must have happened before the drink took place. However, it is also possible that young people are more likely to be aware and be involved in alcohol marketing if they have already taken a drink. If awareness and involvement are taken as the (metric) dependent variables and drink status as the categorical independent variable, then a one-way ANOVA (or an independent samples *t*-test) suggests that the difference in awareness and in involvement between those who have and have not had a proper alcoholic drink is statistically significant ($p = 0.000$). However, remember that the sample is not a random one and could even be seen as a total population – all second-year pupils attending schools in three local authority areas in the west of Scotland. So statistical inference may not appropriate. Even if it were a random sample, then with a sample size of 920, any differences, however tiny, would be statistically significant.

Another way of treating H_1 and H_2 is as a relationship of covariation. Thus for H_1, it could be that those who are aware of alcohol advertising are likely to have had an alcoholic drink and those who are not aware are unlikely to have had an alcoholic drink. If awareness is recoded into two categories, then these two variables can be crosstabulated, as in Figure 5.13. This shows that more of those who had seen alcohol advertising on six or more channels were drinkers compared with those who had seen five or fewer channels (42.5 per cent compared with 26.4 per cent, a percentage difference of 16.1). Cramer's V is 0.169, a low degree of covariation, but, if the sample had been random, would nevertheless have been statistically significant. Crosstabulations and coefficients of association like Cramér's V, however, say nothing about direction, so Figure 15.3 could be equally supportive of the hypothesis that those who have had an alcoholic drink are more likely to be aware of alcohol advertising. Even for a directional coefficient like lambda, the researcher still needs to decide which variable to treat as 'independent'.

			Channels seen into two categories		
			Up to 5 channels seen	Six channels or more seen	Total
Have you ever had a proper alcoholic drink?	Yes	Count	120	198	318
		% within Channels seen into two categories	26.4%	42.5%	34.6%
	No	Count	334	268	602
		% within Channels seen into two categories	73.6%	57.5%	65.4%
Total		Count	454	466	920
		% within Channels seen into two categories	100.0%	100.0%	100.0%

		Value	Approx. Sig.
Nominal by Nominal	Phi	.169	.000
	Cramer's V	.169	.000
N of Valid Cases		920	

Figure 5.13 Drink status by number of channels seen

With H_3 and H_4, drink status is replaced by intentions, which is an ordered category variable.

If awareness and involvement are converted into ordered category variables (using the total scores for these variables and the `Recode` facility on SPSS), then the coefficient gamma can be used to measure the strength of the relationship. However, there are various ways in which the recode can be done. As a general rule, it is better to keep the number of categories low and to make them roughly equal in terms of frequencies. Figure 5.14(a) shows, for example, that just over a third had seen alcohol advertising on up to four channels, almost a third on five or six channels, and the remainder on seven or more. Figure 5.14(b) shows awareness converted into a three-category variable and in Figure 5.15 this is crosstabulated against the likelihood that young people will drink alcohol in the next year (also converted into three categories). The coefficient gamma shows a degree of covariation of 0.187 such that the more channels seen, the more likely young people say they are to drink alcohol in the next year, but the covariation is quite small.

In summary, in terms of a bivariate analysis of the key research hypotheses, the relationships in terms of differences, category clusterings and covariations are all in the expected or predicted direction, but are quite small. Given that the sample is not a random one, saying that these are all, nevertheless, 'statistically significant' is not helpful and not meaningful. Bivariate analysis, however, has its limitations and multivariate procedures may throw further light on the dataset. These procedures are explained in the next chapter.

(a)

		Frequency	Percent	Valid Percent	Cumulative Percent
Valid	0	33	3.6	3.6	3.6
	1	40	4.3	4.3	7.9
	2	56	6.1	6.1	14.0
	3	84	9.1	9.1	23.2
	4	104	11.3	11.3	34.5
	5	137	14.9	14.9	49.3
	6	121	13.2	13.2	62.5
	7	106	11.5	11.5	74.0
	8	83	9.0	9.0	83.0
	9	71	7.7	7.7	90.8
	10	43	4.7	4.7	95.4
	11	24	2.6	2.6	98.0
	12	3	.3	.3	98.4
	13	11	1.2	1.2	99.6
	14	3	.3	.3	99.9
	16	1	.1	.1	100.0
	Total	920	100.0	100.0	

(b)

		Frequency	Percent
Valid	Up to 4 channels	317	34.5
	5 or 6 channels	258	28.0
	7 or more channels	345	37.5
	Total	920	100.0

Figure 5.14 Number of channels on which the ads for alcohol have been seen

Count

		Number of channels seen			
		Up to 4 channels	5 or 6 channels	7 or more channels	Total
Intention to drink alcohol	Definitely or probably not	197	133	163	493
	Not sure, DK or no answer	51	43	64	158
	Definitely or probably yes	69	82	118	269
Total		317	258	345	920

Figure 5.15 Intention to drink by number of channels seen

Chapter summary

There are many different kinds of relationship between two variables, including difference, category clustering, covariation, subset relationships and triangular patterns. Traditional statistical methods can address only the first three of these. Researchers will often begin bivariate analysis with bivariate data display. This might be used in a preliminary diagnostic context in which the researcher runs a number, perhaps a large number, of key crosstabulations to get a feel for the nature of relationships between variables two at a time, or plots scattergrams to check whether relationships between metric variables are roughly linear. In a presentational context, the researcher may use tables and graphs selectively to illustrate key findings in a report.

To obtain summaries of the relationship between two variables it is possible to review the percentage difference between categories in respect of another variable, but this really only works if both variables are binary. It is more effective to use one of the many coefficients that are available. These coefficients are designed to vary from 0 for no association or correlation to +1 for a perfect association or correlation. Where the relationship is negative (and this is possible only for ordinal or metric variables) a maximum value of –1 may be achieved. For two metric variables the standard coefficient is Pearson's r, but for categorical variables there is a large number of coefficients, nine of which are provided in the SPSS `Crosstabs` procedure. Measures appropriate for nominal variables may be based either on the idea of departure from independence or on the notion of proportional reduction in error. Ordinal measures are based on the principle of pair-by-pair comparisons. Each statistic has its own strengths and limitations, so selecting the appropriate statistics to use can be quite complex. For the most part, Cramer's V is probably the best for nominal variables and gamma for ordinal, but if researchers want to take account of which variables are playing an independent role in the research, then lambda may be the preferred statistic.

When testing bivariate hypotheses for statistical significance, the null hypothesis is always that there is no association or correlation. Where both variables are categorical, chi-square is normally used as a test of significance. Where both are metric, the statistical significance of Pearson's r will be generated by SPSS. Where measures are mixed, provided the independent variable is categorical and the dependent variable metric, then analysis of variance may be used. This is based on comparing the variance of the metric variable within the categories with the variance between the categories. Although analysis of variance is commonly used by researchers, the conditions for its proper use are seldom met, particularly when the data were derived from surveys rather than experimental designs.

Testing the null hypothesis only means establishing the probability that the survey result may have come from a population of cases in which the null hypothesis is true. It says nothing about the probability that the null hypothesis actually is true. A statistically 'significant' result does not mean that there is a high level of association or correlation, or even that the difference or the degree of covariation is worth looking at. For large samples of over 300 or so, even very small levels of covariation become statistically significant.

Statistical inference plays a very different role in a piece of research from data summaries. The latter describe features or patterns in the data researchers have before them; statistical inference calculates the probability that these results could have been a random sampling fluctuation. The two procedures are linked in that bigger differences or stronger associations or correlations are more unlikely to have been a random sampling fluctuation; but when samples are large, even quite small differences or weak associations can be statistically significant. In short, statistically significant results are not necessarily important findings. The usefulness of statistical inference for non-experimental datasets has been called into question by many authors and researchers, given that the conditions for its correct use are seldom met in real-life research. It might be better instead to pay more attention to investigating the nature of the relationships between variables, first two at a time and then elaborating this into three or more variables.

Exercises and questions for discussion

1. Bivariate data summary using SPSS is so quick and simple that the temptation must be to crosstabulate everything in sight. Is this a good idea?
2. What result from Cramer's V do you think would count as a 'high' degree of association?

3. Access the alcohol marketing dataset (available at https://study.sagepub.com/kent).

   (i) Crosstabulate brand importance for alcohol by `likeads` and request the coefficients gamma and Cramer's V. Is there any discernible pattern? Try collapsing the table to a three-by-three. Have the coefficients changed?
   (ii) Regroup total importance of brands (`totbrand`) into three categories and crosstabulate against gender, requesting the coefficient Cramer's V. Interpret the results.
   (iii) Recreate Figure 5.15, requesting both gamma and Cramer's V. Can you explain why the two statistics differ?
   (iv) Redo the analysis for (ii) above, but using total involvement.

4. Access the `Trinians` dataset and reread the *Folio* article (see Exercise 6, Chapter 2 for instructions). The researchers have taken one item from the very last question (a semantic differential on the sixth item, left wing ... right wing) and made it the 'outcome' variable to study. However, they have just compared percentages on an item-by-item basis. Try turning the items in Q27 (how often certain protest actions are justified) into a summated rating scale and correlate with the left-wing/right-wing scores. Interpret your answer and compare with the comments in the *Folio* article. Would a test of statistical significance be appropriate?

Further reading

Argyrous, G. (2011) *Statistics for Research with a Guide to SPSS*, 3rd edn. London: Sage.

Chapter 5 covers using crosstabulations to investigate the relationships between variables. Chapter 6 focuses mainly on the coefficient lambda for nominal variables and Chapter 7 on gamma for ordered category variables. Note that Argyrous says this statistic is for 'ranked' data, but not ranked in the sense used in this text. Correlation and regression are considered in Chapter 12.

Diamantopoulos, A. and Schlegelmilch, B. (1997) *Taking the Fear Out of Data Analysis*. London: Dryden Press. Republished by Cengage Learning, 2000.

Chapter 13 is brief, but it explains Cramer's V, Spearman's rho and Pearson's r.

De Vaus, D. (2002) *Analyzing Social Science Data: 50 Key Problems and Data Analysis*. London: Sage.

Part 6 is on how to analyse two variables. Problems 32, 35 and 37 are particularly useful; 39 and 40 can be skipped.

Yang, K. (2010) *Making Sense of Statistical Methods in Social Research*. London: Sage.

Chapter 6 on studying the relationship between two variables is well worth reading. Be warned, however, that for two nominal variables, Yang focuses on the chi-square test for evidence of the existence of a relationship (but presuming a random sample), and for the strength of the relationship recommends the odds ratio.

Suggested answers to the exercises and questions for discussion can be found at the end of this text, pp. 293–321, and on the companion website, (https://study.sagepub.com/kent), which also give links to relevant free online Sage journal articles, PowerPoint slides, an overview of data analysis packages, an introduction to SPSS and weblinks to alternative datasets.

6

MULTIVARIATE ANALYSIS

Learning objectives

In this chapter you will learn about:

- the limitations of bivariate analysis;
- the different types of multivariate analysis and the search for patterns in a dataset;
- the multivariate analysis of categorical variables using three-way and *n*-way tables or log-linear analysis;
- the multivariate analysis of metric variables, and metric and categorical variables mixed, using the dependence techniques of multiple regression, logistic regression and multivariate analysis of variance, and the interdependence techniques of factor analysis, correspondence analysis and multidimensional scaling;
- using SPSS for multivariate analysis;
- how a multivariate analysis of the alcohol marketing dataset suggests that the tendency for those who had seen alcohol advertising on more channels to be more likely to say that they will probably drink alcohol in the next year is stronger for females than it is for males.

Introduction

Bivariate analysis enables researchers to study the patterns of relationships between variables two at a time. However, unfortunately, the real world does not come neatly packaged into isolated pairs of variables, and attempts to account for differences, category clusterings or covariations in a two-by-two fashion will at best be limited, but more likely distorted or even misleading. Multivariate analysis, by contrast, enables the researcher to detect patterns buried in complex, interrelated variables, to explore the degree of influence two or more variables jointly may appear to exert on other variables, or to make predictions of the values of outcome variables based on knowledge of the values of selected independent variables. Although there are simple structures underlying multivariate analysis, the techniques themselves tend to be complex and sophisticated. At the same time, while these techniques have their undoubted strengths, they also have their limitations and they are easily misinterpreted or misused.

The limitations of bivariate analysis

It is perfectly feasible to investigate the relationships between a set of variables two at a time using bivariate analysis, which was explained in the previous chapter. If, for example, there is a single dependent variable or outcome, it is possible to examine the degree of covariation between that variable and all other potential causal factors in the dataset and to say which ones are the most closely associated with the dependent variable. However, there are notable limitations to such analyses that arise at three levels: descriptive, inferential and relational.

At a descriptive level, bivariate analysis may tempt researchers into various degrees of **data dredging**. 'Dredging' normally means casting a net or other mechanical device to trawl whatever is on the sea bottom. In the context of data analysis it is a metaphor for a process of 'trawling' a dataset without specific hypotheses to test, or perhaps even without hunches or specific issues to pursue. In its most extreme form the data are explored in every conceivable way to see what patterns emerge. Researchers may, for example, crosstabulate every variable by every other variable in the dataset in order to see which ones produce the strongest associations. This can easily be achieved using the SPSS `Crosstabs` procedure, but will rapidly produce a lot of tables. If researchers were to crosstabulate each variable by each other variable in a dataset containing, say, 30 variables, they would produce (30 × 29)/2 or 435 tables. Many of the tables would not, of course, be particularly helpful, but some relationships that we might not have expected may well emerge. As a deliberate strategy, however, it is not a particularly efficient one. It also raises the issue of how researchers decided what variables to include in their research in the first place if they had no idea of what patterns they were wishing to study.

Data dredging is an inductive process that may produce some interesting empirical generalizations that require further investigation. There is, however, considerable danger of producing misleading or suspect results (Gorard, 2013: 99). It may be seen as an admission that researchers do not know what they are looking for. However, some patterns *are* discovered by chance, and dredging may turn out to be a fruitful way of producing new insights. It is important to recognize, however, that all the various data dredging procedures are exploratory, and the same data cannot be used to 'test' hypotheses that were derived from such data.

Another limitation of bivariate analysis, also at the descriptive level, is that it gives no indication of joint effects. If, for example, the degree of association between level of customer satisfaction and the perceived quality of the food in a restaurant, both measured as **summated rating scales** and assumed to be metric, gives a Pearson's r coefficient of 0.4, and the degree of association between level of customer satisfaction and the perceived level of service is 0.3, then we might conclude that customer satisfaction is more closely associated with food quality than it is with level of service. However, it is not possible to say from these two separate bivariate associations how much food quality and level of service *jointly* contribute to customer satisfaction. For this, multivariate analysis is required.

At the inferential level, when the researcher is looking at a large number of bivariate associations in a dataset based on a random sample, it becomes increasingly likely that all the statistical random flukes are being picked out, so picking out, say, five tables from 100 bivariate associations and testing for statistical significance with an alpha value of 0.05 will in all probability be picking out the very five tables that will lead the researcher inappropriately to reject the null hypothesis. Putting this another way, the more hypotheses we test using the same variables, the more likely researchers are to believe erroneously that they have found an effect. If, however, it is possible to test a set of hypotheses all at once, then this tendency can be avoided. Multivariate analysis provides exactly this facility.

More important, however, than either the descriptive or the inferential limitations of bivariate analysis are the relational ones. In Chapter 5 it was explained that there is a variety of relationships between two variables that include differences, category clusterings, covariation or asymmetrical patterns. When more than two variables are involved in an analysis a whole new range of complications may arise that bivariate analysis cannot handle. Thus a **conditional relationship** between two variables may be conditional upon, or moderated by, a third variable, for example the relationship between level of customer satisfaction and the perceived quality of the food in a restaurant may be stronger for men than it is for women – the relationship is to some extent conditional upon gender. This type of pattern opens up the way for more sophisticated theorizing, even with only three variables, but as more independent variables are added, each one moderating the original relationship, it quickly becomes very complex.

In a **confounding relationship**, the individual effects of two independent variables on a dependent variable are distorted because the independent variables are themselves related. Where the variables are metric, this kind of relationship is often referred to as **collinearity** (or **multicollinearity** if there are more than two independent variables) and is detected by examining the bivariate correlation between each independent variable. Its impact is to reduce any single independent variable's predictive power by the extent to which it is correlated with the other independent variables. As multicollinearity increases, the unique variance explained by each independent variable decreases. In consequence, the overall prediction increases much more slowly as more independent variables with high multicollinearity are added.

In a **mediating relationship**, an independent variable is seen as having an effect on a dependent variable indirectly through another independent variable. It is in the middle of a causal chain, so it is both dependent on the prior variable and acting as the independent variable to the final outcome (see Figure 6.1).

There can, of course, be several mediating variables in a complex chain. Mediation, furthermore, can be total or partial. In total mediation, the effect can occur only through the mediating variable. With partial mediation, it may be only one pathway among others (see Figure 6.2).

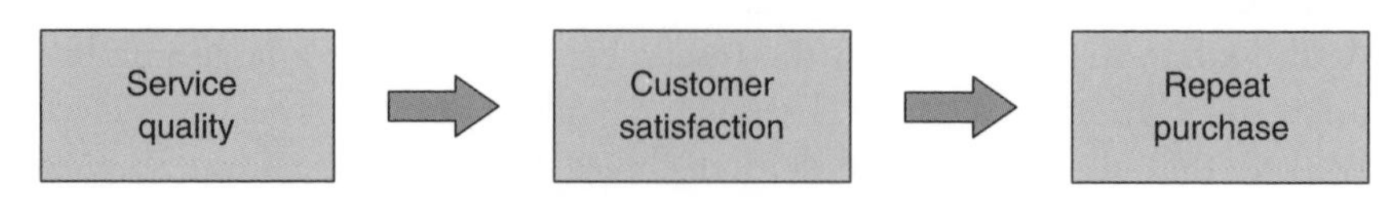

Figure 6.1 A mediating relationship

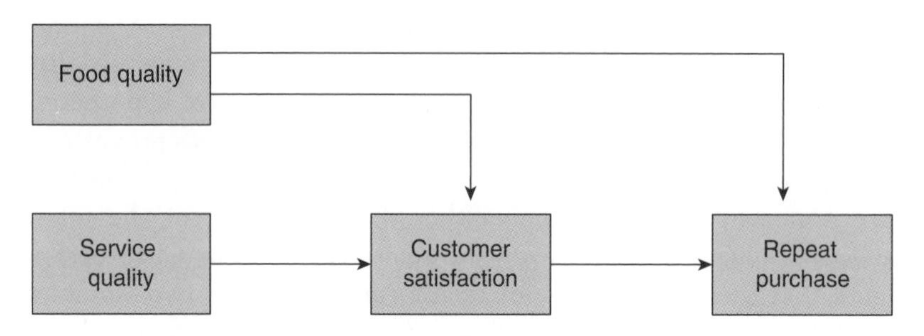

Figure 6.2 A partial mediating relationship

Key points and wider issues

Quite a lot can be achieved by analysing the relationships between variables two at a time, but the researcher should be aware of the limitations of doing so. Thus joint effects may be ignored, there may be an inflated risk of committing what statisticians call Type I errors (falsely rejecting a **null hypothesis**), but, above all, conditional, confounding or mediating relationships may not be unearthed. The process of detecting such relationships is sometimes called **elaboration analysis**. However, as more variables are added to the analysis, the potential for ever more complex relationships grows at an alarming rate. Furthermore, conditional, confounding and mediating effects may all combine in a final pattern. Bivariate strategies cannot even begin to deal with these kinds of complexities.

What is multivariate analysis?

Multivariate analysis is not easy to define; the term is not used consistently by statisticians. Sometimes it is used to refer to the number of dependent variables in an analysis. Thus in an analysis of variance where there is one dependent variable, it may be called 'univariate' analysis of variance, even if there are two or more independent variables. By contrast, multiple regression analysis refers to having more than two independent variables. In general terms, however, we can say that multivariate analysis refers to any simultaneous analysis of three or more variables on a set of cases. Where all the variables to be related are categorical, then it is possible to control, or 'layer' in SPSS terms, the relationship between two variables by a third, fourth, fifth variable, and so on, in the process of **cross-tabulation**. The researcher can use these three-way to *n*-way tables to explore the relationships between three and more than three variables at a time. Unfortunately, any more than three-way tables become very difficult to interpret and unless the sample is very large, the numbers in the cells quickly become very small. An alternative is **log-linear analysis** that looks at interaction effects between a number of categorical variables.

Multivariate techniques for metric variables, or for combinations of metric and categorical variables, may be classified into two broad groups. **Dependence techniques** refer to those procedures where one or more variables are identified as dependent and which are to be predicted or explained by one or more independent variables. By contrast, **interdependence techniques** involve the simultaneous analysis of all the variables in the set, none being identified as either dependent or independent.

It is also important to distinguish between those multivariate techniques that may genuinely be called 'tests' and rely on the cases being analysed constituting a random sample from a population and which use statistical inference to reach a conclusion (in other words, the final output of such procedures is a ***p*-value**), and those techniques that describe one or more features of a dataset, irrespective of whether they constitute a sample or a population.

For dependence techniques, it needs to be emphasized that it is the researcher who decides on the dependent or independent status of variables. The statistics themselves are blind to such allocations. Researchers, however, will often conclude that one variable 'accounts for' a certain percentage of the variability on another variable, but the statistics themselves would also allow the 'accounting for' in the other direction or indeed that both share their variability. The dependent or independent status of the variables comes (or should come) from the research context of the researcher's theoretical ideas, not from the statistics (Spicer, 2005: 15).

Multivariate analysis for categorical variables

Multivariate analysis when all the variables are categorical is, for the most part, limited to using three-way and *n*-way tables, log-linear analysis and correspondence analysis. Correspondence analysis is best considered after looking at its parallel procedure for metric variables, namely factor analysis.

Three-way and *n*-way tables

Where all the variables are categorical, it is possible to conduct a series of three-way analyses. Four-way, five-way, up to any number or '*n*-way' analyses are possible, but they become exceedingly complex and require very large sample sizes as the number of sample splits grows. Such analysis usually begins with a study of the original bivariate relationship. Thus for the alcohol marketing dataset it is plausible that those who have seen ads for alcohol on more channels (and therefore have higher awareness) are more likely to stress the importance of well-known brands for alcohol. In Figure 6.3, number of

			Number of channels seen			
			Up to 4 channels	5 or 6 channels	7 or more channels	Total
Brand importance for alcohol	Important	Count	39	39	75	153
		% within Number of channels seen	14.8%	18.6%	24.9%	19.8%
	Not important	Count	224	171	226	621
		% within Number of channels seen	85.2%	81.4%	75.1%	80.2%
Total		Count	263	210	301	774
		% within Number of channels seen	100.0%	100.0%	100.0%	100.0%

Figure 6.3 Brand importance for alcohol by number of channels on which ads for alcohol have been seen

Count

			Number of channels seen			
Sex of respondent			Up to 4 channels	5 or 6 channels	7 or more channels	Total
Male	Brand importance for alcohol	Important	14	20	38	72
		Not important	94	83	118	295
	Total		108	103	156	367
Female	Brand importance for alcohol	Important	25	19	37	81
		Not important	130	88	108	326
	Total		155	107	145	407
Total	Brand importance for alcohol	Important	39	39	75	153
		Not important	224	171	226	621
	Total		263	210	301	774

Sex of respondent				Value	Approx. Sig.
Male	Nominal by Nominal	Phi		.120	.072
		Cramer's V		.120	.072
	Ordinal by Ordinal	Gamma	Zero-Order	-.249	.019
	N of Valid Cases			367	
Female	Nominal by Nominal	Phi		.106	.102
		Cramer's V		.106	.102
	Ordinal by Ordinal	Gamma	Zero-Order	-.205	.046
	N of Valid Cases			407	
Total	Nominal by Nominal	Phi		.109	.010
		Cramer's V		.109	.010
	Ordinal by Ordinal	Gamma	Zero-Order	-.224	.002
			First-Order Partial	-.224	
	N of Valid Cases			774	

Figure 6.4 Brand importance for alcohol by number of channels on which ads for alcohol have been seen, controlling for gender

channels seen has been recoded into three categories, and importance of brands for alcohol into two categories, resulting in the three-by-two crosstabulation. This shows that a higher proportion of those who have seen such ads on seven or more channels (24.9 per cent) stress the importance of brands than those who have seen ads on fewer channels. As a mirror image, a higher proportion of those who have seen up to four channels say that branding is unimportant. So, there is a degree of covariation here, but does this hold true for both males and females? Figure 6.4 shows a three-way crosstabulation of importance by awareness, controlling (or 'layered' in SPSS terminology) for gender. The lower table shows that whether the coefficients Cramer's V or gamma are used, the degree of association is slightly stronger for the males than for the females. Gamma is probably the better statistic since both variables can be seen as ordered category and it also gives a higher coefficient value. Note that the *p*-value (under `Approx. Sig.`) is much higher for the two partial tables. However, for Cramer's V, the association for both males and females is no longer statistically significant at the 0.05 level (again, only if the sample were a random one). This arises because the number of cases in each of the sub-tables is lower, so even though the value of Cramer's V is higher for males than for the overall table, the *p*-value is now greater than 0.05 and so is not statistically significant. How to produce three-way tables is explained in Box 6.1.

Box 6.1 Three-way analyses using SPSS

Three-way, four-way, up to *n*-way tables may be obtained using the SPSS `Crosstabs` procedure. The control variables are called `layers` in SPSS. The original bivariate table is broken down by each category of the layer variable. To produce Figure 6.4 select `Analyze|Descriptive Statistics| Crosstabs`. Put `Brandal2` into the `Row(s)` box and `Numberseen3` into the `Column(s)` box. Put `Gender` into the `Layer 1 of 1` box. Click on `Statistics` and tick `Phi and Cramer's V` in the `Nominal` box and `Gamma` in the `Ordinal` box. Click on `Continue` then `OK`. Further layers may be added by clicking on the `Next` button.

It is quite possible to layer by more variables, but the sample is getting split into smaller and smaller sub-groups, so it really requires a much larger size of sample. It is probably better, if there are other variables that might act as controls, to do separate three-way analyses for each. Ideally, all the key bivariate relationships that are the focus of the research should be checked in this way against all variables that potentially might affect that relationship. The implications of the results of doing so will need to wait until we have looked at **causal analysis** and the nature of explanation in rather more detail later in Chapter 10.

Log-linear analysis

Three-way and *n*-way analyses compare the two-way relationships for different groups or categories. If, however, a researcher wishes to study the pattern of interactions between three or more categorical variables all treated equally and with no distinction between dependent and independent variables, then log-linear analysis may be used. This takes each possible combination of characteristics and looks at which interaction effects among the variables have a statistically significant impact on the distribution of frequencies. If, for example, there are four variables, besides the one four-way interaction, there will be four three-way interactions and six two-way interactions, making eleven in total, as shown in Table 6.1 for the four variables A, B, C and D. The aim is now to see which of these interactions can be excluded from further consideration in order to arrive at the most parsimonious explanation of the frequencies in the cells.

Table 6.1 Interactions in a set of four variables

Interactions	
Four-way	A×B×C×D
Three-way	A×B×C
	A×B×D
	A×C×D
	B×C×D
Two-way	A×B
	A×C
	A×D
	B×C
	B×D
	C×D

In the four-way interaction, if the variables are binary (or dichotomies) then there will be 2^4 or 16 combinations, each with its own frequency or subset of cases. In log-linear analysis, each frequency is now treated as a kind of 'dependent' variable that has to be 'accounted for' by interactions of variable categories. The goal in log-linear analysis is to find the simplest set of interactions of variables that still models the observed data well. 'Well' in this context means that any differences between observed and expected frequencies are not statistically significant. Log-linear analysis is thus very much an inferential technique that makes extensive use of *p*-values.

The expected frequencies are generated using not the frequencies themselves, but the logarithm of the frequencies to generate a linear model (hence 'log-linear'), whose form is the same as for regression analysis, but with the log of the expected frequency in any cell as the 'outcome' variable. The 'independent' variables are the 11 interaction effects (in the case of a four-variable set) plus the 4 individual effects. A calculation is made for each cell, so there will be 16 of these in a set of four binary variables. In the full or 'saturated' model, all 15 effects (the 11 interaction and 4 individual effects) are included and the expected frequency should correspond with the observed frequency. These calculations are now repeated, systematically removing effects in hierarchical fashion, beginning with the highest order four-way interaction.

For each calculation there will now be a difference between observed and expected frequency whose statistical significance is tested using either the Pearson **chi-square** (which was introduced in Chapter 4), or the **likelihood ratio chi-square**. Instead of squaring the differences between the observed (f_o) and expected frequencies (f_e) and dividing by f_e for each cell, as with the Pearson chi-square, the likelihood chi-square divides f_o by f_e, taking the logarithm of the result and multiplying by f_o for each cell. After summing across all cells the result is multiplied by two. The likelihood ratio chi-square is usually preferred since in the **saturated model** the interaction effects are more likely to add up to zero.

Log-linear analysis begins with the top of the hierarchy, eliminating the four-way interaction, generating expected frequencies and comparing this with the observed frequencies. This is a process called **backward elimination**. If the differences are not statistically significant, the analysis moves down to the next level and tries eliminating in turn the three-way interactions. Any interaction whose elimination produces a statistically significant set of differences will be kept in the model. The most parsimonious overall model consists of the remaining interactions.

The SPSS results of a log-linear analysis using the same variables as for the three-way analysis in the previous section are shown in Figure 6.5. This shows that removing the original three-way interaction (Step 0) produces a chi-square of the difference that is not statistically significant ($p = 0.769$). In the next step, the gender by brand importance for alcohol two-way interaction is not statistically significant and so is removed. The remaining two-way interactions are kept in the model since their removal would result in a statistically significant difference between observed and expected frequencies. The final result at Step 3 suggests that the model that includes the two two-way interactions of gender and number of channels seen, and number of channels seen and brand importance of alcohol, approximates the distribution of frequencies quite well since the final chi-square of the differences between observed and expected frequencies is small (0.661), which is not statistically significant ($p = 0.882$) so the model is a good fit of the data. The original 11 effects have been reduced to 2 without significant loss of model efficiency. How to produce this analysis is explained in Box 6.2.

Data Information

		N
Cases	Valid	774
	Out of Range[a]	0
	Missing	146
	Weighted Valid	774
Categories	Sex of respondent	2
	Number of channels seen	3
	Brand importance for alcohol	2

a. Cases rejected because of out of range factor values.

Step Summary

Step[a]			Effects	Chi-Square[c]	df	Sig.	Number of Iterations
0	Generating Class[b]		Gender*numberseen3*Brandal2	.000	0	.	
	Deleted Effect	1	Gender*numberseen3*Brandal2	.526	2	.769	3
1	Generating Class[b]		Gender*numberseen3, Gender*Brandal2, numberseen3*Brandal2	.526	2	.769	
	Deleted Effect	1	Gender*numberseen3	6.980	2	.030	2
		2	Gender*Brandal2	.135	1	.713	2
		3	numberseen3*Brandal2	9.400	2	.009	2
2	Generating Class[b]		Gender*numberseen3, numberseen3*Brandal2	.661	3	.882	
	Deleted Effect	1	Gender*numberseen3	6.855	2	.032	2
		2	numberseen3*Brandal2	9.275	2	.010	2
3	Generating Class[b]		Gender*numberseen3, numberseen3*Brandal2	.661	3	.882	

a. At each step, the effect with the largest significance level for the Likelihood Ratio Change is deleted, provided the significance level is larger than .050.
b. Statistics are displayed for the best model at each step after step 0.
c. For 'Deleted Effect', this is the change in the Chi-Square after the effect is deleted from the model.

Figure 6.5 A log-linear analysis of the same variables as in Figure 6.4

Box 6.2 Log-linear analysis in SPSS

To produce Figure 6.5, select `Analyze|Loglinear|Model Selection` to obtain the `Model Selection Loglinear Analysis` dialog box. Transfer the variables `Gender`, `Brandal2` and `Numberseen3` into the `Factor(s)` box. SPSS needs to be told what codes have been used to define the categorical variables, so highlight Gender(??), click on Define Range and enter 1 for Minimum and 2 for Maximum (this variable has been coded 1 for male and 2 for female). Click on Continue. Do the same for Brandal2. Numberseen3 has three categories coded 1, 2 and 3, so enter 3 as the maximum value. The default model-building procedure is backwards elimination, which is the one normally used. Click on OK.

There are several components to the output, but focus on the backwards elimination statistics. For a detailed explanation of all the components of the output see Field (2013: Chapter 18).

The analysis shown above, while it illustrates the process of log-linear analysis quite well, does not produce results that are in any way revealing. In fact, it could be argued that the results are less interpretable than those obtained from the three-way table in Figure 6.4. However, with more variables, *n*-way tables become very cumbersome while log-linear analysis can handle many more variables and sort out their interactions. At the same time, remember that log-linear analysis makes extensive use of *p*-values, which, as explained in Chapters 4 and 5, are conditional probabilities and depend on the assumption that the null hypothesis of no effects (or in the case of chi-square that there are no differences between observed and expected frequencies) is true in the population from which the sample of cases was drawn. So, like all statistical inference, not only does it depend on random sampling, but also on these assumptions that all other sources of error can be ignored. There are, in addition, other requirements of the procedure. Tabachnik and Fidell (2001: 223) suggest, for example, that the sample needs to be at least five times the number of cases as there are cells in the analysis. This can quickly mount up with the number of variables; for example, an analysis with four trichotomies would require $3^4 \times 5 = 405$ cases. Furthermore, if there are many cells with very low frequencies or even empty, then estimates of expected frequencies become unstable.

Key points and wider issues

Multivariate analysis of data where all the variables are categorical is limited largely to constructing three-way or *n*-way tables and comparing the various strengths of association in the partial tables. However, the process requires large samples and it becomes unmanageable after two or three layers have been added. An alternative is log-linear analysis, which uses an extension of chi-square to consider the power of every interaction effect and to explain the

(Continued)

(Continued)

distribution of frequencies in every cell in a parsimonious fashion. The result is a list of interactions that are deemed to be primarily responsible for the observed multi-way distribution. The log-linear process is an inferential technique that depends on random sampling, a null hypothesis and the calculation of *p*-values; furthermore, it does not indicate any direction of influence between the variables.

Multivariate analysis for metric variables

Multivariate techniques for metric variables, or for combinations of metric and categorical variables, may be classified into dependence techniques where one or more variables is identified as dependent and which are to be predicted from, accounted by or explained by one or more independent variables. By contrast, interdependence techniques involve the simultaneous analysis of all the variables in the set, none being identified as either dependent or independent.

In similar fashion, the three-way and *n*-way tables in the previous section can be seen as a 'dependence' technique since it means selecting one or more variables as the 'layer' or 'independent' variable and the original two-way table as containing the dependent variables. Log-linear analysis, by contrast, is certainly an interdependence technique in that all variables are treated as having equal status.

Dependence techniques

A correlation between two metric variables can be broken down by two or more categories of a third, categorical, variable. Thus the correlation between age and income can be split between males and females and the two compared. Such a **conditional correlation** can be achieved in SPSS using the `split file` command. If the third variable is metric, then **partial correlation** can be used to calculate the original bivariate correlation with the effects of the third variable removed. This can be achieved in SPSS from `Analyze|Correlate|Partial`. This form of analysis, clearly, is quite limited and it would be more usual to use multiple regression. This is an extension of simple bivariate regression introduced in Chapter 5 to three or more variables.

Multiple regression

Regression analysis is by far the most frequently used data analysis technique; it dominates data analysis in the social, behavioural, educational and health sciences (de Leeuw, 2004). It is a multifunctional technique that may be used for description, for prediction and for inference. It is used to describe the distribution of a metric variable under a number of different conditions, usually combinations of the values of a number of other variables.

Regression analysis in general is nothing more than a study of how particular features of the distribution of a metric variable vary according to different values (which may be metric or categorical) of one or more predictor variables. The particular features of the distributions concerned are in practice directed towards one or more summary statistics, usually the mean and variance. Regression analysis is usually interpreted to mean linear regression, which means imposing a straight line through data. Multiple regression attempts to predict a single dependent variable from two or more independent variables and is an extension of bivariate regression. As we saw in Chapter 5, a linear regression line may be used to make predictions of a dependent variable from a single independent variable. The statistic r^2 indicates how 'good' that line is in making such predictions. In reality, not one but several variables are likely to affect the dependent variable. Thus the level of sales is affected not only by price, but by, for example, advertising expenditure, interest rates and personal disposable income. The formula describing a bivariate regression line is

$$Y = a + bX$$

Multiple regression extends this to

$$Y = a + b_1X_1 + b_2X_2 + b_3X_3 ... + b_nX_n$$

where $X_1, X_2, X_3, ..., X_n$ are the independent variables. The values $b_1, b_2, b_3, ..., b_n$ indicate the rates of change in Y consequent upon a unit change in $X_1, X_2, X_3, ..., X_n$. They are usually referred to as **non-standardized regression coefficients** or **partial slopes**. The calculation for each value of b is made with the degree of correlation between Y and the other variables held constant. This means that combined effects are additive in the sense that if $X_1, X_2, ..., X_n$ were each changed by one unit, the expected change in Y would be the sum of the partial slopes. The magnitude of each partial slope is likely to be less than the bivariate r^2 between each independent variable and the dependent variable since any variation in Y that is shared by the independent variable is taken into account. This model, however, provides an *estimate* for Y, so there is likely to be a degree of error involved. If an error term is added to the equation, then the general form of the model is encompassed.

The product of the analysis is a **multiple R^2**, or **coefficient of multiple determination**, which indicates the percentage of variation in Y associated with the variation in the independent variables. The value of R^2 cannot be less than the highest bivariate r^2 between the dependent variable and each independent variable. If the independent variables are themselves uncorrelated then R^2 will be the sum of bivariate r^2 of each independent variable with the dependent variable.

While R^2 cannot decrease as more independent variables are added to the equation, the amount added by each independent variable will be smaller to the extent that independent variables are themselves correlated.

The value of R^2 for a sample consistently overestimates the population value, and the larger the number of independent values and the smaller the sample size, the worse this bias becomes. Regression computer programs routinely provide an **adjusted R^2**, a deflated estimate that takes into account the number of variables and sample size.

The variables that are entered into the multiple regression equation, along with their partial slopes and the constant, *a*, may be seen as a kind of 'composite' (called a 'variate' by Hair et al., 2010) that is correlated with the dependent variable, *Y*. The partial slopes, however, cannot be used to evaluate the contribution of each factor to *Y* because they will often have been measured in different ways, perhaps in different units. One way of overcoming this problem is to standardize the slopes into what are called **beta coefficients**. If the regression analysis is conducted using standardized scores, which usually means subtracting the means from each score and dividing by the standard deviation to produce **z-scores**, then the slopes are themselves standardized. The use of standardized variables always produces a constant or Y-intercept value of 0, so this adjustment disappears.

In standard multiple regression, all the independent variables are analysed at the same time in one step. However, analytic power can sometimes be extended by conducting a series of regressions, each containing different subsets of the independent variables. The subsets may be chosen by the researcher, in which case the strategy is called **hierarchical regression**, or they may be selected according to some statistical rule, for example including the variables with the largest degree of bivariate correlation first, or the variables that make the largest contribution to R^2 if they are included. This is called **stepwise regression**. The focus is on how the picture changes with each step.

For multiple regression to be legitimately performed, a number of conditions need to be met. First, there must be an adequate number of cases. Tabachnick and Fidell (2001) point out that this is a complex issue because adequacy depends on the alpha level chosen, the number of independent variables, the expected magnitude of relationships, the reliability of measurement, and the frequency distribution of the dependent variable. Green (1991) suggests that if we assume that the cases are a random sample, an alpha level of 0.5, good reliability and a normally distributed dependent variable, the minimum sample size for detecting a medium-sized R^2 is 50 plus eight times the number of independent variables. Thus where there are five independent variables a minimum sample size is 50 + 40 = 90. Larger numbers of cases may be required if some of the other assumptions are not met.

Second, regression analysis assumes that the dependent variable is metric. If it is not, then logistic regression (which is explained in the next section) should be used. The issue, however, is not always clearcut. If the dependent variable is, for example, derived from a five-point rating, then even if the intervals between the points on the scale are assumed to be equivalent, this variable is likely to violate other assumptions made by regression analysis. For the independent variables, more flexibility in scaling is possible; for example, binary variables may be introduced into the regression equation by using them as **dummy variables**. This means assigning the value 1 for cases that possess a characteristic and 0 for cases that do not. If the variable is nominal, it can be reduced to a set of binary variables, each category being represented by a separate binary variable. However, ideally, all or nearly all variables entered into a regression analysis should be metric, otherwise estimates of regression coefficients become unstable.

Third, regression assumes that all metric variables are normally distributed. Normality, however, is a matter of degree, and although there are statistical tests of normality, if these were strictly applied then many variables might be excluded from the analysis. It is often argued, furthermore, that the assumption of normality is a robust one and that its violation except in extreme circumstances is unlikely to affect the calculations, particularly for large samples.

Fourth, multiple regression assumes linearity – that the data are best summarized with a straight line rather than a curved or oscillating one. This is easily checked for bivariate regression, but can be complex for multivariate relationships in multidimensional space. Most researchers will check instead that the errors are normally distributed, a process sometimes called **residual analysis**.

The final assumption is that the independent variables are not themselves highly inter-correlated. Where this is so, there is **multicollinearity**, a term we have already met in the context of confounding relationships between variables. Although these effects are controlled to some extent statistically in the multivariate procedure, there are limits when the independent variables are highly correlated. This makes it difficult to evaluate the relative contributions of the independent variables.

In the alcohol marketing dataset, there are only two continuous metric variables that can be used as 'dependent' variables: age at which young people had their first proper alcoholic drink and total units of alcohol last consumed (check Table 1.1 for a list of the variables). The former has a very narrow distribution (nearly all are aged between 9 and 13) and the latter has a very skewed distribution. Regression requires that the dependent variable must be metric and ideally should have a reasonable variance and a normal or nearly normal distribution.

However, if, for example, the researcher wishes to understand what factors contribute to the amount of alcohol drunk, then despite these limitations the researcher might be tempted to try regressing this against other potential metric factors like total importance of brands, total number of channels seen, total involvement and age at which the first alcoholic drink was taken.

Figure 6.6 shows the SPSS results of a multiple regression of total units of alcohol last consumed against total importance of brands, total number of channels seen, total involvement and age at first alcoholic drink. The adjusted multiple R^2 is very low at 0.095. The standardized beta coefficients, however, show that only total involvement makes some contribution to total units consumed (0.316), and, were the cases a random sample, would be statistically significant.

The bivariate correlation between these two variables is $r = 0.29$ ($r^2 = 0.08$), so is low, but rather higher than the bivariate correlations between total units of alcohol last consumed and the other variables. How to get SPSS to produce Figure 6.6 is explained in Box 6.3.

Model Summary

Model	R	R Square	Adjusted R Square	Std. Error of the Estimate
1	.340[a]	.116	.095	5.596

a. Predictors: (Constant), How old were you when you had your first proper alcoholic drink?, Total involvement, Total importance of brands, Total number of channels seen

Coefficients[a]

Model		Unstandardized Coefficients		Standardized Coefficients	t	Sig.
		B	Std. Error	Beta		
1	(Constant)	-1.940	4.417		-.439	.661
	Total importance of brands	.070	.073	.070	.959	.339
	Total number of channels seen	.050	.183	.022	.274	.784
	Total involvement	1.455	.370	.316	3.933	.000
	How old were you when you had youor first proper alcoholic drink?	.167	.315	.039	.530	.597

a. Dependent Variable: Total units of alcohol last consumed

Figure 6.6 Multiple regression of total units of alcohol last consumed against total importance of brands, total number of channels seen, total involvement and age at first alcoholic drink

---

Box 6.3 Multiple regression using SPSS

To produce Figure 6.6, Select `Analyze|Regression|Linear` to obtain the `Linear Regression` dialog box. Move `Total units of alcohol last consumed` across to the `Dependent` box. Put `Total importance of brands, Total number of channels seen, Total involvement` and age of first alcoholic drink `(initiation)` into the `Independent(s)` box and click on OK.

Key points and wider issues

Multiple regression is an extension of bivariate regression and may be used when there is a clear 'dependent' variable which is metric and is to be predicted from two or more metric independent variables, although a limited number of independent variables may be binary. The researcher should not, however, just throw variables into a regression analysis and hope that findings will emerge: it is not an exploratory technique. There should be sound theoretical reasons for the variables included and, for hierarchical and stepwise regression, for the order in which the variables are entered into the equation. Regression analysis can tell the researcher how well a specified set of variables is able to predict a particular outcome, and which variables in the set are the best predictors when the effects of the other variables are controlled.

The procedure makes a number of assumptions that are seldom met, although statisticians will often argue that the analyses are fairly robust and will not be greatly influenced by, for example, lack of normality. Ideally, the independent variables should each show at least some degree of bivariate correlation with the dependent variable (preferably $r = 0.3$ or above). At the same time the bivariate correlation between each independent variable should not be too high, otherwise there will be **multicollinearity**. Tabachnick and Fidell (2001: 84) suggest that researchers should 'think carefully before including two variables with a bivariate correlation of, say, $r = 0.7$ or more in the same analysis'.

R^2 will always increase as additional independent variables are added to the equation (as long as no pair is perfectly correlated). As the number of variables included approaches the number of cases, so R^2 will approach unity, even if there is no correlation between the dependent variable and each of the independent variables in the population. This is a problem known as 'over-fitting'. To avoid this problem, some statisticians suggest that the number of cases should be 20 times the number of independent variables. Thus the minimum sample size if five independent variables are to be included should be 100.

A further warning relates to the misconception that the beta weight represents the 'worth' of any given variable in predicting the outcome variable. Beta weights, however, will change as independent variables are added to or taken out of the equation. In particular, beta weights can drop dramatically if there is **multicollinearity** between added variables.

Regression techniques have been strongly criticized, for example by Berk (2004: 16), who argues that it is based on a number of judgemental decisions the basis for which go outside the data. These will include the commitment to a linear (straight line) relationship when a non-linear one might be better, the determination of any transformations, for example by taking the logarithm of a curve to make it linear, or the standardization of the slope of the fitted line. However, if these decisions are unexamined, or treated as formally necessary, then the researcher has 'started down the slippery slope towards statistical ritual'. Berk (2004: 203) even goes on to say that 'in the eyes of a growing number of observers, the practice of regression analysis and its extensions is a disaster'. While the primary technical literature on regression analysis and related procedures is usually quite careful about assumptions made by the procedures and other constraints, these 'health warnings' get lost in the translation into textbooks and indeed in research practice itself.

Logistic regression

Multiple regression, as was noted above, assumes that the dependent variable is metric. If it is categorical, then the independent variables cannot be used to predict scores on the dependent variable; what can be predicted, however, is the probability of a case being in a particular category. Pampel (2000) explains that the probability cannot be used directly, but must be indirect using the idea of odds. If the dependent variable is binary (resulting in **binary logistic regression**), and if, for example, 80 cases are in category 1, and 20 cases in category 0, the probability of being in category 1 would be 80/100 = 0.8. Another way of putting this, however, is to say that the odds of being in category 1 would be 80/20 = 4. A case is four times more likely to be in category 1 than in category 0. The next step is to replace the odds with their natural logarithm. These are the **log odds**, sometimes called logits. What this means is that, once again, it is possible to use a linear equation with the predicted log odds as the dependent variable. At the same time, the coefficients of the independent variables are chosen, through log odds, to generate the probability of each case being in a given category.

What logistic regression is testing is the extent to which including the independent variables in the model improves the ability to predict the dependent variable beyond using only the constant in the equation. The efficiency of the prediction is measured by the **log likelihood chi-square**. A logistic regression is performed without the independent variables and again including them. The change in the log likelihood chi-square is a measure of the extent to which all the independent variables jointly contribute to the outcome variable, but does not really offer an interpretable measure of association. Various attempts have been made to develop the equivalent of a pseudo multiple R^2 statistic for logistic regression, but there is little agreement on the appropriate form of such a statistic. SPSS gives the Cox and Snell and the Nagelkerke pseudo R^2 statistics, but the focus more usually changes to the statistical significance of identified changes in the log likelihood chi-square. Sample sizes required for logistic regression are typically larger than for multiple regression. Pampel (2000) suggests that samples under 100 may give misleading results; ideally, 50 cases per independent variable are required.

In the alcohol marketing dataset the obvious binary dependent variable is whether or not respondents had ever had a proper alcoholic drink (drink status). A logistic regression was carried out using this as the dependent variable and total importance of brands, total number of channels seen and total involvement as the independent variables. Selected results from SPSS are shown in Figure 6.7. The `Model Summary` gives the final value of the log likelihood chi-square and the two pseudo R^2 equivalents. These show a very low degree of correlation. The `Variables in the Equation` table tells us the estimates for the coefficients of the predictors in the model. The coefficients B are the equivalent of the slope in a regression equation, but tell us the change in the log odds of the outcome variable associated with a one-unit change in the predictor variable. The crucial statistic is the Wald statistic, which has a chi-square distribution and tells us whether the coefficient for that predictor is significantly different from zero. Both total involvement and total importance of brands are statistically significant (or would be if a random sample). However, only total involvement has a beta coefficient (0.346) high enough to suggest that it is a moderate predictor of drink status. Note that this is fairly similar to the standardized beta coefficient in the multiple regression predicting total units of alcohol last consumed. How to undertake logistic regression using SPSS is explained in Box 6.4.

Model Summary

Step	-2 Log likelihood	Cox & Snell R Square	Nagelkerke R Square
1	665.168[a]	.054	.073

a. Estimation terminated at iteration number 4 because parameter estimates changed by less than .001.

Variables in the Equation

		B	S.E.	Wald	df	Sig.	Exp(B)
Step 1[a]	Totbrand	.035	.015	5.311	1	.021	1.036
	totalinvolve	.346	.096	13.016	1	.000	1.413
	Totalseen	.019	.040	.241	1	.624	1.020
	Constant	-2.126	.466	20.817	1	.000	.119

a. Variable(s) entered on step 1: Totbrand, totalinvolve, Totalseen.

Figure 6.7 Logistic regression of drink status against total importance of brands, total number of channels seen and total involvement

Box 6.4 Logistic regression using SPSS

To obtain Figure 6.7, select `Analyze|Regression|Binary Logistic`. This will give you the main `Logistic Regression` dialog box. Transfer the variable `Drinkstatus` into the `Dependent` box. Transfer `Totbrands`, `Totalinvolve` and `Totalseen` into the `Covariates` box. There are various methods of logistic regression, including stepwise procedures, but the default is what SPSS calls `Entry`. This is the forced entry method: all the covariates are placed into the regression model in one block. Click on `OK`.

Key points and wider issues

Logistic regression is normally used instead of multiple regression when the dependent variable is binary. If the dependent variable has three or more categories, then multinomial regression may be used, but discriminant analysis,

(Continued)

(Continued)

which is explained in Chapter 8, may be preferable. Independent variables may be metric or binary. The procedure makes fewer assumptions than multiple regression (e.g. there is no assumption about normal distributions). However, the sample size required is greater, particularly where there are unequal frequencies in the categories and there is considerable missing data. Logistic regression can be used descriptively by just reporting the R^2 equivalents, but if the Wald statistic is used to assess each coefficient, then it is inferential and assumes random sampling.

Multivariate analysis of variance

Analysis of variance (ANOVA) was introduced in Chapter 5. It is essentially an inferential technique that tests the null hypothesis that the groups come from populations in which there is no difference between the means of categories or groups. It was explained in Chapter 5 that ANOVA recognizes that there will be variations in individual scores within groups as well as differences between groups. The logic of ANOVA is that it compares the variance of scores between the groups with the variance within the groups. If the variance between groups is larger, we conclude that the groups differ; if not, we conclude that the results are not statistically significant.

ANOVA is, in fact, a whole family of related procedures, but they all involve comparing means for groups or categories of cases. In short, the dependent variable or variables are metric and the independent variables are categorical (the opposite of logistic regression). Which particular technique is appropriate depends on two things. First, it depends on the number of (categorical) independent variables. When there is only one such variable, it is called 'one-way' ANOVA. One-way ANOVA can be thought of as a bivariate technique – there is one independent variable and one dependent variable. 'Two-way' ANOVA is used when there are two independent variables, and so can be thought of as a multivariate technique since a total of three variables is involved.

The appropriate ANOVA technique depends, second, on the number of (metric) dependent variables. ANOVA assumes that there is only one dependent variable. MANOVA (the multivariate analysis of variance) is an extension of ANOVA to accommodate more than one dependent variable. It measures the differences for two or more metric dependent variables on a set of categorical variables acting as independent variables. A further extension to analysis of variance is to introduce **covariates**. These are uncontrolled metric independent variables that create extraneous (nuisance) variation in the dependent variable, and whose effect can be removed through the use of regression-like procedures. This process is usually called **analysis of covariance (ANCOVA)** or **multivariate analysis of covariance (MANCOVA)** where there are several dependent variables. The process allows for more sensitive tests of treatment effects.

Even if there is just one independent variable, once there are more than two categories there are many comparisons between means that are being made.

Thus with three categories, then besides the three pairwise differences, there are also differences between subsets, for example between one group and the average of the other two groups. Once there is more than one independent variable an effect may be found both for each variable or for the variables operating jointly – an **interaction effect**. This requires what is called a factorial design in which every category of each variable is combined, so that, for example, two independent variables each with three categories will produce nine different means, all of which are to be compared, both with themselves and with the overall grand mean and the mean for each category.

For two-way ANOVA, MANOVA, ANCOVA and MANCOVA, SPSS uses what is called the **general linear model (GLM)**. This combines analysis of variance with regression analysis. A univariate GLM means that there is a single metric dependent variable and one or more **factors** (independent variables) which will normally be categorical, but metric variables may be included as

Between-Subjects Factors

		Value Label	N
Sex of respondent	1	Male	132
	2	Female	168
Does your father drink alcohol nowadays?	0	No, not sure, no father	99
	1	Yes	201
Have you ever had a proper alcoholic drink?	0	No	4
	1	Yes	296

Tests of Between-Subjects Effects

Dependent Variable:Total units of alcohol last consumed

Source	Type III Sum of Squares	df	Mean Square	F	Sig.
Corrected Model	66.029[a]	6	11.005	.394	.883
Intercept	472.910	1	472.910	16.931	.000
Gender	32.064	1	32.064	1.148	.285
Fdrinks	10.086	1	10.086	.361	.548
Drinkstatus	16.428	1	16.428	.588	.444
Gender * Fdrinks	.739	1	.739	.026	.871
Gender * Drinkstatus	39.820	1	39.820	1.426	.233
Fdrinks * Drinkstatus	12.915	1	12.915	.462	.497
Gender * Fdrinks * Drinkstatus	.000	0	.	.	.
Error	8183.881	293	27.931		
Total	14552.910	300			
Corrected Total	8249.910	299			

a. R Squared = .008 (Adjusted R Squared = -.012)

Figure 6.8 Multivariate analysis of variance of total units of alcohol last consumed against gender, father drinks and drink status

covariates. The univariate GLM model is used for two-way, *n*-way and ANCOVA. To obtain MANOVA or MANCOVA then a multivariate GLM is required. This allows for several (metric) dependent variables to be entered. The SPSS output is similar to that for univariate GLM except that **Wilks's lambda** is used to examine the contribution of each main and interaction effect.

An ANOVA was carried out on the alcohol marketing data, taking total units of alcohol last consumed as the metric dependent variable (as with the multiple regression) but with the categories gender, whether father drinks alcohol and drink status as the factors.

The results are shown in Figure 6.8. The overall model is not statistically significant ($p = 0.883$) and the adjusted R^2 is tiny (0.012). None of the factors nor any of the factor combinations are statistically significant. The variance within the groupings is greater than the variance between them.

Box 6.5 Multivariate analysis of variance using SPSS

To obtain Figure 6.8, select `Analyze|General Linear Model|Univariate` to obtain the `Univariate` dialog box. Put `totunits` into the `Dependent Variable` box. Transfer `Gender`, `Fdrinks` and `Drinkstatus` to the `Fixed Factor(s)` box. Click on `OK`.

Key points and wider issues

There are many situations in social research where researchers have, or assume that they have, a metric dependent variable, like the results of deploying summated rating scales, and want to check whether there are any differences in mean scores between categories of respondents in a survey that are not due to random sampling error. Analysis of variance consists of a number of parametric, inferential techniques for testing the null hypotheses that discovered differences between means on one or more metric dependent variables relating to the categories of one or more categorical independent variables are a result of chance random sampling fluctuations. The tests compare the variation in scores (as measured by the variance, the square of the standard deviation) within the groups with the variation between the groups. The sampling distribution of the ratio of the former to the latter (the *F*-ratio) is then used to determine the probability that the null hypothesis is true. When more than two groups are involved, a statistically significant result indicates that the means of at least two of the groups do differ. Which particular means are involved needs to be determined using further statistical tests. The aim of the analysis is to determine which independent variables (factors) and which interactions between those factors account for a significant proportion of the overall variance in a variable.

The simplest analysis of variance is a one-way ANOVA for a binary independent variable, in which case it is the equivalent of an independent samples *t*-test. One-way ANOVA may be thought of as a bivariate technique and is found on the `Compare Means` menu in SPSS. Two-way ANOVA, *n*-way ANOVA and MANOVA are multivariate techniques and may be undertaken using the SPSS `General Linear Model` procedure in which covariates may be added.

Analysis of variance is often used in circumstances that technically do not justify its use. All the various procedures assume that:

- the respondents whose results are being compared are a random sample;
- the values of the dependent variables are continuous metric and normally distributed within each group;
- the variance within the groups is approximately equal;
- the groups or categories are 'independent' samples.

Analysis of variance is designed essentially for experimental designs and in this context is extremely useful. When used for the analysis of survey data, however, many of these assumptions are not met and it is probably an overused, and frequently misused, technique.

Multivariate analysis for metric variables: interdependence techniques

Interdependence techniques involve the simultaneous analysis of all the variables in the set currently being analysed, none being identified as either dependent or independent. The main techniques include factor analysis, cluster analysis, multidimensional scaling and correspondence analysis. Cluster analysis is explained in Chapter 8 as a quantitative case-based technique. Multidimensional scaling and correspondence analysis are considered very briefly after factor analysis.

Factor analysis

The primary purpose of factor analysis is to define the underlying structure in a large set of metric variables that refer to a common theme or topic, for example a set of attitude rating items. Factor analysis recognizes that when many variables are being measured, some of them may be measuring different aspects of the same phenomenon and hence will be interrelated. It systematically reviews the correlation between each variable forming part of the analysis and all the other variables, and groups together those that are highly inter-correlated with one another, and not correlated with variables in another group. The groups identify **factors** that are in effect 'higher order' or **latent variables**. This helps to eliminate redundancy where, for example, two or more variables may be measuring the same construct. The factors themselves are a weighted composite of all the variables that are seen to be inter-correlated. They are 'latent' in the sense that they are not directly observable, but each variable has a **factor loading** which is the correlation between the variable and the factor with which it is most closely associated. The effect, and advantage, of factor analysis is to reduce a large number of variables to a more manageable set of factors that themselves are not correlated.

Factor analysis begins by calculating a correlation matrix – a table of the value of **Pearson's r** for each variable with each other variable. If there are, for example, just five variables, then the correlation matrix might look like that in Table 6.2. From visual inspection it is clear that variables 4 and 5 are highly correlated and both are negatively correlated with variable 3. Variables 1 and 2 are also correlated, but neither is correlated with variables 4 or 5. A factor analysis might produce a 'solution' like Table 6.3. Variables 3, 4 and 5 combine to define the first factor and the second factor is most highly correlated with variables 1 and 2.

Table 6.2 A correlation matrix

Variable	1	2	3	4	5
1	1.00	0.61	0.47	–0.02	–0.10
2		1.00	0.33	0.19	0.32
3			1.00	–0.83	–0.77
4				1.00	0.93
5					1.00

Table 6.3 Factor loading on two factors

Variable	Factor 1	Factor 2
1	–0.25	0.72
2	0.06	0.87
3	–0.94	0.33
4	0.94	0.21
5	0.95	0.26

The extraction of factors can be achieved in a variety of ways. SPSS offers six different methods, but the default is **principal components analysis**. 'Component' is only another term for 'factor'. With this procedure both the amount of variance to be accounted for and the number of components to be extracted equal the number of variables. The first factor will always explain the largest proportion of the overall variance, the next factor the next largest that is not explained by the first factor, and so on. Each variable is correlated with (loads on) each factor.

To calculate the proportion of the total variance explained by each factor, square the loadings of the variables on that factor and sum the result to give the **eigenvalue**. If this is now divided by the number of variables, this gives the proportion of the total variance explained by that factor. Factors with an eigenvalue of less than one are usually dropped. The remaining factors are then rotated to make their meaning clearer. There are various ways in which this can be done. SPSS gives five methods, but the most common one is **varimax**. This keeps the factor vectors at right angles so that they are unrelated to one another.

There are problems associated with factor analysis. First, it is possible to generate several solutions from a set of variables. Second, a subjective decision needs to be made as to how many factors to accept. Third, the grouping has to make intuitive sense. Thus if variables 1–5 above were consumer reactions to a new product, then variables 4 and 5 might be two questions that tap the 'value-for-money' factor, and variables 1–3 are different aspects of 'benefits-derived-from-use'. Factor analysis will always produce a solution; whether it is a good or helpful one is another matter. There may not, in fact, be any factors underlying the variables.

The uses of factor analysis can broadly be classified into exploratory or confirmatory. **Exploratory factor analysis** may itself be undertaken for different purposes, for example for data reduction – replacing a large set of variables with a smaller number of factors, for testing the characteristics for one or more measuring instruments, or for explaining the patterning in the data. **Confirmatory factor analysis**, by contrast, is used to test the probability that a particular hypothesized factor structure is supported by the data. More than one model can be tested to discover which might provide the best fit. Fit is sometimes determined by using the likelihood ratio chi-square test. This compares the difference between the correlation matrix of the original data with the correlation matrix produced by the model. The bigger the difference, the larger chi-square will be and the more likely it is that the difference will not be due to random sampling error.

Confirmatory factor analysis is considered to be one of a family of techniques that estimate multiple and interrelated dependence relationships called **structural equation modelling**. This uses separate relationships for each set of dependent variables. It combines factor analysis with multiple regression in a series of structural equations. It is not available on SPSS, but the company has another program for this purpose called AMOS (see www.spss.com/amos). More commonly used, however, is a program called LISREL, which is available from www.ssicentral.com.

There is no large set of metric variables in the alcohol marketing dataset, but for the purposes of seeing the SPSS factor analysis output, a factor analysis on the nine items rating the importance of branding for a range of different products was carried out. These variables, you may recall, are five-point rating scales which, for the purpose of this exercise, are treated as metric variables. The result, using principal components analysis and varimax rotation, is shown in Figure 6.9. Three factors have been extracted, showing, unsurprisingly, that chocolates and sweets, fizzy drinks and crisps all load highly on the first factor; trainers, clothes and to a lesser extent perfume or aftershave load on the second; and cigarettes and alcohol on the third. Another part of the output from SPSS (Figure 6.10) shows that over 72 per cent of the variance is explained using the varimax rotation. How to do this in SPSS is explained in Box 6.6.

Rotated Component Matrix[a]

	Component		
	1	2	3
Brand importance for chocolate or sweets	.838	.167	.139
Brand importance for fizzy drinks	.865	.148	.100
Brand importance for crisps	.819	.124	.143
Brand importance for trainers	.192	.859	.047
Brand importance for clothes	.162	.888	-.025
Brand importance for magazines	.487	.321	.137
Brand importance for perfume/Aftershave	.164	.697	.196
Brand importance for cigarettes	.181	.062	.933
Brand importance for alcohol	.164	.118	.933

Extraction Method: Principal Component Analysis.
Rotation Method: Varimax with Kaiser Normalization.

a. Rotation converged in 5 iterations.

Figure 6.9 Factor analysis of nine items relating to brand importance

Total Variance Explained

Component	Initial Eigenvalues			Extraction Sums of Squared Loadings			Rotation Sums of Squared Loadings		
	Total	% of Variance	Cumulative %	Total	% of Variance	Cumulative %	Total	% of Variance	Cumulative %
1	3.725	41.389	41.389	3.725	41.389	41.389	2.508	27.865	27.865
2	1.575	17.496	58.885	1.575	17.496	58.885	2.198	24.419	52.283
3	1.256	13.958	72.843	1.256	13.958	72.843	1.850	20.559	72.843
4	.733	8.141	80.984						
5	.589	6.546	87.530						
6	.415	4.612	92.142						
7	.290	3.219	95.361						
8	.250	2.778	98.139						
9	.167	1.861	100.000						

Extraction Method: Principal Component Analysis.

Figure 6.10 Explained variance

Box 6.6 Factor analysis using SPSS

To obtain Figures 6.9 and 6.10, select `Analyze|Dimension Reduction|Factor` to produce the `Factor Analysis` dialog box. Transfer the nine brand importance items to the `Variables` box. Click on `Rotation` and select `Varimax`, then `Continue` and `OK`.

Key points and wider issues

Factor analysis is a descriptive data reduction technique that makes sense only when there is a large set of metric variables or items measuring a given concept, but which might contain latent sub-dimensions. It also makes sense only if there are reasonable correlations between the variables. Tabachnick and Fidell (2001) recommend an inspection of the correlation matrix for evidence of correlations greater than 0.3. If few are above this level then factor analysis is not appropriate. The technique also requires a reasonably large sample size – at least 150. According to Hair et al. (2010) there should be five times as many cases as variables. The analysis consists of three steps: first, a correlation matrix of the original variables is computed; second, a few factors are extracted from the correlation matrix; and third, the factors are rotated to maximize the correlation of each variable with one of the factors.

Sometimes factor analysis is used where, as with the brand importance items in the alcohol marketing dataset, the dimensions are fairly obvious, or in some cases just reconfirm the dimensions that researchers have used to construct the total set of items. This can be quite helpful if the researcher wants to confirm that items on a questionnaire do in fact cohere in the way intended. Alternatively, the results may be used to reduce the number of variables that are entered into another form of analysis. Here, the rotated sums of squares can be used as replacement variables in a regression equation or converted into set memberships for fuzzy set analysis (see Chapter 7). A third possibility is that the factors are used in a theoretical sense to 'explain' the patterning in the variable set.

Other interdependence techniques

Multidimensional scaling, correspondence analysis and cluster analysis may, along with factor analysis, be considered as interdependence techniques. Multidimensional scaling (MDS) refers to a series of techniques that help researchers to identify latent dimensions relating to respondents' evaluations of objects that are then positioned in a spatial map. It is often used in marketing to identify the relative position of competing brands or shops as perceived by customers, and to uncover key dimensions underlying customers' evaluations. It could also be used to position schools or hospitals or to assess cultural differences between distinct groups of people. It seeks to infer underlying dimensions from a series of similarity or preference judgements provided by respondents about objects within a given set. Data on perceptions may be gathered directly or they may be derived from attribute ratings. In direct approaches respondents are asked to judge how similar or dissimilar various brands or other stimuli are, using their own criteria. They may be asked, for example, for each pair of brands to rate them on a five- or seven-point scale from very dissimilar to very similar. Alternatively, respondents may be asked to rank-order all the possible pairs from the most similar to the least similar.

The focus in MDS is not on the objects themselves, but rather on how the individual perceives them. There is no set of variables that is determined by the researcher; rather the perceptual dimensions of comparison are inferred from overall measures of similarity between objects. This, say Hair et al. (2010), is a bit like providing the dependent variable (similarity among objects) and figuring out what the independent variables (the perceptual dimensions) must be. In this sense, the researcher can never be really sure what variables the respondent is using to make the comparisons.

Correspondence analysis has objectives similar to MDS, that is to create perceptual maps, but uses researcher-specified categorical variables. Variables are crosstabulated against objects or other variables and the relative sizes of the frequencies in the cells are used to see how close categories 'correspond'. The statistic chi-square is used to standardize the frequencies and from which to create perceptual maps. From these, underlying dimensions may be detected. It is a kind of factor analysis, but using categorical variables.

Going into detail on MDS and correspondence analysis is beyond the scope of this text. To follow up on these techniques see the further reading at the end of this chapter. Correspondence analysis is available on SPSS, but not MDS. Cluster analysis is explained in Chapter 8.

Implications of this chapter for the alcohol marketing dataset

Chapter 4 explained that the original research hypotheses could be broken down into a series of bivariate hypotheses; for example, the more young people are aware of alcohol marketing, the more likely they say they are to drink alcohol in the next year. Awareness is measured by the number of channels on which respondents claim to have seen alcohol advertised. If this is recoded into three categories, intention to drink alcohol is also recoded into three categories and awareness is taken as the independent variable, then the two-way crosstabulation with column percentages and measures of association are as in Figure 6.11. This shows that a higher percentage of those who had seen alcohol marketing on seven or more channels would probably or definitely drink alcohol in the next year than those who had seen four or fewer channels. The reverse is true for those who said they probably or definitely would not drink alcohol. The statistic gamma shows a low degree of association (0.187), which would nevertheless be statistically significance if the 920 cases were a random sample. However, at the very least, these bivariate associations should be checked against other variables in a series of three-way crosstabulations. Does the relationship hold for both males and females, for example? Figure 6.12 shows that the relationship is stronger for the females than for the males. Both the percentage difference and

% within Number of channels seen

		Number of channels seen			
		Up to 4 channels	5 or 6 channels	7 or more channels	Total
Intention to drink alcohol	Definitely or probably not	62.1%	51.6%	47.2%	53.6%
	Not sure, DK or no answer	16.1%	16.7%	18.6%	17.2%
	Definitely or probably yes	21.8%	31.8%	34.2%	29.2%
Total		100.0%	100.0%	100.0%	100.0%

		Value	Approx. Sig.
Nominal by Nominal	Phi	.137	.002
	Cramer's V	.097	.002
Ordinal by Ordinal	Gamma	.187	.000
N of Valid Cases		920	

Figure 6.11 Number of channels seen by intention to drink alcohol

% within Number of channels seen

			Number of channels seen			
Sex of respondent			Up to 4 channels	5 or 6 channels	7 or more channels	Total
Male	Intention to drink alcohol	Definitely or probably not	62.2%	58.7%	48.9%	55.7%
		Not sure, DK or no answer	13.4%	11.9%	18.3%	15.0%
		Definitely or probably yes	24.4%	29.4%	32.8%	29.3%
	Total		100.0%	100.0%	100.0%	100.0%
Female	Intention to drink alcohol	Definitely or probably not	62.1%	44.7%	45.5%	51.7%
		Not sure, DK or no answer	17.9%	21.2%	18.8%	19.1%
		Definitely or probably yes	20.0%	34.1%	35.8%	29.2%
	Total		100.0%	100.0%	100.0%	100.0%
Total	Intention to drink alcohol	Definitely or probably not	62.1%	51.6%	47.2%	53.6%
		Not sure, DK or no answer	16.1%	16.7%	18.6%	17.2%
		Definitely or probably yes	21.8%	31.8%	34.2%	29.2%
	Total		100.0%	100.0%	100.0%	100.0%

Sex of respondent				Value	Approx. Sig.
Male	Nominal by Nominal	Phi		.125	.146
		Cramer's V		.089	.146
	Ordinal by Ordinal	Gamma	Zero-Order	.158	.022
	N of Valid Cases			433	
Female	Nominal by Nominal	Phi		.181	.003
		Cramer's V		.128	.003
	Ordinal by Ordinal	Gamma	Zero-Order	.226	.000
	N of Valid Cases			487	
Total	Nominal by Nominal	Phi		.137	.002
		Cramer's V		.097	.002
	Ordinal by Ordinal	Gamma	Zero-Order	.187	.000
			First-Order Partial	.197	
	N of Valid Cases			920	

Figure 6.12 Number of channels seen by intention to drink alcohol and gender

the value of gamma are larger for the females. So, are females more likely to be more strongly affected by alcohol advertising? Ideally, all bivariate crosstabulations should be checked in this way against a range of categorical variables that the researcher feels may affect the relationship.

Three-way and *n*-way analyses do not, however, show the interactions between variables at the same level. For this, log-linear analysis would be needed, as illustrated in Box 6.2. In addition, given that the variables are categorical, correspondence analysis would also be a possibility.

In terms of the metric variables in the alcohol marketing dataset, there are only two continuous metric variables that can be used as 'dependent' variables using dependence techniques: age at which young people had their first proper alcoholic drink; and total units of alcohol last consumed. Neither is particularly suited to multiple regression, but despite this a regression of total units of alcohol last consumed against total importance of brands, total number of channels seen, total involvement and age at first alcoholic drink shows a very low adjusted multiple R^2. Only total involvement is a moderate predictor of total units consumed.

Logistic regression allows for a binary dependent variable such as whether or not respondents had ever had a proper alcoholic drink (drink status). Once again, among the metric variables, only total involvement showed any degree of contribution to drink status.

Analysis of variance takes one or more metric variables as dependent with one or more independent categorical variables. The technique assumes that a random sample has been taken so, strictly speaking, is not appropriate to the alcohol marketing dataset. However, for the purpose of illustration, a random sample is assumed. An analysis of variance taking total units of alcohol last consumed as the metric dependent variable (as with the multiple regression) but with the categories gender, whether father drinks alcohol and drink status as the independent variables, shows that neither the overall model nor any of the individual variables are statistically significant. The variance within the groupings is greater than the variance between them.

In terms of interdependence techniques, only factor analysis was applied to the alcohol marketing dataset. Ideally, there should be a large set of metric variables whose structure is not obvious or pre-planned, but for the purposes of illustration, a factor analysis on the nine items rating the importance of branding for a range of different products was carried out. Three factors that were rather unsurprising and related to groups of products were extracted.

Chapter summary

Multivariate analysis is the simultaneous analysis of three or more variables on a set of cases. It can overcome some of the limitations of bivariate analysis, for example the joint effects of several variables operating together can be assessed, the risk of committing Type I errors (falsely rejecting a null hypothesis) is minimized, while conditional, confounding or mediating relationships can be detected. How this is done is a topic taken up in Chapter 10.

Where all the variables to be related are categorical, then it is possible to control the relationship between two variables by a third, fourth, fifth variable, and so on, in the process of three-way and *n*-way tabular analysis. The researcher can use these tables to explore the strength of relationships between three and more than three variables at a time. An alternative is log-linear analysis that looks at interaction effects between a number of categorical variables.

Multivariate techniques for metric variables, or for combinations of metric and categorical variables, may be classified into dependence and interdependence procedures. The former depend on the researcher being able to establish the status of variables as dependent or independent and include multiple regression, logistic regression, discriminant analysis and multivariate analysis of variance. Interdependence techniques include factor analysis, cluster analysis, correspondence analysis and multidimensional scaling. Not all of these are considered in this chapter. Cluster analysis and discriminant analysis are explained in Chapter 8. Correspondence analysis and multidimensional scaling are only briefly looked at.

One message that should come over clearly in this chapter is that not all the techniques can be applied or sensibly applied to a given dataset. If the dataset is not a random sample, then any techniques that use or rely on statistical inference must be treated with extreme caution or best not used at all. Dependence techniques are appropriate only when there is a sufficiently developed theoretical model that indicates which variables can sensibly be treated as dependent and which ones as independent. Within that framework, the researcher needs to be clear which of each is categorical and which is metric.

Exercises and questions for discussion

1. This chapter checked the bivariate association between seeing ads for alcohol on more channels (and therefore having higher awareness) and stress on the importance of well-known brands for alcohol against gender, and found in a three-way crosstabulation that it was a little greater for males than for females. From the alcohol marketing dataset, using three-way crosstabulation on SPSS, check whether the same is true of the four hypotheses listed in Chapter 5 under implications of the chapter for the alcohol marketing dataset. Pick out one or two categorical variables other than gender that you think might affect these relationships and try a three-way crosstabulation with all four hypotheses.
2. This chapter ran a multiple regression of total number of alcohol units last consumed against total importance of brands, total number of channels seen, total involvement and age at which the first alcoholic drink was taken. The resulting multiple R^2 was very low. Using SPSS, check out the bivariate correlations between total units consumed and the other variables.
3. How can the status of any variable as 'dependent' or 'independent' be established?
4. The appropriate use of regression-based techniques depends on a number of assumptions being met. Given that these are seldom met in their entirety, or not at all, to what extent has the use of regression been, in the words of Berk (2004: 203), a 'disaster'?

Further reading

De Vaus, D. (2002) *Analyzing Social Science Data: 50 Key Problems and Data Analysis*. London: Sage.

Part Seven of this book is about how to carry out multivariate analysis. In particular, look at Problem 41 which reviews what de Vaus calls 'elaboration analysis'. This looks at what might happen to a relationship between two variables when a third 'test' variable is introduced. Problems 46–50 on multiple regression are well worth a read.

Field, A. (2013) *Discovering Statistics Using IBM SPSS Statistics*, 4th edn. London: Sage.

This covers multiple regression and logistic regression in Chapters 7 and 8 in detail and with explanations of how to use SPSS and interpret the outputs. Similarly with factor analysis (Chapter 17) and log-linear analysis (Chapter 18). There is probably far too much on analysis of variance (Chapters 9–14) which he treats as a form of regression, but section 10.2.3 explaining why is worth a read.

Hair, J., Black, W., Babin, B. and Anderson, R. (2010) *Multivariate Data Analysis: A Global Perspective*, 7th edn. Upper Saddle River, NJ: Pearson Education.

This book has been around for over 30 years and is now in its seventh edition. It is regarded by many as the best introduction to multivariate analysis. If you do nothing else, look at Chapter 1, which gives an overview of the whole area including a neat summary of each technique. There is, however, little on SPSS.

Spicer, J. (2005) *Making Sense of Multivariate Data Analysis*. Thousand Oaks, CA: Sage.

This is a concise, conceptual introduction to multivariate data analysis. There are few symbols and few equations. It covers multiple regression, logistic regression, discriminant analysis, multiple analysis of variance, factor analysis and log-linear analysis. Pay particular attention to the section in each chapter on the trustworthiness of each technique. There is no coverage, however, of SPSS or any other software.

Suggested answers to the exercises and questions for discussion can be found at the end of this text, pp. 293–321, and on the companion website, (https://study.sagepub.com/kent), which also give links to relevant free online Sage journal articles, PowerPoint slides, an overview of data analysis packages, an introduction to SPSS and weblinks to alternative datasets.

III

CASE-BASED ANALYSES

7

SET-THEORETIC METHODS AND CONFIGURATIONAL DATA ANALYSIS

Learning objectives

In this chapter you will learn:

- that set-theoretic methods – and configurational data analysis in particular – offer an alternative to deploying the variable-based statistical analyses explained in Part Two of this book;
- the various ways in which set memberships may be assessed;
- the key different kinds of set relationship;
- the general principles of configurational data analysis and how it focuses on which conditions may be necessary and which conditions or combinations of conditions may be sufficient for an outcome to happen;
- the principles behind and how to use one particular piece of software for analysing both crisp and fuzzy sets;
- that a fuzzy set analysis of the alcohol marketing data suggests that *not* having siblings who drink, *not* being involved in alcohol marketing, and *not* liking ads for alcohol is sufficient in nearly all cases to ensure that the person thinks that they will *not* drink alcohol in the next year.

Introduction

Part Two of this text has considered the more traditional and commonly used variable-based data analysis procedures that focus on the distributions of values down the columns in a data matrix. Part Three turns to approaches to data analyses that are still quantitative, but focus on cases and the arrangement of values across the rows for each case. These case-based procedures are less well known and often not part of any traditional statistics course, but are increasingly being used by researchers. Chapter 1 introduced the distinction between properties of cases as variables and properties as set memberships. This chapter develops the idea of sets and set-theoretic methods which analyse social reality through the notion of set memberships and their relationships. Relations between social phenomena are seen in terms of relationships of intersection, union, negation and subset which are then interpreted in terms of sufficiency and necessity. The chapter begins by examining how set membership is assessed, the different forms of set relationship, and then turns to one particular set-theoretic method: **configurational data analysis** (CDA). Chapter 8 then looks at other case-based approaches.

The assessment of set membership

In Chapter 1 it was suggested that sets may be seen as containers to which cases may or may not belong (or belong only to a certain extent). Sets are based on notions of inclusion and exclusion according to boundaries that are defined in an absolute sense, that is they are not relative to characteristics of other cases in the dataset, as with variable-based analyses. Chapter 1 also explained that sets may be crisp or fuzzy; the former are identical to binary variables and are given a set membership value of either [1] or [0] for membership and non-membership, while the latter are an extension of crisp sets to allow for degrees of set membership in the interval between [1] and [0] with a crossover value of [0.5] for maximum ambiguity between membership and non-membership.

The assessment of crisp set membership may be a simple matter of direct observation (e.g. whether or not a brand is on the shelf) and making a record (a value of [1] for yes and [0] for no). Assessment may also be a matter of translation from recorded values of variables. A binary variable is already, by definition, also a crisp set and simply has to be given a value of [1] for membership (or presence of a characteristic) and [0] for non-membership. Nominal and ordered category variables will need a rule that clarifies how the translation is to take place. A dichotomy (Is it an A or a B?) may require two crisp sets: A/not A and B/not B. Three or more categories may require several crisp sets recording the absence or presence of each category. An ordered category measure is more difficult to translate into a crisp set. A set of five Likert categories from 'Strongly agree' to 'Strongly disagree' could, for example, be converted into a crisp set by taking the set 'Agree' as including both 'Strongly agree' and 'Agree' categories, while the remaining three categories are taken as non-membership. Ordered category variables are, however, usually better translated into fuzzy sets (see below).

Where the variables to be translated into set memberships are metric, it makes little sense to deliberately create crisp sets. This would entail selecting a value that is the dividing line between membership and non-membership of a set (e.g. a level of GDP per capita that divides 'rich' and 'poor' countries). The temptation must be to make this value the average value. However, this is only a relative position that is data driven. It is not an absolute value, but rather depends on the values of other cases. It is not a result of any substantive meaning, theoretical knowledge or empirical evidence about what set membership (e.g. being a 'rich' country) entails. Choosing threshold values on any criteria, however, is always going to be somewhat arbitrary and the results of any analyses based on such data will be sensitive – perhaps very sensitive – to different thresholds. It also means that a lot of information on variation is being thrown away. Despite this, some researchers nevertheless deliberately create crisp sets in order to be able to apply crisp set analysis. Rihoux and Ragin (2009) in fact devote several pages of their book to giving guidance on how to create crisp sets.

Fuzzy sets must at a minimum have three values, usually [1], [0.5] and [0], but these could be other values like [0.8], [0.6] and [0.4] if cases are neither fully in nor fully out of a set, and there are no cases for which it is not possible to say whether they are in or out. There does not need to be a value of [0.5]; indeed it is better if no cases are given this value because, as will be explained later, values of exactly [0.5] are excluded from certain parts of fuzzy set analysis. The intervals between the fuzzy values, furthermore, do not need to be equal. It is a case of what makes sense, given the nature of the property and the researcher's understanding of it. Fuzzy sets may have any number of values; these may be fixed by the researcher or they may be continuous. Fixed values may be labelled as in the five-value set in Figure 7.1 or they may not.

Fuzzy sets are binary and metric at the same time. Cases with membership values above [0.5] are more 'in' than 'out' of a set, and those with values below [0.5] are more 'out' than 'in'. This characteristic comes in very useful at a later stage, as we will see, when determining how many cases 'belong' to a given configuration of conditions.

Three-value sets are not particularly helpful for analysis purposes. Ideally fuzzy sets should be as fine-grained as possible, but it depends on the nature and understanding of the property being assessed. Fuzzy sets may be created directly from a questionnaire if respondents are asked questions in fuzzy set format, for example 'On a scale of 0–10, would you say you were a "heavy" drinker?' If the respondent gives a rating of, say, 6 out of 10, then this can be given a fuzzy value of [0.6] and so on. Such self-defined set memberships, however, run counter to the idea that set memberships should, ideally, be determined in a process of calibration using criteria that are defined in an absolute sense and reflecting the researcher's detailed understanding of the properties and the cases involved.

It is more usual to create fuzzy sets from properties already measured as ordered category or metric variables. Thus a five-value measure can take the highest or most extreme category as full membership, with the remaining values assigned fuzzy membership values as, for example, in Figure 7.1 for a Likert-scale item. However, this is a somewhat mechanical approach that may or may not 'make sense' in terms of set memberships. It is also likely to produce too many values of exactly [0.5]. For attitude scaling techniques and for other forms of respondent rating this is still based on self-defined evaluation that runs counter to the idea that set memberships should, ideally, be determined in a process of researcher calibration.

Fully in	Mostly in	Neither in nor out	Mostly out	Fully out
1.0	0.75	0.5	0.25	0.0
Strongly agree	Agree	Neither agree nor disagree	Disagree	Strongly disagree

Figure 7.1 Fuzzy set membership – a five-value set

For metric variables there are several fuzzy set assessment options. It might be tempting simply to transform the entire range of the metric variable into the [0] to [1] interval, preserving the relative distances from one case to the next. This can be achieved, for example, by dividing the metric value for each case by the highest value among all the cases. This, however, would make the fuzzy values obtained void of any substantive meaning. Determining membership of the set 'Rich countries' by using Gross Domestic Product (GDP) per capita as an indicator sounds like a sensible idea, but simply using the highest GDP per capita as a basis for determining the extent to which a country is 'rich' will make the cut-off between membership and non-membership of the category 'rich country' a GDP per capita simply of half that of the richest country.

Ragin (2008) suggests that not all the variation in a metric variable is necessarily relevant to a given fuzzy set. If a researcher, on the basis of extensive knowledge about standards of living throughout the world, suggests that any country with a GDP per capita (at purchasing power parity, which takes account of the cost of living and inflation) of over $30,000 is a 'rich' country, then all countries above this figure have full membership, and the fact that Qatar, according to International Monetary Fund estimates, has a GDP per capita of nearly $100,000 is not relevant to set membership. Similarly, any country with less than $2,000 (Senegal and the Gambia have just below this figure and the Democratic Republic of the Congo at the bottom of the list has $349) may be considered as fully out of the set. The point of maximum ambiguity is probably near a crossover value of about $15,000, which would just include countries like Malaysia, Botswana, Lebanon and Uruguay as 'rich'. Note that this value is not an average nor a value of the middle-ranking countries (which would be about $8,000), but a value that appears to separate rich from not-rich countries. Ragin suggests a method, using these three anchor points, to convert the metric scale to fuzzy set values, ignoring irrelevant variation. The method uses estimates of the log of the odds of full membership as an intermediate step to providing fuzzy set values. The software fsQCA, which is explained later in this chapter, does these computations for the researcher using a transformation that is S-shaped rather than linear. The selection of the threshold values and the crossover value are crucial to this method and will not only affect the intermediate fuzzy values between [1] and [0], but also determine the number of full and non-memberships.

An alternative is to treat membership and non-membership as ideal types such that few if any cases will attain full membership, for example, of the category 'fully bureaucratic organization'. The highest fuzzy set value for membership of the set bureaucratic organization may thus be [0.8] and the lowest [0.2]. All organizations will have some bureaucratic characteristics and none will have them all.

Set membership assessment for the alcohol marketing study

The key research hypotheses for this study are that the more young people are aware of and involved in alcohol marketing, the more likely they are to have consumed alcohol, and the more likely they are to think that they will drink alcohol in the next year. The two key outputs thus relate to drinking behaviour. Drinking status is assessed by asking whether they have ever had a proper alcoholic drink, not just a sip. This, clearly, is a binary variable that is readily converted into a crisp set with 'Yes' = [1] and 'No' = [0]. As an output this cannot be used for fuzzy set analysis, so crisp set analysis would be required. Future drinking intention is assessed by asking about the likelihood that they would drink alcohol during the next year – 'Definitely not', 'Probably not', 'Probably yes' and 'Definitely yes'. They are also given a 'Not sure' option. To convert this into a fuzzy set requires the creation of a category to which respondents could have a degree of membership, for example degree of membership of the category 'intention to drink alcohol in the next year'. 'Definitely yes' can be given a full membership value of [1] and 'Definitely no' a value of [0]. 'Probably yes' could be converted into [0.7], 'Probably not' into [0.3] with 'Not sure' located at the crossover point of [0.5].

To measure awareness, respondents are asked if they have seen any adverts for alcohol in any of 15 channels, for example television, cinema, newspapers, websites or sponsorships. Responses are recorded into 'Yes', 'No' and 'Don't know'. A summated scale of 0–15 'yeses' may be converted into the degree of membership of the category 'aware of many channels'. Since over 90 per cent of respondents claim to have seen nine or fewer channels, nine or more may be given full membership and fewer channels given fuzzy values as follows: 8 [0.9], 7 [0.8], 6 [0.7], 5 [0.6], 4 [0.4], 3 [0.3], 2 [0.2], 1 [0.1] and 0 [0]. Note that [0.5] is avoided. Ragin's procedure explained above is probably not appropriate since this measure is discrete rather than continuous metric. To measure involvement in alcohol marketing, pupils are asked whether they have, for example, received free samples of alcohol products, free gifts showing alcohol brand logos or promotional mail or email. This may be converted into degree of membership of the category 'involved in alcohol marketing'. Over 95 per cent say they have been involved in three or fewer of the eight types of involvement, so three or more may be given full membership, with 2 [0.8], 1 [0.6] and 0 [0]. The rationale here is partly data driven since 43 per cent claim no involvement, so any involvement is given a fuzzy value of over [0.5]. The remaining fuzzy and crisp sets used in the alcohol marketing study are outlined in Table 1.2 (p. 35).

Key points and wider issues

In order to apply set-theoretic methods and configurational data analysis, it is necessary to create a dataset in which properties of cases are recorded as set memberships rather than as variables. Sets are qualities that cases may or may not possess or may possess to a certain extent. The assessment of set memberships is according to criteria that are determined by the researcher and external to the dataset, so they are absolute rather than relative to the properties of other cases in that dataset. These criteria, furthermore, need to make sense according to the concepts, theories or models that are being applied to the research and which are at the same time consistent with the empirical evidence about what set membership entails, for example what it means to be a 'rich' or 'democratic' country or a young person with the 'intent to drink alcohol in the next year'.

Set memberships may be crisp or fuzzy. Both may be created either directly from observation or, more usually, indirectly from transformation of variables into set memberships. Binary variables readily translate into crisp sets. Nominal variables may require several sets, one for each category. Ordered category and metric variables are best converted into fuzzy sets. There is no one 'correct' way in which this should be done, but whatever procedure is chosen it does need to make sense in theoretical terms. It also needs to be kept in mind that the eventual results may be very sensitive to the choices made and that it may be prudent at a later stage in the analysis to reconsider these choices. The values of ranked variables are, by definition, relative to one another, so are difficult to transform into fuzzy sets unless they are degrees of membership of the category 'highly ranked'.

Treating properties of cases as set memberships rather than as variables probably requires a lot more thinking about concepts, theories and measurement procedures. It is only too easy for researchers when undertaking variable-based analysis just to accept the highest and lowest values for each variable or the frequencies of categories as the 'range' of variation without considering the substantive meaning of the values achieved.

Set relationships

Thinking in the social sciences – indeed in everyday language – commonly implies notions of sets and set relationships. If, for example, a theory states that public organizations are bureaucratic and dysfunctional, then implicitly this is a statement about the way two sets overlap. Organizations in general may be bureaucratic but not dysfunctional, dysfunctional but not bureaucratic, or neither, but public organizations, according to the theory, are in the intersection of being both. If, alternatively, the theory states that large organizations are bureaucratic, the claim implies that large organizations are a subset of all bureaucratic organizations. The claim is asymmetrical: all large organizations are bureaucratic, but not all bureaucratic organizations are large. Unfortunately, researchers in the variable-based tradition will translate the claim into a symmetrical **bivariate hypothesis** that focuses on the degree of correlation between organizational size and degree of bureaucracy: large organizations are bureaucratic and by implication small ones are not.

The relationships between two or more sets can be divided into four main types:

- intersection;
- union;
- negation;
- subset.

Two sets are said to intersect where cases combine the properties of both sets; similarly for three or more sets. Where sets are crisp, cases need to be members of each set to be members of the intersection. The shaded area in Figure 7.2 shows the cases that are members of both Set X and Set Y. Note that in set theory the symbol '*' means 'and' but is sometimes omitted, so 'X and Y' becomes simply 'XY'. The symbol '~' means 'not'.

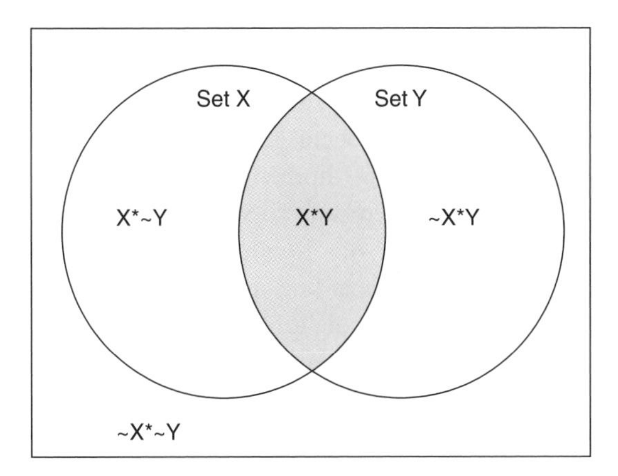

Figure 7.2 Intersection: crisp sets

Where sets are fuzzy, rather than requiring for intersection that cases have to have membership values of over [0.5] in each condition, it makes more sense to take account of the degree of membership on each condition by taking the membership value for the intersection as the *minimum* of the individual set memberships. If a country has a membership value of [0.8] in the set 'wealthy country', [0.6] in the set 'industrialized' and [0.3] in the set 'democratic', its membership value of the intersection wealthy AND industrialized AND democratic is

[0.3] – it is more 'out' than 'in'. Membership of the intersection is the 'weakest link' in the chain. This is very different from a variable-based approach that would take the average of the values and come up with [0.57], which is more 'in' than 'out'.

The set relationship of union describes logical alternatives. In crisp sets this is realized if at least one of the conditions is present – cases need to be members of only one of the sets. Where sets are fuzzy, the membership value for the union is the *maximum* of the individual set memberships. The fuzzy membership value of the union wealthy OR industrialized OR democratic for the example in the previous paragraph is therefore [0.8].

A concept that is related to both intersection and union is a measure of the degree of overlap between two or more sets. Borgna (2013) suggests a measure of 'set coincidence', which for crisp sets is the number of cases in the intersection of two or more sets divided by the number of cases in the union of the same sets. For fuzzy sets, intersection corresponds to taking the minimum membership values, and union corresponds to the maximum values for each case. Set coincidence is then the sum of the minimum values divided by the sum of the maximum values. Both indexes vary from 0, the absence of coincidence, to 1, perfect set coincidence, in which case the two sets in effect constitute the same set.

Negation is the membership value given to the absence of membership. For crisp sets, negation switches membership values from [1] to [0] or [0] to [1]. For fuzzy sets it is [1] minus the fuzzy set value, so the negation of [0.8] is [0.2]. Strictly speaking, negation is not a relationship between two or more sets, but of values within a set. Intersection, union, coincidence and negation may be seen as descriptive tools that enable the researcher simply to analyse patterns of overlap and non-overlap in set memberships in a set of cases. In their research on homicide in the USA, Miethe and Regoeczi (2004), for example, used binary set memberships to study over 400,000 homicide situations. Differences in homicide situations were evaluated according to variation in 'offense' elements (e.g. type of weapon, motive, number of offenders, location of the crime), victim characteristics, and offender attributes (e.g. age, gender, race). In all, 15 variables were used, but there was a degree of 'forcing' of some properties into binary crisp sets, so that 'motive' was classified as instrumental or expressive, victim–offender relationships into family/intimate and other, and weapon used into gun and other. Some dummy variables were also created, for example age became under 20, yes/no; 20–29, yes/no; and so on. Miethe and Regoeczi profiled the combinations of characteristics in each homicide situation; they did not analyse causal relationships. The cases used in the study were all homicides and the authors were asking what they had in common. The authors were able to conclude, for example, that most homicides between 1976 and 1998 were characterized by combinations of intra-racial and intra-age group shootings of strangers by a single adult male offender in the course of arguments occurring in large urban areas.

Borgna (2013) argues that intersection, union and coincidence have never received proper attention within the framework of set-relational research, which is essentially focused on **subset relationships** and the analysis of causality. A subset relationship exists when members of one set are totally contained within a wider set. The existence of a subset relationship does not, however, by itself demonstrate that being a member of the subset has any influence on, explains or causes membership of the superset – or vice versa. Furthermore, a subset by itself demonstrates neither sufficiency nor necessity. However, if two properties are allocated to condition and outcome, then if the condition is a subset of the outcome, this indicates logical sufficiency, as illustrated in Figure 7.3 for crisp sets. Thus if all students who study hard get good examination results, then the former is logically sufficient to attain the latter. The situation can also be illustrated in a two-by-two table, as in Table 7.1. All students who studied hard obtained good exam results and there were no cases of studying hard and not getting good results. The condition, while sufficient, is not necessary since there were 50 students who did not study hard but still got good results. This is the equivalent of saying that all Xs are Ys, but not all Ys are Xs. It is asymmetrical and non-reversible.

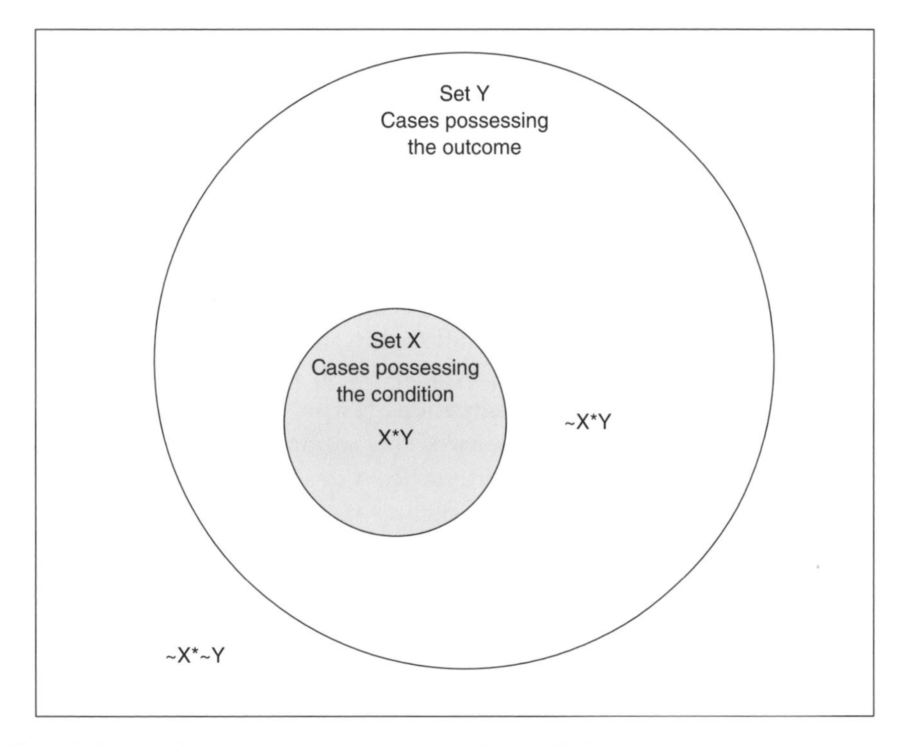

Figure 7.3 A subset relationship: crisp set logical sufficiency

It will be explained in a later section on establishing logical necessity that if, alternatively, the outcome is a subset of the condition, this indicates logical necessity. The use of the word 'logical' implies that a subset relationship exists and that memberships of the property designated as the condition and memberships of the property designated as the outcome are consistent with the pattern needed for causal sufficiency (or necessity) even if that does not 'make sense'.

Table 7.1 A sufficient condition

Good exam performance (Y)	Studied hard (X_1): X_1	Studied hard (X_1): Not X_1
Y	100	50
Not Y	0	100

The sense-making comes from a **theory** that suggests that a causal interpretation is at least plausible. If, for example, researchers hypothesize that religious fundamentalists are politically conservative, they are implying that the former is a subset of the latter and that, furthermore, their fundamentalism is a cause of their conservatism. Causality implies that there is a degree of efficacy or 'forcing' involved and that there is a given temporal sequence.

Social scientists using variable-based analysis, when using the term 'cause', do not usually distinguish between sufficiency and necessity. This is largely because variable-based data analyses cannot, or at least cannot without difficulty or only in certain limited situations (as in the two-by-two table in Table 7.1), distinguish between the two. This argument is taken up in more detail in Chapter 10. For the moment, suffice it to say that, for the most part, researchers in the variable-based data analysis tradition tend to focus on conditions or actions that make a fixed contribution to an outcome. The relationship is seen as symmetrical, focusing on patterns of either covariation or category clustering (see Chapter 5). Usually, only one pathway or recipe for an outcome is considered, so no consideration is given to the possibility that a condition may be sufficient but not necessary or necessary but not sufficient.

If set memberships are crisp, then, given a plausible theoretical argument, a condition may be considered causally sufficient if, whenever it is present across a set of cases, the outcome is also present in those cases. It is equivalent to the simple 'If X then Y' statement; in terms of set relations we can say that, in these circumstances, X is a subset of Y – the condition is a subset of the outcome. Note that this says or implies nothing about the outcome Y if cases are not members of X. The statement 'If X then Y' does not imply, as it would with variable-based analyses, 'If not X then not Y'.

If the sets are fuzzy, sufficiency requires that, for each case, membership on X (the condition) is less than (or equal to) its membership on Y (the outcome). This means that, once again, X is a subset of Y since Y is always greater than X. The degree of membership of X ensures a minimum degree of membership of Y. However, even if the degree of membership of X is sufficient to ensure a degree of membership of Y, if there are other ways on which a high value of Y can come about, then X may not be necessary for Y, so high membership of Y may be accompanied by a wide range of values on X. The situation is illustrated in Figure 7.4. All cases are in the upper left triangle. Note that while this XY plot looks similar to a scatterplot used in variable-based analyses, the diagonal is not a regression line. It is not intended to summarize the pattern of the plots, but rather represents all those points where memberships of X and Y are exactly equal.

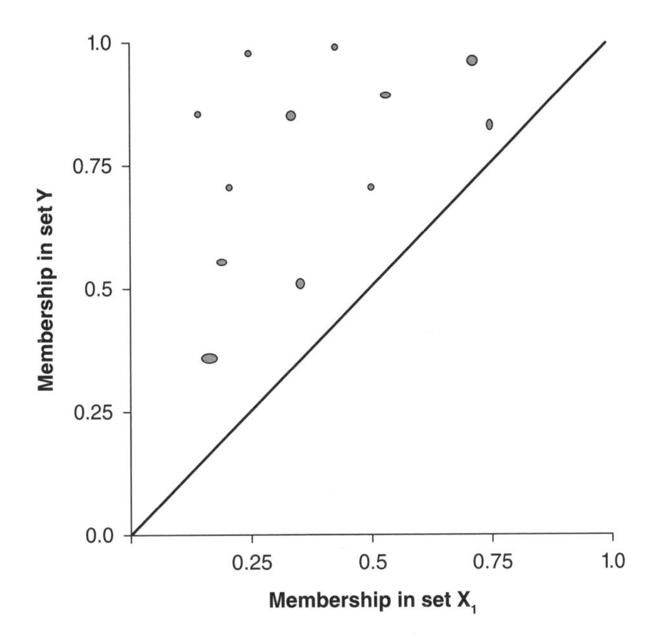

Figure 7.4 A fuzzy set sufficient but not necessary condition

Conditions are more likely to be sufficient when in combination. Fuzzy set membership of the combination of conditions – the intersection – can be determined, as explained above, by taking the minimum fuzzy value for each separate condition. This can then be used to assess the fuzzy subset relationships by comparing values on the causal combination with membership values on the outcome.

In the real world, perfect subset relationships are not common; there are usually errors, exceptions, inconsistencies or contradictions. Figure 7.5 shows a less than perfect subset relationship for crisp sets. Any cases that have the condition X but not the outcome (X~Y) are inconsistent with the subset relationship, but there are relatively few of them compared with those that are both X and Y

(X*Y). With fuzzy sets, any cases that are below the diagonal in Figure 7.4 would be inconsistent with a subset relationship.

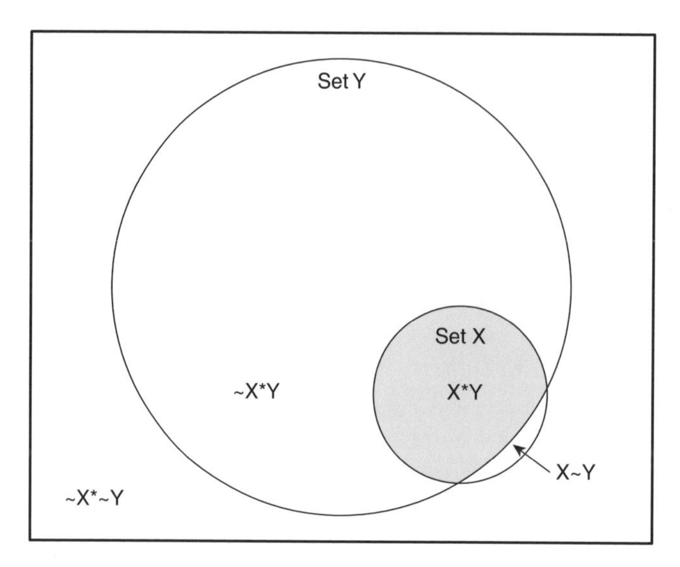

Figure 7.5 A subset relationship: crisp set logical sufficiency with some inconsistency

The logic behind establishing a **necessary condition** can be seen as the mirror image for establishing **sufficient conditions**. A condition may be considered necessary if, whenever the outcome Y is present, the condition X is also present. It is equivalent to the simple 'If Y then X' statement. Y cannot be achieved without X. In terms of subset relations, Y is now a subset of X – the outcome is a subset of the condition. The procedures for establishing logical necessity are considered in detail in a later section in this chapter.

Key points and wider issues

The relationships between sets are very different from the kinds of relationships traditionally considered by researchers using variable-based analyses and which were explained in Chapter 5. The key set relationships for the purpose of configurational data analysis are those of intersection, union, negation and subset. Intersection involves the overlap of sets such that some or many cases are members of both or all sets. It is indicated by the Boolean operator 'and', which is usually symbolized with an asterisk linking the sets (this can be confusing because the asterisk in variable-based analyses means multiplication). Membership of an intersection for fuzzy sets is a matter of degree and its assessment entails taking the lowest fuzzy set value among the intersecting sets.

Union describes logical alternatives among sets and is indicated by the Boolean operator 'or', which is usually symbolized with a plus sign linking sets (this is also confusing since this symbol means addition in variable-based analyses). Membership of a union of fuzzy sets is also a matter of degree and its assessment entails taking the highest fuzzy set value among the alternative sets.

Negation means converting the negation of membership into the membership itself, such that, for example, the negation of 'democratic country' is 'non-democratic country' and countries may now be considered as full or non-members of the latter category. For fuzzy sets, set membership of the negation is one minus the fuzzy value of that membership.

The main set-theoretic relationship for configurational data analysis is that of the subset. A subset relationship exists when members of one set are totally contained within a wider set. If one set is considered to be the outcome that the researcher is investigating and the other set a condition (or configuration of conditions with its own intersection value), then if the condition is a subset of the outcome, this indicates logical sufficiency. If set memberships are crisp, then, given a plausible theoretical argument, a condition may be considered causally sufficient if, whenever it is present across a set of cases, the outcome is also present in those cases. If the sets are fuzzy, sufficiency requires that, for each case, membership of the condition is less than (or equal to) its membership of the outcome. This means that the condition is a subset of the outcome since membership of the outcome is always greater than membership of the condition.

Alternatively, if the outcome is a subset of the condition, this indicates logical necessity. The relationships between the outcome (Y) and the condition (X, which also stands for a configuration) are asymmetrical for both sufficiency and necessity. For sufficiency, all Xs are Ys, but not all Ys are Xs. For necessity, all Ys are Xs, but not all Xs are Ys. The subset relationship is, as we will see in the next section, key to configurational data analysis.

Configurational data analysis

In the late 1980s Charles Ragin and programmer Kriss Drass implemented **set-theoretic methods** in a computer package they called Qualitative Comparative Analysis (QCA) which operated only on binary (they called them 'crisp') sets. Since then, QCA has been modified, extended and improved several times. Unfortunately, the acronym QCA now often gets used as a generic term that encompasses not only the original binary set procedures, but a whole 'family' of variants including its application to multi-value sets and to fuzzy sets. It is also used outside the context originally envisaged by Ragin to include applications in large-*n* (large populations or samples) studies that do not entail comparative analysis, but data derived from standard survey research where cases are at the micro, individual level rather than the macro level of cultures, societies or nation-states. QCA is, furthermore, not a 'qualitative' method in the sense understood by most qualitative researchers in the social sciences. It does not analyse speech, text or images; rather, 'qualitative' is used in the engineering sense of 'quality' as opposed to quantity and refers to categorical rather than numeric data. Categories still have frequencies, however, while, for fuzzy sets, there are also numeric values that relate to degrees of membership of categories. It is, in short, distinctly a quantitative method. Interestingly, Rihoux and Ragin (2009), in their recent book on the subject, use the term 'configurational comparative methods' to avoid its necessary connection with qualitative research, but it is still unambiguously tied to the 'comparative' mast. Accordingly, in this text, these procedures will be called 'configurational data analysis' or CDA for short. CDA can then be seen as a particular realization of the wider 'set-theoretic' methods.

The 'home' of much work in CDA is to be found in the COMPASSS website at www.compasss.org. COMPASSS stands for COMParative methods for the Advancement of Systematic cross-case analysis and Small-*n* Studies. COMPASSS is a research group bringing together scholars and practitioners who share a common interest in theoretical, methodological and practical advancements in a systematic comparative case-based approach to research which stresses the use of a configurational logic, the existence of complex causality and the importance of a careful construction of research populations. There are over 70 working papers that can be downloaded for free along with the software and accompanying manuals. There is also a bibliographical database with references to over 2,000 published and unpublished papers.

Whereas variable-based approaches to data analysis focus on the distribution of categories or metric values within variables down the columns of a data matrix across a set of cases, for example the distribution of ages across a set of survey respondents, CDA sees each case as a particular arrangement or configuration of set memberships across the rows. Only cases with exactly the same configuration may be considered to be identical, for example cases that are male, aged 76 and claiming to be 'very dissatisfied' with a service. A change in any one characteristic potentially makes it a different kind of case. The researcher can now study diversity – the various combinations of characteristics that tend to be found together in practice, and the combinations that are unusual or lack empirical existence entirely. It is possible, furthermore, to go on to think about which configurations give rise to an outcome that is of interest to, or under investigation by, the researcher (e.g. making a complaint about the service) and which ones do not.

According to Schneider and Wagemann (2012: 8–9), CDA distinguishes itself from other set-theoretic methods by combining three essential features. First, it is necessarily causal and requires that one property (assessed in terms of set membership) be singled out as the 'outcome' while the other properties as set memberships are all potential causal conditions. Second, it makes use of truth tables that enable researchers to study all possible configurations from two or more conditions and to assess the extent to which they are subsets of the outcome, that is logically sufficient. Third, CDA makes use of the principles of Boolean minimization by which the results are expressed in a more parsimonious but logically equivalent manner.

The fuzzy set analysis software

The particular software that is the focus of this chapter goes by the name of Fuzzy Set Qualitative Comparative Analysis or fsQCA for short. It was developed and is still being developed by Charles Ragin at the University of Arizona. It is freeware, so to download a copy just Google `fsqca` and choose `fs/QCA Software`. As an exercise, try following Box 7.1, which invites you to enter the data in Figure 7.6.[1] These data show the results of a study into the conditions relevant to the survival or breakdown of democracy in 18 countries in interwar Europe. The outcome of interest is the degree of membership in the set of countries with democracies that survived the many political upheavals of this period, (`survived`). Degree of membership in the set of countries experiencing democratic breakdown (`breakdown`) is simply the negation of `survived`.[2] The potential causal conditions are degree of membership in the sets of developed countries (`developed`), urbanized countries (`urban`), industrialized countries (`industrial`), literate countries (`literate`) and politically stable countries (`stable`).

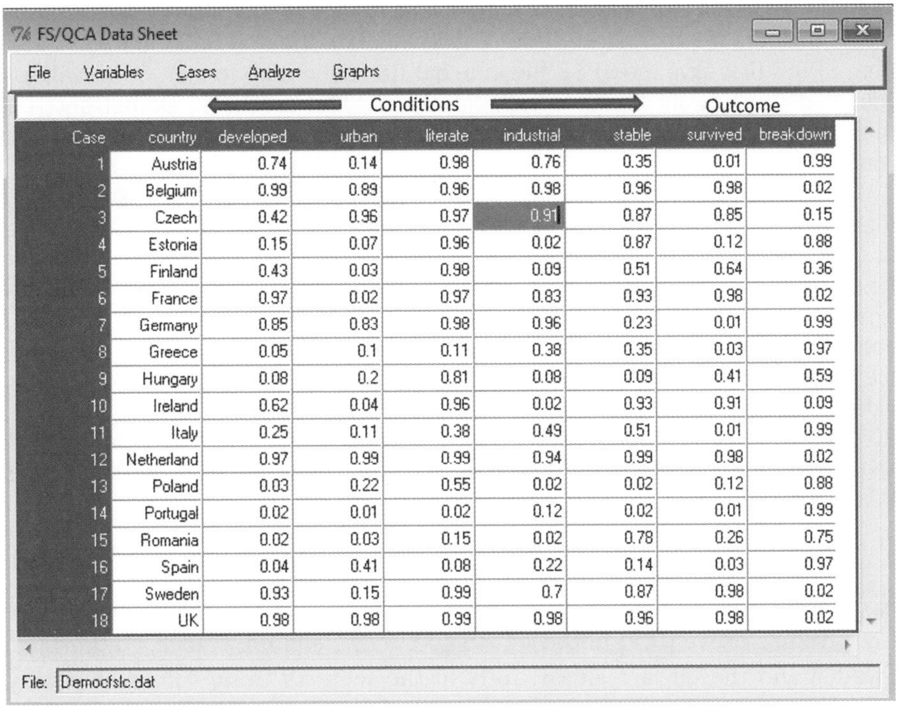

FS/QCA Data Sheet

File Variables Cases Analyze Graphs

		Conditions					Outcome	
Case	country	developed	urban	literate	industrial	stable	survived	breakdown
1	Austria	0.74	0.14	0.98	0.76	0.35	0.01	0.99
2	Belgium	0.99	0.89	0.96	0.98	0.96	0.98	0.02
3	Czech	0.42	0.96	0.97	0.91	0.87	0.85	0.15
4	Estonia	0.15	0.07	0.96	0.02	0.87	0.12	0.88
5	Finland	0.43	0.03	0.98	0.09	0.51	0.64	0.36
6	France	0.97	0.02	0.97	0.83	0.93	0.98	0.02
7	Germany	0.85	0.83	0.98	0.96	0.23	0.01	0.99
8	Greece	0.05	0.1	0.11	0.38	0.35	0.03	0.97
9	Hungary	0.08	0.2	0.81	0.08	0.09	0.41	0.59
10	Ireland	0.62	0.04	0.96	0.02	0.93	0.91	0.09
11	Italy	0.25	0.11	0.38	0.49	0.51	0.01	0.99
12	Netherland	0.97	0.99	0.99	0.94	0.99	0.98	0.02
13	Poland	0.03	0.22	0.55	0.02	0.02	0.12	0.88
14	Portugal	0.02	0.01	0.02	0.12	0.02	0.01	0.99
15	Romania	0.02	0.03	0.15	0.02	0.78	0.26	0.75
16	Spain	0.04	0.41	0.08	0.22	0.14	0.03	0.97
17	Sweden	0.93	0.15	0.99	0.7	0.87	0.98	0.02
18	UK	0.98	0.98	0.99	0.98	0.96	0.98	0.02

File: Democfslc.dat

Figure 7.6 The survival of democracy in interwar Europe: fuzzy set memberships

---

Box 7.1 Using fsQCA: entering data

To enter the data from Figure 7.6, download fsQCA. Just Google `fsQCA` and select `fs/QCA software`. Download `fsQCA 2.0`. Open it, select `File|New` and the `Add New Variables` window will open. Note that the software uses

(Continued)

(Continued)

the language of 'variables' even though, as explained earlier in this chapter, the properties are set memberships. Type in the variable names, clicking on `Add` after each one. `Country` should be `String` under `Variable Type` and the rest should be `Free Numeric`. Note that variable names are not case sensitive, they cannot exceed 15 characters and spaces or any non-alphanumeric characters like hyphens cannot be used. When all variables have been entered, click on `OK` and enter `18` as the number of cases. Click on `OK` and the `Data Sheet` will open. Enter the data from Figure 7.6. It is easiest to do this by column since pressing `Enter` will move the highlight to the cell underneath. There is no automatic backup save procedure, so before closing the data sheet, you need to save it. Use `File|Save As` and select `Tab` under `Set File Type`. Be sure to browse to where you want the file to be saved. When you next log on to fsQCA, select `File|Open|Data`.

Data can be either entered directly into fsQCA, as above, or imported from other sources including SPSS, Excel and a range of other programs or formats. Editing facilities in fsQCA are somewhat limited and in practice it may be advisable to create the data matrix first in SPSS, Excel, Minitab, Stata or SAS. To import data from SPSS, first save the SPSS file in a .dat (tab delimited) format. SPSS will ask you whether you want to 'Write variable names to spreadsheet'. Do not uncheck this option. In fsQCA select `File|Open|Data` and browse, if necessary, to your .dat file.

To get fsQCA to convert a metric variable into a fuzzy set, select `Variables|Compute`. Name your new variable in the `Target Variable` box. In the `Functions` box, select `Calibrate` and put it into the `Expression` box by clicking on the up arrow against `Functions`. Edit the expression to include the name of the old metric variable and the upper, crossover and lower values, for example `Calibrate(age,65,40,20)`. In this example, anybody aged 65 or over will be given a full membership value of [1] of the category 'old', anybody aged 40 will have the crossover value of [0.5] and anybody 20 or younger will be given a value of [0] for non-membership of that category. Click on `OK`. Check in the `Data Sheet` that you now have a new fuzzy set that ranges from [0] to [1].

A descriptive analysis of Figure 7.6 shows that 8 of the 18 countries have fuzzy values above [0.5] in the set `survived`. Belgium, France, Netherland, Sweden and the UK are almost fully in the set with fuzzy values of [0.98]. All of these are developed, literate and industrial, but France is not urban. Ireland, Czech and Finland have lower values on survival. Ireland is developed and literate, but not urban or industrial; Czech is urban and literate; and Finland is literate. So what characterizes all eight countries that survived is that they are all literate. Of the ten countries that suffered breakdown, four did not lack literacy, but all except Germany were not urban. It is possible, then, that literacy may be a necessary but not sufficient condition for the survival of democracy.

It is possible to analyse both sufficiency and necessity from the data matrix using the idea of subset relationships explained earlier in this chapter. For sufficiency, the researcher can check whether, for each case, membership of a condition is less than or equal to membership of the outcome. This then needs to be repeated for the negation of the condition. Then for pairs of conditions, the negation of the pairs, then for three conditions, and so on, up to all five conditions. From this analysis it would be possible to build up a set of statements about which configurations may be sufficient for the outcome to occur. Such an analysis, however, would be tedious and cumbersome. Much more efficient is to use truth tables, which are explained in the next section.

The truth table

Truth tables are at the core of CDA and they shift the focus from empirical cases to configurations of set memberships (Schneider and Wagemann, 2012: Chapter 4). They also focus on the analysis of causal sufficiency; establishing causal necessity is a separate procedure which is considered later in this chapter. At first glance truth tables look like a data matrix, but the rows now represent each of the logically possible and combinations (intersections) between conditions (see Table 7.2). All the values in a truth table relate to crisp sets having the value of [1] for the presence of a characteristic and [0] for its absence.

The number of logically possible configurations may be calculated by taking two to the power of the number of conditions, or 2^k. Thus if a set of cases is characterized by just three binary (presence/absence) characteristics, then there will be 2^3 or eight theoretically possible combinations. With four binary sets, this goes up to 16 and with ten there are 1,024. In reality, not all these configurations are likely to materialize empirically. Configurations for which there are no or insufficient instances are referred to as 'logical remainders' in CDA and give rise to the phenomenon of '**limited diversity**' (Ragin, 1987: 104ff.). Handling logical remainders is considered later in this chapter.

Table 7.2 A truth table

Configuration	Dissatisfied	Male	Old	Frequency	Complained
1	1	1	1	30	1
2	1	1	0	15	1
3	1	0	0	5	1
4	1	0	1	12	0
5	0	1	1	2	0
6	0	1	0	5	0
7	0	0	1	0	NA
8	0	0	0	14	0

1=factor/outcome present 0=factor/outcome absent NA= not applicable

Each case in the table is assigned to the row that corresponds with its configuration. It can belong to only one truth table row, but rows may contain more than one case. Where there are few cases, each case with an identical configuration may be named. Where there are many cases, simply the frequency may be recorded, as in Table 7.2. Where sets are fuzzy, most cases will have partial membership in most rows and seldom full membership in any of the rows. However, remember that fuzzy sets are binary and metric at the same time, so fuzzy set values above [0.5] are more 'in' the set and can be given a crisp value of [1], and fuzzy values below [0.5] a crisp value of [0]. Any fuzzy set values of exactly [0.5] are excluded from the truth table since it cannot be determined whether they are more 'in' or 'out' of the set. As with crisp sets, it is now possible to assign cases uniquely to one row. Fuzzy set membership values of exactly [0.5] are best avoided wherever possible because, although they are included in fsQCA calculations of logical sufficiency, cases where any membership value in the configuration is exactly [0.5] are not included in the frequency column in the truth table. If there are many such cases, descriptive analyses of diversity may be misleading. Box 7.2 discusses some of the issues involved in avoiding such values.

Box 7.2 Avoiding fuzzy set values of [0.5]

If the original variables being converted into fuzzy values are metric, then it is unlikely that there will be more than a few membership values that end up as exactly [0.5]. However, if there are few cases it may be important, for the purpose of descriptive analysis, to keep all cases in the truth table. If it makes sense to do so, a fuzzy set value of [0.5] can be 'nudged' up or down to [0.51] or [0.49], for example, on the basis of knowledge about individual cases. For ordered categories, if there is an uneven number of categories with a middle neutral value such as with Likert-type items, then there could be many fuzzy values of [0.5]. It may make sense, however, to contrast those who 'agree' with a statement (including those who 'strongly agree' and those who 'agree') with those who do not, so the 'neither agree nor disagree' category can be given a fuzzy set value of 0.4 and will join those who are indicated as 'out' of the category 'agree' with a binary value of 0 in the truth table.

Besides listing all possible configurations, truth tables also indicate the outcome for each one. If the outcome is a crisp set, then its presence or absence is noted with a value of [1] or [0]. Table 7.2, for example, shows the number of cases that possess each logically possible combination of membership of the categories 'dissatisfied with the service received', being 'male' and being 'old', plus whether or not the outcome of interest – made a complaint in this example – occurred. A value of [1] shows the presence of a characteristic and [0] its absence. Thus there are 30 cases that combine being dissatisfied, male and old; 15 that combine being dissatisfied, male, but not old; 5 that combine being dissatisfied, female and not old; and so on. There were no instances of the configuration not dissatisfied, female and old (configuration 7), so it is not possible to say whether or not the outcome occurred.

At a descriptive level it is possible to summarize the table by saying that configuration 1 was the most common (30 out of the total 83 cases – the sum of the frequencies – or 36 per cent) while configurations 1 and 2 together accounted for over half (53 per cent) of the cases. We can also say that there were 12 older women who, although dissatisfied, did not complain. At an interpretive level, these results can be compared with what might be expected from theory about complaining behaviour, for example that people will complain only if they are dissatisfied and males are rather more likely to complain than females. Also, there were no cases of older females being dissatisfied; this might be a 'surprising' finding and might benefit from some further analysis.

With fuzzy sets it is, similarly, possible to study diversity at a descriptive level. From Figure 7.7 it can be seen that there are three countries (top row) that are developed, urban, industrial, literate and stable (having values of more than [0.5] on each) and three (bottom row) that have none of these characteristics. From Figure 7.6 it is possible to see that the three countries in the first group are Belgium, Netherland and the UK, and in the second group, Greece, Portugal and Spain. In all there are 10 configurations that exist and 22 that have no empirical existence. These have been deleted from the table since the outcome of such configurations, were they to exist, is unknown. In the terminology of fsQCA they are 'remainders'. Where there are few cases or an intermediate (small-*n*) number, remainders are usually those with no empirical existence. For datasets with larger numbers of cases such as the alcohol marketing data with 920 cases, it may be advisable to exclude configurations for which there are relatively few instances, for example frequencies of five or fewer.

The next element in the process of data analysis (see Figure 1.8 in Chapter 1) is looking at patterns of relationships between variables. From Table 7.2, for example, it is now possible to examine the subset relationships between the outcome being present and the presence or absence of various combinations of conditions. Each row in the truth table is, in effect, a statement of logical sufficiency. For configurations 1, 2 and 3, each combination of conditions is logically sufficient for complaint behaviour (i.e. membership of that configuration is a perfect subset of the outcome), so there is more than one pathway to an outcome. Reality is, of course, seldom that neat. If, for example, 3 of the 30 cases with all three conditions did not complain, then there is inconsistency. The row, in short, contains cases that contradict the statement of logical sufficiency. The proportion of such cases that do not display the outcome can be taken as a measure of inconsistency, in this case 3/30 = 0.1. Alternatively, a measure of consistency would be 0.9. Ragin (2006) argues that consistency scores below 0.75 make it difficult on substantive grounds to maintain that a subset relationship exists.

For fuzzy sets, the outcome could be converted into crisp sets, as for the conditions, but this would not take into account the fuzzy value relationships for

developed	urban	literate	industrial	stable	number	survived	raw consist.
1	1	1	1	1	3	1	0.884337
1	0	1	0	1	1	1	0.77381
1	0	1	1	1	2	0	0.725352
0	1	1	1	1	1	0	0.675497
0	0	1	0	1	2	0	0.508197
0	0	1	0	0	2	0	0.506173
1	1	1	1	0	1	0	0.392857
1	0	1	1	0	1	0	0.37931
0	0	0	0	1	2	0	0.311628
0	0	0	0	0	3	0	0.225543

Figure 7.7 The completed truth table

logical sufficiency. The solution is to make the values of [1] or [0] in the column for the outcome property in the truth table refer not to the presence or absence of an outcome, but to a judgement by the researcher that the fuzzy membership values for that configuration for all or nearly all the cases are less than or equal to the fuzzy membership of the outcome. In other words, there is, again, a subset relationship that is acceptably consistent with the notion of logical sufficiency. Recall from the explanation of fuzzy set intersection earlier in the chapter that the fuzzy set membership value for a configuration is the minimum value in that configuration. Thus in Figure 7.6, for Czech the fuzzy set membership value of the configuration `developed` and `urban` and `industrial` and `literate` and `stable` is [0.42] – the lowest value, which is for `developed`. Note that this value is less than the fuzzy set membership of the outcome, which is [0.85] for `survived`. This is also true for all cases except Austria, Germany, Greece and Spain. Figure 7.8 shows the relationships as an XY plot. Notice that all countries except these four are in the upper left triangle, giving a high consistency score on sufficiency. Figure 7.7 shows this to be 0.884. This figure can also be seen in Figure 7.8 to two decimal places at the top of the Y-axis. The degree of membership of this configuration is, for the most part, sufficient to ensure a degree of membership of the outcome. Fuzzy values on the outcome are, for the most part, higher than or equal to fuzzy values on the configuration. Note that this XY plot is only one of 32 plots that could be created from 2^5 configurations, for example plots where one of the conditions is absent.

While the assessment of the distribution of cases across configurations in a truth table excludes cases where the fuzzy membership value is exactly [0.5], the assessment of the extent to which a given configuration may be consistent with the proposition of logical sufficiency includes every case. The assessment also incorporates membership of every configuration, including the ones that have no empirical existence (since all cases are members of each configuration to some degree).

Each case thus has a membership value on each condition, on each configuration and on the outcome, so the calculation is thus done entirely with fuzzy

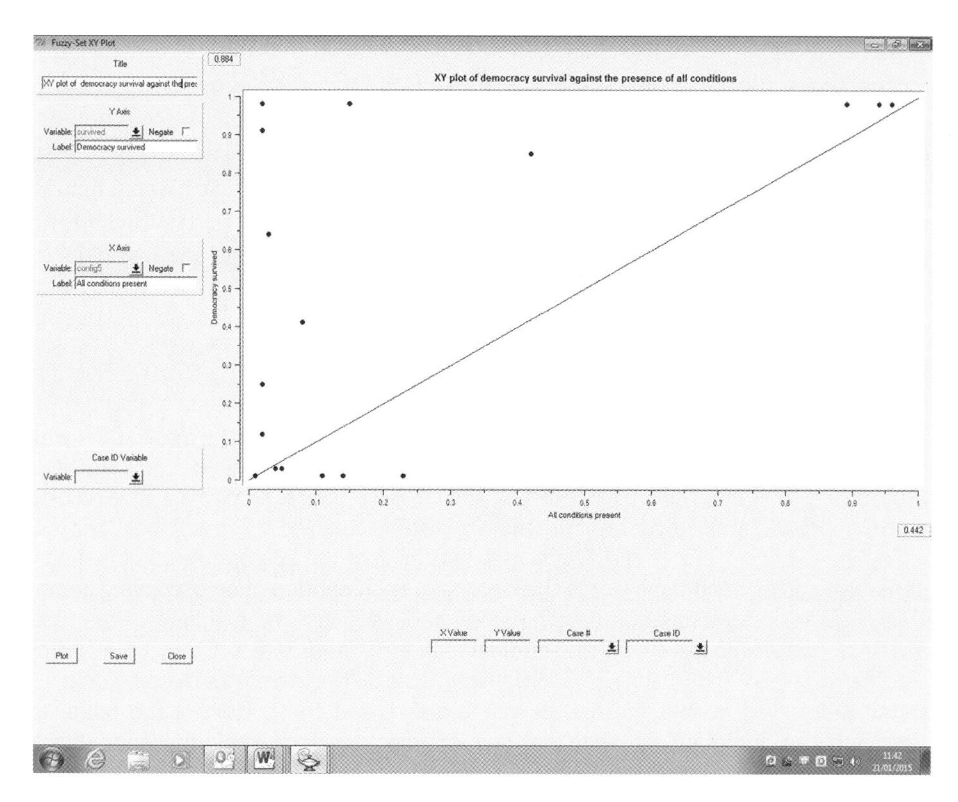

Figure 7.8 An XY plot of membership scores for democracy survived and configuration of all five causal conditions present

values. The fsQCA software calculates an index of consistency for each configuration. It takes the lesser of the two fuzzy set membership values for the configuration and for the outcome for each case and sums the total across cases. This is then divided by the total of the configuration membership values across cases. The calculation is summarized in the formula

$$\text{Consistency} = \frac{\sum \left(\min\ (X,Y)\right)}{\sum X}$$

If the membership value of the configuration is always less than the value of the outcome, the numerator and the denominator will be identical, and the result will be unity. If outcome values drop below configuration values for some of the cases, then inconsistency appears, but is weighted by how far they are below the diagonal (the size of the difference). The result shows the extent to which fuzzy membership of the configuration is a subset of the fuzzy membership of the outcome. This means that the membership value on the outcome is, to an acceptable degree, consistently higher than the membership value of the causal combination. The index of consistency is lowered by the number of cases that show inconsistency, the degree of that inconsistency, and the extent to which inconsistent cases have a large membership value on the causal configuration.

Consistency scores of less than 0.75 or even 0.8 mean that there is considerable inconsistency. Ideally, scores should be above 0.9. Figure 7.7 shows a truth table completed in fsQCA for the democracy survival data. Box 7.3 shows you how to do this. Only the top configuration has a consistency above 0.8. The next is at least above 0.75. The rest have too much inconsistency. So, only the first two configurations may be interpreted as consistent with being, for the most part, sufficient for democracy to survive. These are given a crisp set value of [1] and the rest [0].

Box 7.3 Using fsQCA: constructing a truth table

Select `Analyze|Fuzzy Sets|Truth Table Algorithm`. FsQCA will prepare an `Elimination Table` and give you a `Select Variables` box. Put `survived` into the `Outcome` box and `developed`, `urban`, `literate`, `industrial` and `stable` into the `Causal Conditions` box. Click on `Run`. The `Edit Truth Table` screen lists each of the 2^k possible combinations of the five causal conditions (32 in this analysis), each configuration occupying a row. To exclude configurations that have no empirical cases, put the highlight in the first row that has no cases (0 in the number column), and select `Edit/Delete current row to last row`. Note: Check that the `number` column is sorted before you do this so that all remainders are at the bottom of the table. If not, put the highlight in the top row of the `number` column and select `Sort|Descending`.

To see the range of consistency scores, it is best to sort the `consist` column into descending values. Put the highlight in the top box under `raw consist` and select `Sort|Descending`. The truth table can now be completed. Enter `[1]` under `survived` for the first two rows and a zero for all the rows below. There is, unfortunately, no way to print off a truth table directly from fsQCA. However, in the `Edit Truth Table` window, you can save the table (after editing) as a .csv file. Select `File|Save As CSV File`. This file can then be opened in Excel and copied into Word or PowerPoint.

Each row of the truth table is a statement of sufficiency, so there are two configurations deemed to be logically sufficient to ensure the survival of democracy:

developed*urban*literate*industrial*stable ⟶ survived

or

developed*~urban*literate*~industrial*stable ⟶ survived

The asterisk (*) denotes the Boolean operator 'and' while (~) denotes 'not', so, for example, the second equation reads: being developed, not urban, literate, not industrial and stable are, to an acceptable level of consistency, sufficient to ensure the outcome survived.

The truth table may result in several statements of this kind. The next stage in the analysis tries to simplify these in a more parsimonious solution that is entirely consistent with the data but eliminates logically redundant conditions. The process is usually referred to as Boolean logical minimization. This uses the general rule that if two truth table rows are both linked to an outcome but differ on only one condition, then that condition can be considered logically redundant. Thus configurations 1 and 2 in Table 7.2 are identical except that the outcome occurs whether or not the case is old, so the sufficient conditions reduce to being dissatisfied and male. Several such minimizations may be possible by taking all pairwise comparisons of expressions. The end products of the logical minimization process through pairwise comparisons might be further reduced because some of these expressions may themselves be logically redundant. The end result will be fewer solutions containing fewer conditions, but each will be logically equivalent to the original expressions; these solutions express the same information contained in the truth table, they do not contradict each other and they do not contradict the information contained in the truth table.

Minimization is likely to be limited or even not possible in two main circumstances. First, if there are many truth table rows lacking sufficient empirical evidence (remainders) and these are excluded from subsequent analysis. Second, if only one or two statements in the truth table reach an acceptable level of raw consistency no minimization may be possible. In both these situations, the fsQCA solution statements will be identical to the rows in the truth table that meet an acceptable level of raw consistency.

To produce solutions to the minimization process, fsQCA offers the researcher a number of options: a standard analysis and an analysis that allows the researcher to specify a number of constraints on the minimization process. The `Standard Analysis` generates three solutions: the most complex, which excludes remainders; the most parsimonious, which incorporates all remainders that can provide a simplifying assumption; and an intermediate solution somewhere between the two that allows researchers to include their directional expectations about each causal condition. This will ask the researcher, for each causal condition, whether its presence or absence is expected to contribute to the outcome. If, for example, the researcher, on the basis of substantive knowledge about the cases or on the basis of other evidence, suspects that the presence rather than the absence of a cause contributes to an outcome, then this can be indicated in the `Intermediate Solution` window – conversely with its absence (see Box 7.4).

In a normal Boolean minimization the configurations A*B*C or A*B*~C (where '~' means 'not') would, if there are instances of both configurations resulting in an outcome, be reduced to A*B since the absence or presence of C seems to make no difference to the outcome. If, however, there are no instances of A*B*~C, then the parsimonious solution will make an assumption that had such a configuration materialized, the outcome would still have come about. The software, in fact, will make whatever assumption about the outcome that results in minimization. Suppose, however, that empirical evidence and/or theory suggests that the presence of C is likely to contribute to the outcome, then the assumption that the outcome would occur in the absence of C becomes implausible, so the minimization is not made. In short, indicating that the outcome is linked to the presence rather than the absence of this condition limits the use of this configuration as a simplifying assumption.

When diversity is limited, there will be many remainders, and excluding them from the analysis by using the complex solution will result in little or no Boolean simplification. Likewise, a parsimonious solution, which makes all remainders available as simplifying assumptions (and which ignores any absent or present conditions entered in the intermediate solution), can be unrealistic, over-simple and may make assumptions that are untenable. Intermediate solutions strike a balance between parsimony and complexity by incorporating the researcher's substantive and theoretical knowledge. In general, intermediate solutions are preferred since they are the most interpretable and incorporate only those assumptions that can plausibly be made.

In the current example, it would seem reasonable to assume that being developed, urban, literate, industrial and stable would all contribute to the survival of democracy. The main output from an fsQCA analysis is a list of different configurations (which may be interpreted as causal combinations) that have met specified criteria of sufficiency for the outcome to occur. Figure 7.9 shows the fsQCA output from the standard analysis. The complex solution results in no minimization and the solution just repeats the first two rows from the truth table in Figure 7.7. The intermediate solution shows that two configurations are

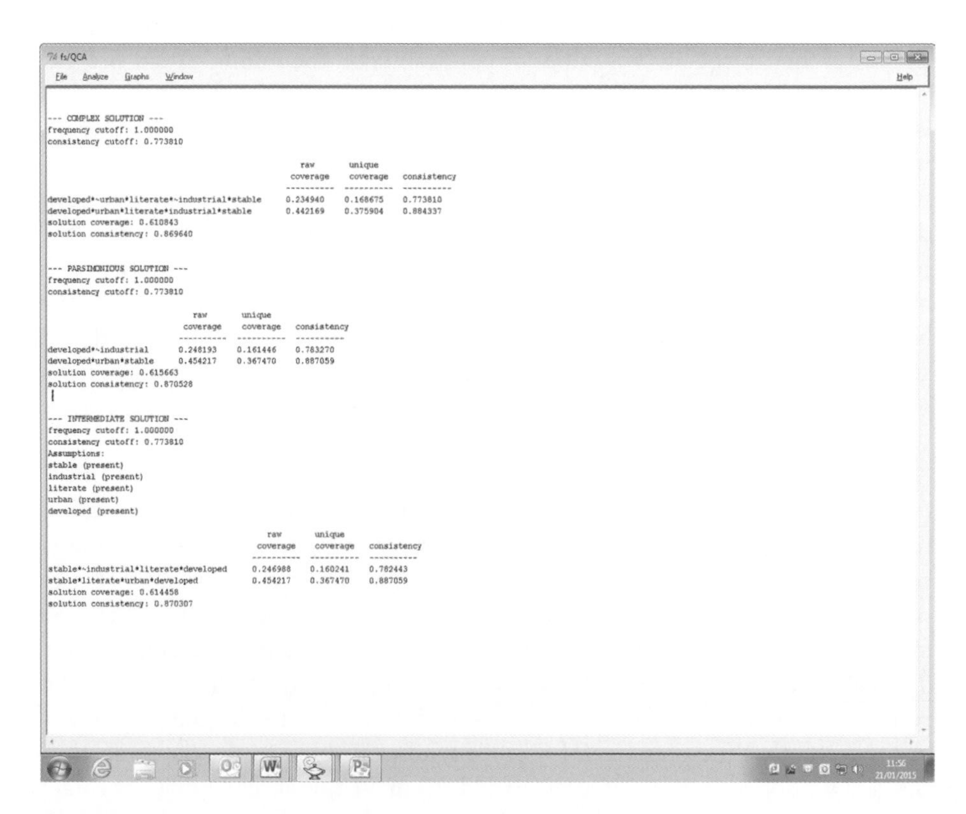

Figure 7.9 fsQCA complex, parsimonious and intermediate solutions

above the consistency cut-off selected in the truth table. The configuration stable*literate*urban*developed has the highest consistency of 0.887. The symbol [*] in Boolean terminology means 'and', so that among the countries studied, the causal combination of being developed, urban, literate and stable was sufficient to ensure the survival of democracy. The 'missing' causal factor is industrial. From Figure 7.7 you can see that democracy survived whether or not countries were industrial, so this factor can be dropped from the equation. The other configuration shows that if countries were not industrial, then for democracy to survive, it needs to be combined with being stable, literate and developed. The parsimonious solution is almost certainly making too many untenable simplifying assumptions. Box 7.4 explains how to obtain results for an analysis of logical sufficiency from fsQCA.

Box 7.4 Using fsQCA: obtaining the results

Once the truth table has been completed (see Box 7.3), in the Edit Truth Table screen click on Standard Analysis to obtain the Intermediate Solution screen. Click on all five radio buttons as Present, and on OK. (If the Prime Implicants screen appears, just click on Mark All and OK.) The truth table analysis gives three solutions: the complex, the parsimonious and the intermediate. Figure 7.9 should be reproduced. The solution itself can be directly printed by selecting File|Print. Parts of the output not required for presentation purposes can be selected and deleted before printing. There is no Edit|Copy procedure. Required outputs can be selected; whether or not they can be copied depends on the set-up of your mouse. An alternative is to use the Print Screen on your keyboard and paste into Word or PowerPoint.

To obtain an XY plot, select Graphs|Fuzzy|XY Plot from the Data Sheet. In the Fuzzy-Set XY Plot screen select your outcome variable from the Y-Axis drop-down list, for example survived, and a causal condition variable from the X-Axis drop-down list. You can also put in a title and a label for each axis. Click on Plot. The XY plot cannot be printed directly from fsQCA. It needs first to be saved as a PostScript file (.ps). Click on the Save button and select PostScript file. Open it in a program that reads .ps files. Alternatively, hit the Print Screen button on your keyboard and paste into Word or PowerPoint.

To obtain membership values for a causal combination, it is necessary to create a new variable that records membership values for that configuration. From the Data Sheet select Variables|Compute. In the Target Variable box, give your new variable a name, for example config5. In the Functions box, select fuzzyand(x,...) and click on the up arrow against Functions to put this expression in the Expression box. Now select each of the five causal conditions in turn, clicking on the up arrow against Variables, which will put them into the expression itself. Each condition needs to be separated by a comma. Click on OK and check your Data Sheet. config5 should appear as the last variable and have given you the minimum set membership of each of the five causal conditions, for example [0.14] for Austria (this has a membership value of [0.14] for urban). You can now use config5 in an XY plot as in Figure 7.8.

If `breakdown` is substituted for `survival` (and this is only 1 – `survival`), then a different analysis emerges. In the truth table, there are now six configurations with a consistency of over 0.8. The results are not just a mirror image of the analysis for democracy survival, but show that the conditions leading to the breakdown of democracy are very different from those that are sufficient for its survival. Thus, being developed, literate, industrial, but unstable is sufficient to produce democratic breakdown. However, an alternative pathway is being not developed, not urban and not industrial. Finding sufficient conditions for the breakdown of democracy is thus more effective than looking for conditions for its survival. In general, it is always good practice to analyse the negative of an outcome (Y and ~Y). However, the researcher needs to be aware of the possibility that because consistencies are never perfect, it is possible that a configuration can be sufficient both for the occurrence of an outcome and for its non-occurrence. How to handle such a **contradiction** is explained in a later section.

Solutions offered by fsQCA may be evaluated against two main criteria: consistency and coverage. Consistency and coverage between them evaluate the degree of empirical support for hypotheses specifying causal sufficiency (and, using a separate analysis, for causal necessity). Consistency for each solution statement is measured in the same way as for `raw consistency` in the truth table, but on configurations that have been minimized. It assesses the degree to which cases sharing a given minimized configuration are a subset of cases displaying the outcome. Consistency across all solutions measures the degree to which membership across all the solutions is a subset of membership of the outcome. The *maximum* of each case's membership across the solution configurations is compared with membership of the outcome.

Coverage is the proportion of cases that are covered by a configuration that has an acceptable level of consistency: that is, it forms a subset relationship – there is little point in considering coverage for configurations that cannot be considered subsets of the outcome. There are two main kinds of coverage: matrix coverage and outcome coverage. **Matrix coverage** is the number of cases in a subset relationship as a proportion of the total number of cases in the data matrix (which, as explained in Chapter 1, may be a sample, random or non-random, a census of the research population or an incomplete census). For crisp sets this is the number of cases that are both members of a condition (or configuration) and members of the outcome (i.e. they are in a subset relationship) divided by the total number of cases. In Table 7.2, for example, there are 30 cases with membership of the first configuration out of a total of 83 cases, giving a matrix coverage of 0.36; the next configuration has a coverage of 0.18. Coverage across all sufficiency statements is 45/83 or 0.54. For fuzzy sets, matrix coverage is the number of cases included in a configuration that meets an acceptable level of consistency divided by the total number of cases.

Outcome coverage is the number of cases in a subset relationship as a proportion of the number of cases that are instances of the outcome. This is a useful measure when there is equifinality, that is, the outcome may result from two or more configurations, each sufficient but not necessary for the outcome. A configuration (which may be interpreted as a causal combination) that covers or accounts for only a small proportion of the instances of the outcome is not as empirically important as one that covers a large proportion (Ragin, 2006).

For crisp sets, outcome coverage can be measured by taking the number of cases that are both members of a condition (or configuration), X, and also members of the outcome, Y, divided by the number of cases that are members of the outcome (which includes both those that are both X and Y plus those who are Y but not X) as in Figure 7.3 (p. 193) for a perfect subset relationship and Figure 7.5 (p. 196) where there is some (but an acceptable degree of) inconsistency. For fuzzy sets, Ragin (2006) suggests a parallel calculation, taking a measure of the intersection as the sum of the lesser of *X* and *Y*, divided by the sum of the *Y*-values:

$$\text{Outcome coverage} = \frac{\Sigma(\min(X,Y))}{\Sigma Y}$$

Outcome coverage is a function of the relationship in size between subset and superset; matrix coverage is a function of the relationship in size between subset and the sample or population used in the research. Both are measures of the 'importance' of a subset relationship. Ideally, researchers would like their findings to cover a significant proportion of both the cases in the data matrix and the number of cases that manifest an outcome. However, where either matrix coverage is very high or even total or where outcome coverage is very low, then this is a situation that Schneider and Wagemann (2012) call 'skewed' set memberships. Both kinds of coverage may, as with consistency, be calculated in three main contexts: on the 'raw' configurations, on the minimized solution statements, and across solutions. Where there is equifinality, outcome coverages may overlap, so Ragin distinguishes in the fsQCA software between coverage that may be shared with other acceptably consistent subset relationships (which, unfortunately, he calls 'raw' coverage) and unique coverage which measures the degree to which memberships in the outcome are explained by the conditions in the configuration minus their overlap with other configurations, that is, uniquely due to that pathway.

Researchers, of course, are looking for high consistency and high coverage. In reality, consistent statements may cover relatively few cases. Consistency and coverage often work against one another so that, in fuzzy set analysis, if the researcher raises the raw consistency cut-off, coverage is likely to be lower, and vice versa. Coverage is relevant only for configurations that are acceptable subsets of the outcome. Consistency, like statistical significance in variable-based analyses, signals whether an empirical connection merits the close attention of the researcher. If the hypothesized subset relation is not consistent with the proposition of causal sufficiency, then the researcher's theory is not supported. Coverage, like the degree of association or correlation in variable-based analyses, indicates the empirical importance of a set-theoretic connection.

The analysis of logical necessity

By focusing on the truth table, fsQCA puts the emphasis on the analysis of sufficiency. The truth table is purely an analysis of sufficiency, not necessity. Drawing conclusions about necessity from a truth table can be very misleading. The focus on truth tables has, unfortunately, often been to the detriment of the analysis of logical and causal necessity. In fsQCA there is no such thing as a necessary configuration; if conditions are necessary, then they are necessary individually. Necessary conditions are analysed separately in an analysis that does not use truth tables.

Logical necessity requires that one property has been selected as the outcome, and that a subset relationship exists such that the outcome is a subset of the condition, as in Figure 7.10 for crisp sets (compare this with Figure 7.3).

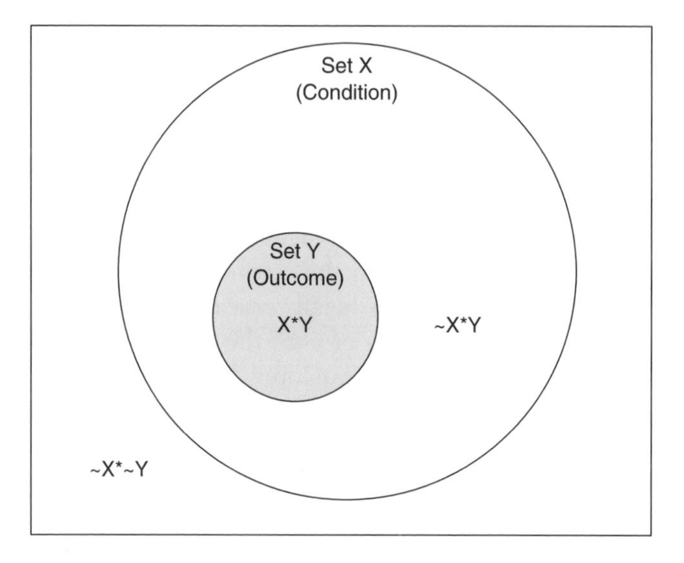

Figure 7.10 A subset relationship: crisp set logical necessity

Table 7.3 A necessary condition

		Has a higher degree(X_1)	
		X_1	Not X_1
University lecturer (Y)	Y	100	0
	Not Y	50	100

All cases with the outcome manifest that condition, for example all university lecturers have a higher degree but not all those with higher degrees are university lecturers. Having a higher degree is a necessary but not sufficient condition for being a university lecturer. The situation can also be seen in a two-by-two crosstabulation (Table 7.3). There are no university lecturers who do not have a higher degree – the top right box is empty.

For fuzzy sets, a necessary but not sufficient condition requires that for each case the membership value on the condition is greater than (or equal to) the fuzzy set value on the outcome. This means that Y is a subset of X. Graphically, all cases fall below the main diagonal as in Figure 7.11. High membership on Y presupposes high membership on X – but low membership on Y may be accompanied by high or low values on X. High X may be necessary but not sufficient for Y. The degree of membership of X in effect sets a ceiling on the degree of membership of Y.

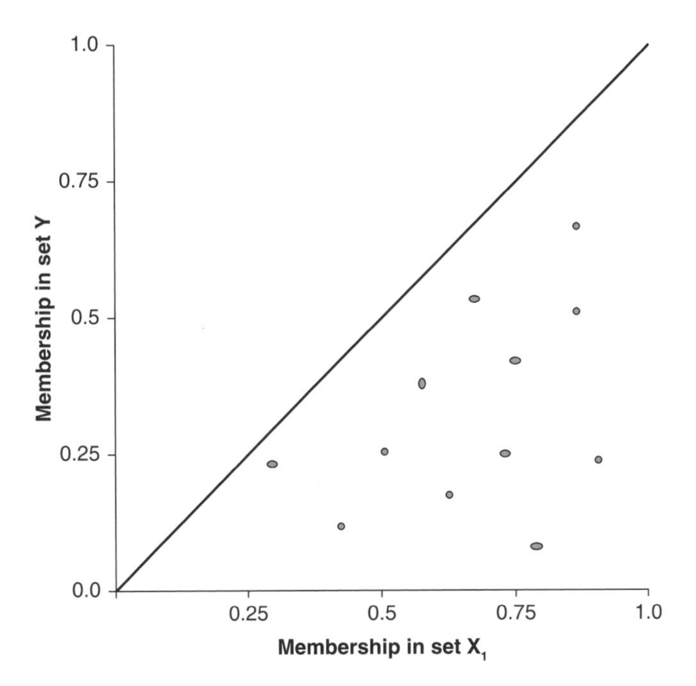

Figure 7.11 A fuzzy set necessary but not sufficient condition

Whereas it is possible for a condition not to be sufficient on its own but to be sufficient in combination with other conditions, it is not possible for a condition to be not necessary on its own but necessary in combination, so necessary conditions are always treated singly.

Logical necessity might imply causality, but does not demonstrate it. It only indicates that a causal interpretation may be appropriate. For *causal* necessity, it still needs to make theoretical sense as with causal sufficiency. The treatment of necessary conditions in fsQCA has been the subject of some discussion. For example, should the analysis of necessary conditions be carried out before focusing on sufficiency? Should necessary conditions be included in the analysis of sufficiency? These points are taken up later in a section on standards of good practice in fsQCA.

The analysis of necessary conditions in fsQCA is a separate procedure that looks at which *individual* conditions may be necessary or mostly necessary for the outcome to occur. Box 7.5 explains the procedure. An analysis of necessary conditions for the survival of democracy shows that being literate and being stable each have a consistency of over 0.9 (see Figure 7.12), so are probably necessary conditions. The analysis for breakdown is less clear and shows that the condition of being not developed and not urban are separately consistent with necessity with a score of over 0.8. Note that fsQCA produces equivalent measures of consistency and coverage for the analysis of necessity. The calculations are not explained here, but see Schneider and Wagemann (2012: 139–47) for details.

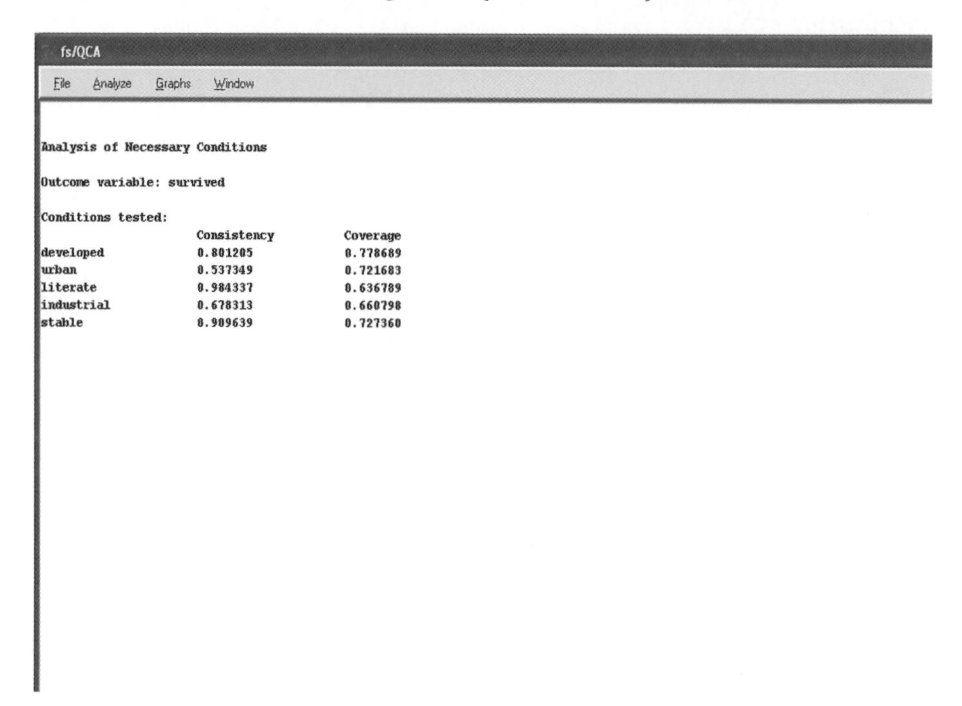

Analysis of Necessary Conditions

Outcome variable: survived

Conditions tested:

	Consistency	Coverage
developed	0.801205	0.778689
urban	0.537349	0.721683
literate	0.984337	0.636789
industrial	0.678313	0.660798
stable	0.909639	0.727360

Figure 7.12 fsQCA analysis of logical necessity

Box 7.5 The analysis of necessary conditions in fsQCA

From the `Data Sheet`, select `Analyze|Necessary Conditions`. In the `Select Conditions` window select either `survived` or `breakdown` as your outcome, then add the conditions from the `Add Condition` box. You can put in both the condition and its negation, but for the most part put whatever you think might be necessary, for example `~stable` for `breakdown` or `stable` for `survived`. In the output, consistency and coverage are listed for each condition entered.

Key points and wider issues

Configurational data analysis (CDA) requires the creation of a data matrix in which the properties of cases are expressed as set memberships, that one property is singled out as the outcome under investigation, and that the remaining properties are conditions that are potentially causally relevant. The focus of this chapter is on the analysis of fuzzy sets and the use of one particular software package, fsQCA, which stands for fuzzy set Qualitative Comparative Analysis. This makes use of truth tables that enable researchers to study all possible configurations from two or more conditions and also to assess the extent to which any of the configurations may be accepted as subsets of the outcome, that is, logically sufficient. Where there are several such subsets, the principles of Boolean minimization are then deployed so that the results can be expressed in a more parsimonious but logically equivalent manner. The manner in which this is achieved depends on assumptions that are made about what an outcome would have been for configurations that do not exist empirically in the dataset. The analysis of logical necessity is a separate procedure that does not use truth tables.

The extent to which subsets exist is measured by an index of raw consistency, while the results are evaluated by using measures of causal statement consistency on the minimized solutions and by measures of coverage. It is important that the results are evaluated not only for the outcome, but also for its negation, since the conditions that are sufficient or necessary for the former may be different for those that may be sufficient or necessary for the outcome not to occur.

Pitfalls in the analysis and interpretation of fsQCA findings

Researchers need to be aware of potential dangers in the analysis of set memberships using fsQCA, dangers that may remain hidden unless specific attention is paid to them. These dangers arise from three main sources: skewed set memberships, lack of diversity and contradiction.

Skewed set memberships

Membership in a set is skewed if all or nearly all cases hold very high or very low membership in a set – either in the outcome or in a condition. Skewness can arise in two main contexts: very high matrix coverage when either all or nearly all the cases manifest the outcome or the condition; or very low outcome coverage when the condition never or very seldom occurs. It often happens, for example, that all or nearly all the cases in a dataset manifest the outcome or have high memberships of the outcome. In this situation, all or nearly all conditions or configurations readily become sufficient for the outcome to happen. They are sufficient, but in a trivial sense. Figure 7.13 shows a **trivial sufficient condition** for crisp sets; as the size of the superset Y_1 expands to cover the matrix of cases (the rectangle Y_2), it becomes increasingly trivial. Figure 7.14 shows the parallel situation for fuzzy sets. All cases have full or nearly full membership of set Y.

Alternatively, there may be very few cases that possess a condition; it may be a perfect subset relationship, but not particularly relevant to the analysis. Figure 7.15 shows this for crisp sets where subset X shrinks to near non-existence, and Figure 7.16 for fuzzy sets. All the cases have non-membership or very low memberships of set X.

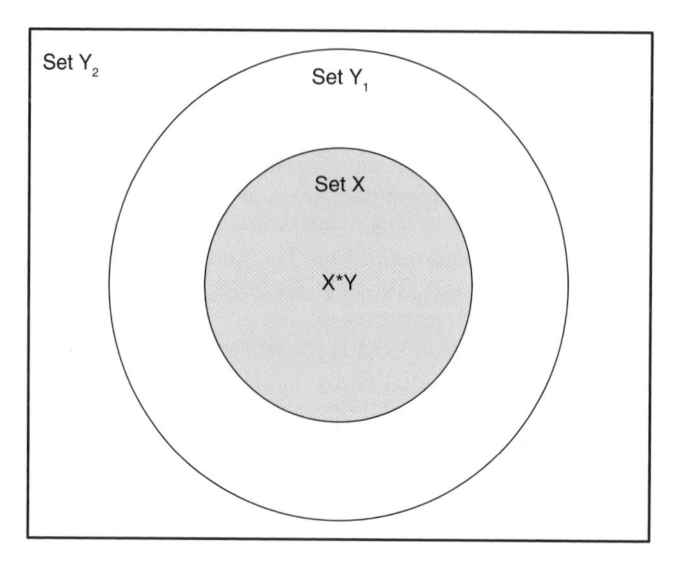

Figure 7.13 A subset relationship: trivial crisp set logical sufficiency

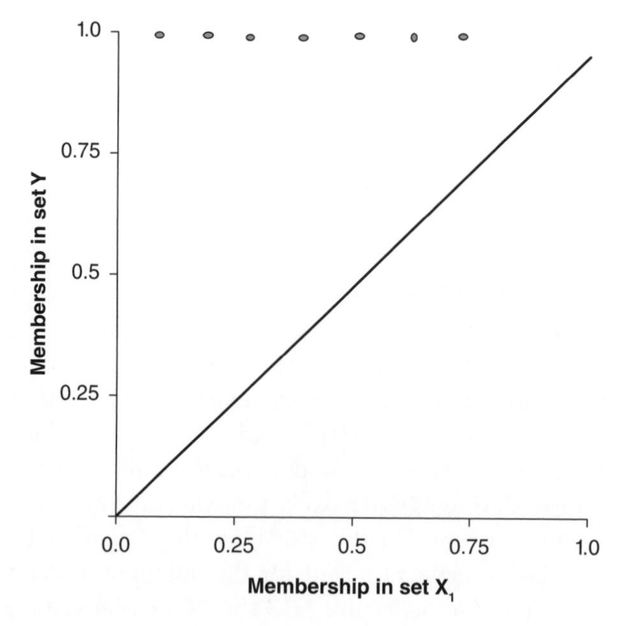

Figure 7.14 A trivial fuzzy set sufficient but not necessary condition

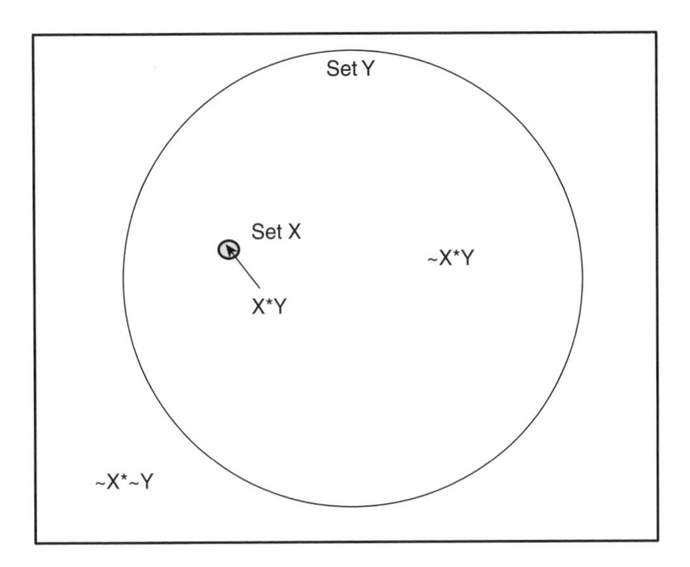

Figure 7.15 A subset relationship: irrelevant crisp set logical sufficiency

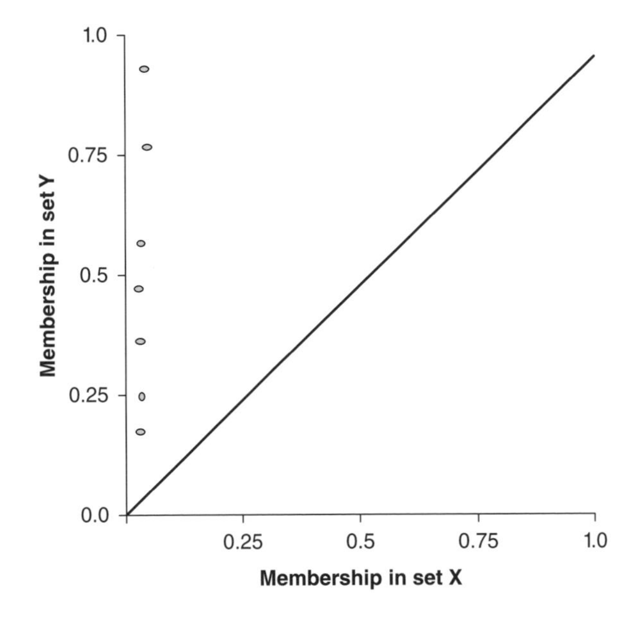

Figure 7.16 An irrelevant fuzzy set sufficient but not necessary condition

For necessary conditions, subset and superset are reversed. A **trivial necessary condition** exists when the condition always or nearly always occurs, so it readily becomes necessary, whatever the outcome. In Figure 7.17, Y is always a subset of X since X expands to cover the matrix of cases. For fuzzy sets, all cases have high membership of set X even though they are all in the lower right hand triangle in a perfect subset relationship (Figure 7.18).

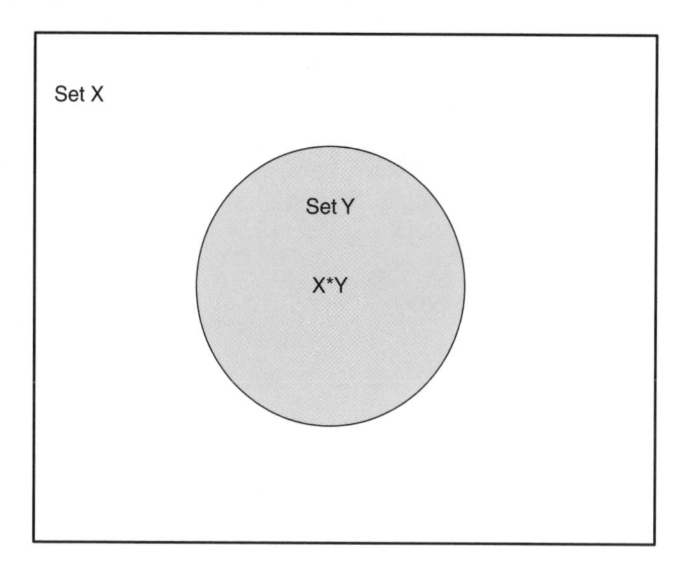

Figure 7.17 A subset relationship: trivial crisp set logical necessity

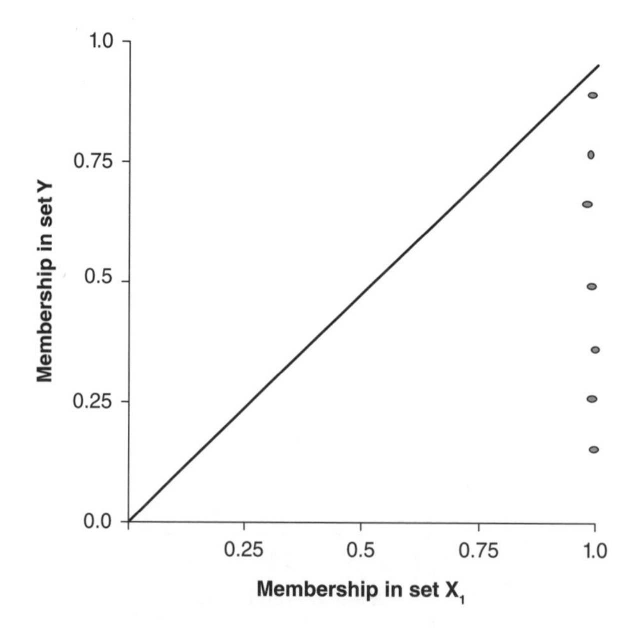

Figure 7.18 A trivial fuzzy set necessary but not sufficient condition

An irrelevant necessary condition is one in which the outcome Y never or seldom occurs, as in Figure 7.19 for crisp sets and Figure 7.20 for fuzzy sets.

Skewed set memberships, in short, tend towards triviality when matrix coverage expands to cover the entire matrix and towards irrelevance when outcome coverage is very low. Skewness is a matter of degree and measures of its extent would be helpful. Ragin's measure for fuzzy set coverage in fsQCA explained

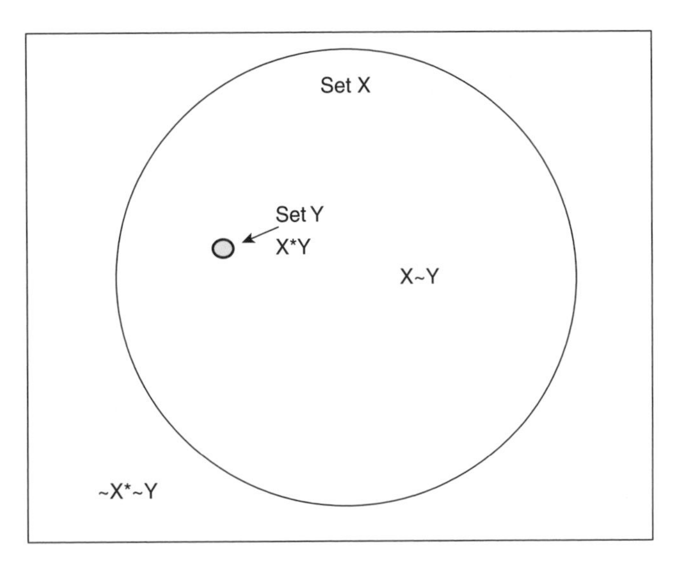

Figure 7.19 A subset relationship: irrelevant crisp set logical necessity

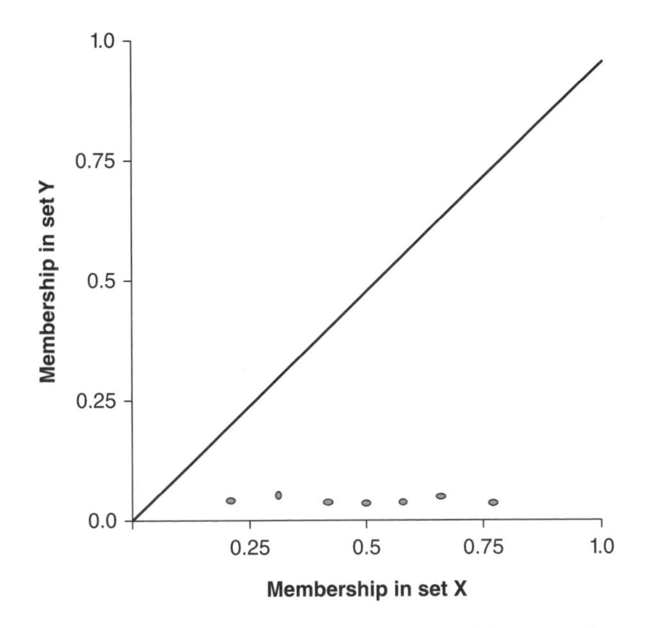

Figure 7.20 An irrelevant fuzzy set necessary but not sufficient condition

on p. 211 relates only to outcome rather than matrix coverage and is hence a measure of relevance rather than triviality. Maximum outcome coverage is achieved when a condition is both necessary and sufficient. For crisp sets, this means that set X and set Y totally coincide; for fuzzy sets, the cases all line up along the diagonal. Minimum coverage is when either X or Y never occurs or has zero fuzzy set memberships.

Lack of diversity

Diversity is limited when many possible configurations do not have an empirical existence (or for large-*n* datasets there are not enough cases in a configuration to be able to draw conclusions). Each configuration that fits into this category is a 'logical remainder' in fsQCA terminology. In short, a logical remainder is a row in a truth table without enough cases in it. (Remember that each case can fit into only one row, whether the sets are crisp or fuzzy.) The more logical remainders there are in relation to the logically possible rows, the greater the lack of diversity.

Schneider and Wagemann (2012: Chapter 6) argue that limited diversity may be a result of three main sources. First, the number of truth table rows may simply outnumber the cases at hand. Thus if there are five conditions, this will produce 32 rows and there may be only, say, 20 cases, so some rows have to be empty. Second, some combinations are not logically possible, like a pregnant male, or just very unlikely. Third, in reality, conditions tend to cluster together, so even if there are many cases, there will still be some configurations that, although theoretically possible, just do not exist.

A study of diversity and lack of diversity may be profitable in itself, but for fsQCA there are some serious decisions that need to be made about what to do with remainders. You may recall from the explanation of the standard analysis in fsQCA above that remainders may be (a) excluded from the analysis, (b) included as simplifying assumptions or (c) limited by directional expectations.

Where there is reasonable diversity and the number of cases greatly exceeds the number of truth table rows, then excluding remainders from the analysis may be a sensible approach and some minimizations may be possible. This will have the added advantage that no simplifying assumptions have been made, tenable or untenable. However, where there is lack of diversity or relatively few cases (or both) this approach will lack any parsimony and the solutions may just repeat truth table configurations that meet acceptable levels of consistency. Where a researcher's theory is not well developed or there are few directional expectations, then the parsimonious solution that takes advantage of any simplifying assumption may be sensible. Where diversity is limited and directional expectations are well developed, then the intermediate solution is the obvious approach. Any analysis of assumptions that are made in the absence of empirical evidence is usually called 'counterfactual analysis'. Box 7.6 discusses counterfactual analysis and the different kinds of counterfactual assumptions that may be made. It should be clear from Box 7.6 that fsQCA considers only those counterfactuals that are simplifying and, as a subset of these, those that are directional. This means that some directional expectations that are not simplifying are excluded from the analysis, as are any simplifying assumptions that are not directional.

Box 7.6 Counterfactual analysis

All causal statements make assumptions about what would have occurred if the causal condition or set of conditions had not manifested themselves. Thus, if we say that A causes B, we are making the assumption that, in the absence of A, B will not happen. Where data are historical, it is, of course, not possible to rerun history to see what would have happened if the causal event had not occurred. Thus the assertion that the assassination of Archduke Franz Ferdinand in Sarajevo caused the outbreak of the First World War hinges on the assumption that if the archduke had survived, there would have been no world war. Such an assumption about an unobserved event is a 'counterfactual'. Probing the historical and logical consistency of counterfactuals is the essence of counterfactual analysis; for example, were there forces that would have led to war even in the absence of the assassination?

Causal statements, in short, are based on comparing an outcome when a causal event or condition did occur (the 'factual') with an outcome it is assumed would have happened if a causal event or condition did not occur (the 'counterfactual'). For a single event, the counterfactual is, by definition, unobserved. For situations where there are multiple events (as in repeated experiments) or multiple cases (as in a survey) then it is possible to compare outcomes for situations when causal factors did occur with outcomes where they did not. However, we are still making the assumption that, after controlling for confounding factors as far as possible, cases or events are reasonably comparable to allow for assessment of causal effect. We still cannot know what the outcome *would* have been for each individual case if the causal factor had not been present. Counterfactuals are, by definition, unobserved.

In experimental designs in which treatments are manipulated and cases are randomly assigned to treatment and control groups, these counterfactual assumptions are probably quite reasonable and justifiable. In survey research and in macro-comparative research, it is not possible to assign cases randomly to treatment and control groups or to apply treatments in a discriminatory manner. Here, the counterfactual assumptions that are made can be crucial to any conclusions about causality. According to King and Zeng (2007), every quantitative analysis is making a counterfactual prediction and each one needs to be evaluated. Emmenberger (2010) argues that qualitative researchers normally use counterfactuals to assess causality for singular events and have developed criteria for assessing them; however, these same criteria, he suggests, can be used for quantitative research to identify erroneous causal inferences and help in formulating more realistic hypotheses.

The counterfactual assumptions being made are clearest when an argument is being put forward for causal necessity. If an event or condition is necessary for an outcome to occur then, by definition, this entails the counterfactual that without that event or condition, the outcome will not happen or, historically, would not have happened, even if that event or condition was not by itself sufficient. Goertz and Levy

(Continued)

(Continued)

(2007) call these necessary condition counterfactuals. Sufficient condition counterfactuals are rather more complicated. Issues of causality and how it is established are taken up in more detail in Chapter 10.

There are different kinds of counterfactual. The distinction has already been made between simplifying and non-simplifying counterfactuals. The `Standard Analysis` allows only for the former. Within this subset, Ragin and Sonnett (2004) distinguish between 'easy' and 'difficult' counterfactuals. Easy counterfactuals are those that are in line with both the empirical evidence and existing theoretical knowledge on the effect of a single condition on the outcome and emerge in the `Standard Analysis` as directional expectations. Difficult counterfactuals may be in line with the empirical evidence at hand, but not with theoretical expectation. These are best left as `Absent or Present` in the `Standard Analysis` and will be excluded from the truth table solution.

Contradiction

Inconsistency is a matter of degree and, in the context of fuzzy set analysis, arises as cases that are exceptions to the subset pattern. Contradiction, by contrast, consists of two statements that say the opposite of one another. It is not a matter of degree; either the contradiction exists or it does not. An example is to claim that a condition is necessary and at the same time to allow a configuration as sufficient that includes the absence of that condition. This can happen if a condition that is found to be necessary is subsequently included in the truth table analysis of sufficiency. If, furthermore, a separate analysis of necessary conditions is not carried out, necessary conditions may be hidden because they get minimized away by accepting the absence of a necessary condition in order to advance parsimony. In principle, this could happen only for the parsimonious solution in the `Standard Analysis`, but unless the researcher is aware that a condition is necessary, indicating its presence as contributing to the outcome in the intermediate solution may be overlooked. It is also possible that even with no assumptions about remainders, the acceptance of even a small amount of inconsistency at the truth table stage may still allow as a sufficient condition a configuration that contains the absence of a necessary condition. The remedy, suggest Schneider and Wagemann (2012), is to exclude remainder rows that might provide the basis for such contradictions before the `Standard Analysis` is carried out. Thus remainder rows that include the absence of a necessary condition can be deleted.

A second form of contradiction can arise when a case's membership on the condition can be less than its membership of both the outcome and its negation. The result can be a condition that is sufficient for both the occurrence and the non-occurrence of the outcome. It is quite possible, in fact, for a truth table row to reach an acceptable degree of consistency for both the outcome and its negation. Ragin has developed a measure of such contradiction that he calls the PRI, which stands for Proportional Reduction in Inconsistency, and which is produced automatically in fsQCA on the truth table output. The bigger the difference in consistency scores for Y and ~Y, the larger the PRI and the more confidence the researcher can have in selecting either Y or ~Y as the proper subset relationship. PRI is thus comparing the results of the outcome with its negation without the researcher having to rerun a separate analysis for the negation. If the contradiction cannot be resolved, for example because the PRI score is low, then the researcher may need to consider (a) adding (or excluding) conditions that increase the PRI score, (b) changing the definition or assessment procedures for the outcome or (c) changing the specification of relevant cases to include in the analysis.

Requirements and limitations

In order to use fsQCA it is necessary that:

- there is a clear outcome that the research is designed to investigate;
- the outcome can be assessed in terms of fuzzy set memberships;
- each condition and the outcome need to be thought of in terms of a category or 'quality' to which cases may or may not belong, or belong to a certain extent, and the assessment of which is related to theoretical and empirical knowledge external to the data;
- there are hypotheses or at least justifiable hunches about the underlying causal structure for that outcome;
- the researcher wishes to interpret the results in terms of necessary and sufficient conditions;
- it is expected that the causal structure is complex, that is equifinal (there are different pathways to an outcome) and conjunctural (conditions are often sufficient only in combination);
- the number of cases is at least 15, preferably over 30 or so.

In terms of limitations, while fsQCA can handle a mixture of crisp and fuzzy set conditions, if the outcome is crisp, then it is necessary to use crisp set QCA – which is also available in the fsQCA software. However, to undertake crisp set analysis, all the conditions need to be crisp sets.

The results from fsQCA are very sensitive to the decisions and assumptions the researchers will have made in the process of data construction and data analysis. These will include the ways in which set memberships have been assessed, the number of conditions that are included in a particular run of fsQCA, the thresholds chosen both for defining frequencies of configurations that will count as remainders and for the acceptable level of raw consistency, and the assumptions made about the inclusion of remainders in the analysis.

The output from fsQCA is not easy to interpret. There is usually no single result such as would emerge from a multiple regression analysis; instead there may be several different causal expressions, each with its own level of consistency and its own level of coverage. The role played by any particular condition may vary considerably depending on which other conditions it is combined with.

Standards of good practice

While CDA might be seen as an alternative to the more traditional variable-based approaches, it might alternatively be regarded as an addition to the bag of data analysis tools available to the researcher. The two approaches can be seen as complementary, each having its own strengths and weaknesses, so one standard of good practice is to consider both as potential solutions to achieving research objectives and perhaps even to apply both to a dataset. What each is good and not so good at doing is taken up in more detail in Chapter 9. It is worth emphasizing here, though, that CDA focuses on establishing, separately, causal sufficiency and causal necessity in circumstances where causal complexity is likely. To apply CDA simply because there are not enough cases for variable-based analyses is not good practice.

Once it has been decided to use CDA, careful thought must be given to the definition and assessment of set memberships for all conditions. Fuzzy set memberships should be as fine-grained as possible, taking particular care in identifying, clarifying and justifying which values constitute full membership, and where the crossover value of [0.5] should be drawn, but at the same time trying to avoid giving cases exactly that value.

The number of conditions used in an analysis run needs to be kept to a moderate level, ideally three or more but no more than about ten. In fact, beyond about six conditions, issues of lack of diversity and having too many remainders can become a problem. An alternative is to divide up the variables into two or more groups and run each separately. Schneider and Wagemann (2012: 253–5) suggest a two-stage approach, first running a truth table with contextual or 'remote' conditions which are then minimized. A second run then includes these along with more immediate or proximate conditions or events. The number of conditions can sometimes be reduced by developing higher-order, more abstract constructs such as might result from exploratory factor analysis, but it needs to be remembered that each needs to be characterized as a category to which cases may belong to various degrees that can be defined in absolute rather than relative terms and which also relate clearly to theory and empirical evidence.

It has already been explained that results can be very sensitive to decisions made about frequency thresholds, acceptable levels of raw consistency and what to do about remainders. So good practice would involve undertaking a sensitivity analysis that shows the effect of different decisions. If truth table solutions and analyses of necessary conditions remain largely unaffected, then the findings are more robust.

It is always good practice to undertake analyses for the negative of the outcome as well as for the outcome itself and to check for contradictions, triviality and irrelevance. Consistency and coverage of the truth table solutions both need to be considered. It is usually a good idea to undertake analyses of necessity first, then watch for remainders in the analysis of sufficiency that include the absence of any conditions that are clearly necessary.

Key points and wider issues

There are a number of hidden dangers in the use and interpretation of the results from fsQCA. These include situations where set memberships are heavily skewed such that all or nearly all cases hold very high or very low membership in a set – either in the outcome or in a condition. Skewness can arise in two main contexts: very high matrix coverage when either all or nearly all the cases in the matrix manifest the outcome or the condition; or very low outcome coverage when the condition never or very seldom occurs. Skewed set memberships tend towards triviality when matrix coverage expands to cover the matrix of cases and towards irrelevance when outcome coverage is very low.

Further dangers arise when there is a severe lack of diversity, that is, there are many configurations that, while perfectly feasible, do not exist or, for larger datasets, exist in insufficient numbers. The researcher needs to make decisions about what to do with these remainders – to exclude them from the analysis, to allow the software to make whatever simplifying assumptions add to parsimonious results, or to make counterfactual assumptions based on theory or empirical knowledge or understanding of cases. Lack of diversity is a matter of degree, but the more it exists, the more the results are contingent upon these decisions and the greater the possibilities for contradiction to arise.

Contradiction consists of two statements that say the opposite of one another, for example, claiming that a configuration is sufficient for an outcome to occur and also sufficient for its negation – for it not to occur – or claiming that a condition is necessary and at the same time allowing a configuration as sufficient that includes the absence of that condition. For fuzzy set analysis where a degree of inconsistency is generally acceptable, and particularly when there is a severe lack of diversity, the software may produce such contradictions. As far as possible, contradictions need to be resolved before the analysis is carried out, for example by excluding remainder rows that might provide the basis for such contradictions.

In order to use fsQCA, there needs to be a clear outcome that can be assessed in terms of fuzzy set memberships; each condition and the outcome need to be thought of in terms of a category or 'quality' to which cases may or may not belong or belong to a certain extent and the assessment of which is related to theoretical and empirical knowledge external to the data; there must be hypotheses or at least justifiable hunches about the underlying causal structure for that outcome; the researcher must have a desire to interpret the results in terms of necessary and sufficient conditions; it is expected that the causal structure is complex and that the number of cases is at least 15, preferably over 30 or so.

Once these conditions are met, it is good practice to consider the following:

- be clear about why fsQCA is being used and how it complements (or contradicts) the results of variable-based analyses;
- fuzzy set memberships should be as fine-grained as possible, taking particular care in identifying, clarifying and justifying which values constitute full membership, and where the crossover value of [0.5] should be drawn, but at the same time trying to avoid giving cases exactly that value;

(Continued)

(Continued)

- the number of conditions used in an analysis run should be between about three and eight;
- sensitivity analyses should be carried out showing the effects of different decisions, particularly on chosen levels of acceptable consistency and what to do about remainders;
- analyses of the negation of the outcome should be undertaken;
- coverage of the results needs to be checked for triviality and lack of relevance;
- careful thought needs to be given to the presentation of results to an audience who may not be familiar with fuzzy set analysis. Issues of presentation are taken up in Chapter 10.

Fuzzy set analysis and time

fsQCA is a cross-sectional data analysis tool that treats the data matrix as a single period of time. The order in which conditions are linked through either logical AND or OR does not matter. A*B is equivalent to B*A. However, the plausibility of causal claims often depends crucially on the sequence or temporal order of conditions or events. Besides the general idea that outcomes should come 'after' causal conditions in some sense, it may well be that, for example, A followed by B is more likely to result in the outcome than B followed by A. It may be possible to investigate change or trends generally by running separate fsQCA analyses at different points of time. Alternatively, changes could be measured and incorporated as set membership values in the data matrix. More formally, Caren and Panofsky (2005) and Ragin and Strand (2008) have developed a temporal QCA called tQCA. This incorporates information on the sequence of events so that, for example, A/B reads 'A then B'. This considerably expands the number of logically possible combinations, so that, for example, even with just two conditions there are eight logically possible sequences including ~A/B, A/~B, and so on. For an explanation of how tQCA copes with this, see Schneider and Wagemann (2012: 263–4). tQCA is available in the R Package QCA3 (see below) or in fsQCA version 2.5. Remember, however, that tQCA handles only crisp sets. There is as yet no fuzzy set version.

Other fuzzy set analysis software

Software solutions for fuzzy set analysis tend to be provided free of charge. Their developers have invested a lot of time and effort into programming them, so if you make use of a program, please acknowledge this by citing it in your work.

Tosmana is not, strictly speaking, for analysing fuzzy sets. However, it allows the user to go beyond binary variable crisp sets to include nominal variables with three or more categories. It was developed by Lasse Cronqvist, who calls it multi-value QCA or mvQCA in short. It can be downloaded for free at www.tosmana.net. It covers csQCA as well as mvQCA, giving both complex and parsimonious solutions. Its main advantage is that cases remain identifiable throughout the analysis. However, there are no intermediate solutions and no analysis of necessary conditions.

R Package QCA3 is the second package for the R environment for statistical computing and graphics. It can be installed from within R or downloaded from http://cran.r-project.org. It was released in 2009 by Ronggui Huang. It includes csQCA, mvQCA, fsQCA and tQCA, but there is no automated procedure for intermediate solutions.

Stata Package 'fuzzy' is a package for the commercial statistical software Stata. It has been developed with a strong orientation towards the incorporation of probabilistic procedures. The authors are Kyle Longest and Stephen Vaisey. It requires the commercial software Stata and has no intermediate solutions.

Generally, programs with a graphical user interface (GUI) are more user-friendly, but limited in their capabilities (fsQCA, Tosmana). Command line interface (CLI) software, in contrast, has a much steeper learning curve because it requires programming, but it is usually also more flexible and powerful (R Package QCA3, Stata Package 'fuzzy').

Implications of this chapter for the alcohol marketing dataset

The background and key variables used in the alcohol marketing study were explained in Chapter 1. Chapters 3–6 explored the results of applying variable-based analyses to the dataset. For the purpose of this chapter, a number of key variables were converted into fuzzy sets. How these were assessed was explained earlier in this chapter. A list of the properties as set memberships is given in Table 1.2.

Figure 7.21 shows the fuzzy set data for the first 30 or so cases. A preliminary analysis of logically necessary conditions for intention to drink alcohol in the next year shows that none of the conditions meet an acceptable level of consistency; similarly for the negation of intention (would definitely not drink alcohol in the next year). Accordingly, there is no rationale for excluding any conditions for the analysis or sufficiency, nor, indeed, for deleting any remainder rows that entail the absence of any of the conditions.

Figure 7.22 shows the completed truth table for the analysis of sufficiency. A frequency threshold of six was chosen, so that configurations with fewer than six cases were treated as remainders; 16 of the possible $2^5 = 32$ rows were thus treated. This threshold was chosen because only 10 per cent of cases were thereby excluded. A minimum consistency score of 0.75 was chosen (there is a sudden drop after this point) leaving six configurations that are consistent with the idea of their being sufficient to ensure the outcome of intending to drink alcohol in the next year. These configurations, however, cover only 72

FS/QCA Data Sheet

File Variables Cases Analyze Graphs

Case	fsaware	csdrinkstatus	fsinvolve	fslikeads	fslikeschool	cssibsdrink	fsintention
1	1	0	1	0	1	1	0
2	0.4	0	0.6	0.7	0.5	0	0.3
3	0.4	1	0	0.5	0.3	0	0.3
4	0.1	0	0	0	0.3		0
5	0.9	1	0	0.5	0.5	0	0.3
6	1	0	0.6	0.5	0.7	1	0.7
7	1	0	0.6	0.5	0.7	1	0.3
8	1	0	0	0.5	0.7	0	0.3
9	1	1	1	0.7	0.5	0	0.7
10	0.3	1	0	0.5	0.7		1
11	0.7	1	0.6	0.5	1	1	0.7
12	0.8	0	0	0.5	1	0	0
13	0.3	1	0	0.5	0.7		0.7
14	0.6	0	0	0.5	1	0	0.3
15	0.9	0	0.8	0.3	1	1	0.3
16	0.4	1	1	0.3	0.7	1	0
17	1	0	0	0	0.3	1	0.3
18	0.4	0	0	0.5	0.5		0.5
19	0.9	0	0	0	1	0	0
20	1	1	0.8	0	0.5	0	0.7
21	0.7	0	0.6	0.3	0	0	0.3
22	0.7	1	0.6	0.3	0.3	0	0.7
23	1	1	0.6	0	0.7	0	0
24	0.4	0	0.6	0.5	0.7	1	0
25	0.6	0	0	0	0.7	0	0.5
26	1	0	0.8	0.5	0.3		0.3
27	0.7	1	0	0.3	0.3	1	0.7
28	0.4	0	0.6	0.5	0.3	0	0
29	1	0	0.8	0.3	0.3		0.7
30	0.7	1	0.6	0.5	0	1	1
31	0.2	1	0.6	0.5	0.3	1	1
32	0.6	0	0	0	0.5	0	0.3
33	0.3	0	0.6	0.3	0.3	1	0.3
34	0.7	0	0	0.3	1	0	0

Figure 7.21 The fuzzy set data

Edit Truth Table

File Edit Sort

fsaware	fsinvolve	fslikeads	fslikeschool	cssibsdrink	number	fsintention	raw consist.	PRI consist.	product
1	1	1	0	1	12	1	0.887234	0.707182	0.627436
1	1	0	0	1	18	1	0.835443	0.600877	0.501999
1	1	1	1	1	7	1	0.830928	0.600000	0.498556
1	1	0	1	1	14	1	0.797599	0.535433	0.427061
1	0	0	1	1	12	1	0.787965	0.451852	0.356043
1	1	1	0	0	9	1	0.761017	0.350231	0.266531
0	0	0	1	1	16	0	0.704160	0.321555	0.226426
1	1	0	0	0	18	0	0.685534	0.260355	0.178482
1	1	1	1	0	14	0	0.678160	0.253333	0.171800
1	0	0	0	0	9	0	0.635513	0.124688	0.079241
0	0	0	0	0	6	0	0.611247	0.091429	0.055886
0	0	1	1	0	7	0	0.594086	0.087613	0.052050
1	1	0	1	0	57	0	0.546153	0.161634	0.088277
0	1	0	1	0	19	0	0.532491	0.088028	0.046874
1	0	0	1	0	45	0	0.513851	0.104593	0.053745
0	0	0	1	0	49	0	0.471048	0.058496	0.027554

Figure 7.22 The completed truth table

out of the 920 cases or 7.8 per cent. The three most common configurations (with frequencies of 57, 45 and 49) have low levels of consistency.

A standard analysis was conducted, making the assumptions listed in Figure 7.23. Thus the presence of four of the conditions was assumed to contribute to intention to drink alcohol in the next year, but the absence of liking school was assumed to do so. Figure 7.23 shows the intermediate solution, which makes only those assumptions allowed by the researcher. The highest level of consistency arises from the combination of not liking school (~ indicates the absence of a condition), liking alcohol ads, being involved in alcohol advertising and being aware of alcohol advertising. An alternative configuration is being aware of alcohol advertising and having siblings who drink, but the level of consistency has fallen considerably. Selecting a higher level of consistency, for example 0.8, which leaves the top three configurations as acceptable logical sufficiency statements (Figure 7.24), produces a better result and shows that it is the combination of liking alcohol ads, being involved in alcohol ads and having siblings who drink that is important. Liking or disliking school can be omitted since both expressions are consistent and differ only in membership of the category liking school (Figure 7.24). Note, however, that the level of coverage has declined considerably by raising the level of consistency.

More striking and much clearer in the analysis, and possibly of more relevance to policy issues, was to analyse the negation of intention to drink alcohol, so that *not* having siblings who drink, *not* being involved in alcohol marketing and *not* liking ads are sufficient in nearly all cases to ensure that the person thinks that they will *not* drink alcohol in the next year. Note that the last two analyses were run without entering drink status as a potential 'causal' factor.

The results from a fuzzy set analysis of the alcohol marketing data are very different from the results obtained from the variable-based analyses that were carried out in Chapters 3–5. Several different causal expressions meet the

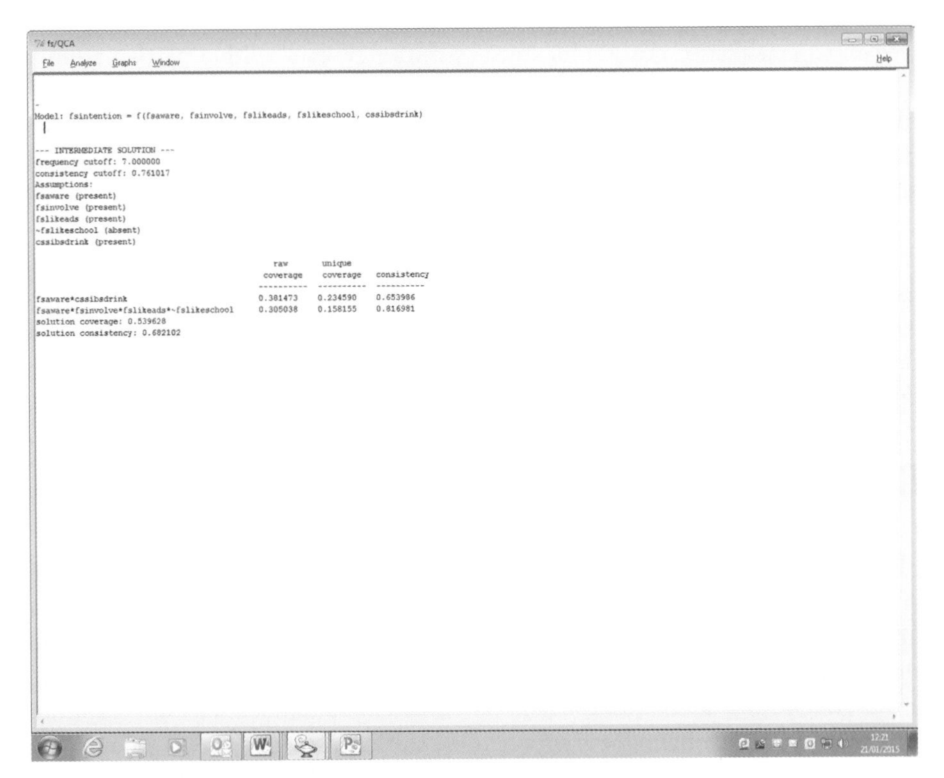

Figure 7.23 fsQCA intermediate solution

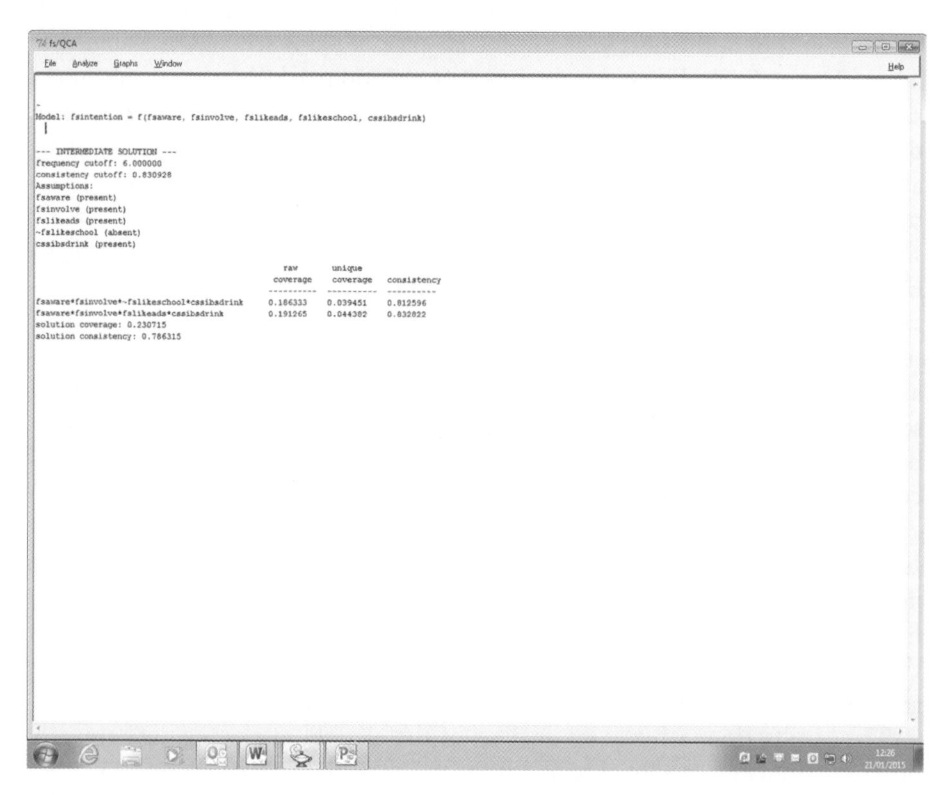

Figure 7.24 fsQCA intermediate solution with higher consistency cut-off

researcher-established criteria for causal sufficiency, each with a particular level of consistency and level of coverage. In variable-based analyses, the 'importance' of discovered findings is indicated by the degree of covariation between variables; such analyses, however, do not address issues of causality, necessity or sufficiency. Fuzzy set analyses, by contrast, focus mainly on combinations of conditions that appear to meet criteria of causal sufficiency and individual conditions that appear to be necessary. There is no measure of covariation between variables; rather the 'importance' of the results is assessed by measures of coverage. Chapter 9 considers in detail the comparisons, advantages and limitations of both variable-based and case-based approaches to the analyses of datasets in general and directly compares the results from the alcohol marketing dataset.

Chapter summary

Configurational analysis is one of a number of alternatives to deploying the variable-based statistical analyses which were explained in Part Two of this book. Analyses that use only binary variables can focus simply on case membership or non-membership of categories of potential causal conditions and outcomes, and then investigate which combinations of category memberships (configurations) are prevalent in a set of cases, which potential configurations have no empirical existence, which configurations appear to be sufficient to give rise to an outcome and which individual conditions appear to be necessary for an outcome to happen.

Fuzzy set analysis extends binary set configurational analysis by allowing for degrees of category membership, but not only is it still possible to evaluate which potentially causal configurations may be sufficient but not necessary for an outcome to occur, or which individual condition may be necessary but perhaps not sufficient, but also, as pointed out by Ragin (2000), the analysis in fact becomes sharper and more demanding.

The processes of data analysis outlined in Chapter 1 apply as much to configurational analysis as they do to variable-based analyses. Data still need to be prepared in various ways Data records need to be checked for usability and edited for consistency and accuracy. Each property then needs to be give a set membership value. Variables that are already in binary form should be given a value of [1] for the presence of a characteristic and [0] for its absence. Nominal variables will need to be converted into binary form, each category being given a value of [1] for membership or [0] for non-membership. Ordered categories or rankings will usually be converted into a single fuzzy set. Metric variables will need an algorithm to define an upper value beyond which full category membership is being proposed, a lower value below which non-membership is proposed and a crossover value of maximum ambiguity.

Data description will take the form of examining the nature of diversity among the configurations while interpretation will be needed to link empirical diversity with theoretical expectations. Relating variables together takes the form of selecting an outcome which is the focus of the investigation, then looking at subset relationships to establish which single conditions may be causally necessary and which conditions or configurations of conditions may be causally sufficient. Evaluation is then required of causal statements that meet an acceptable level of raw consistency. For sufficiency statements where there is **equifinality** (there are several pathways to an outcome), they are first minimized and then consistency and coverage are recalculated on the minimized statements. Levels of consistency and coverage will need to be tested for sensitivity to different frequency cut-offs, for different raw consistency cut-offs, for different assumptions about remainders and for the negation of the outcome. Researchers also need to check for triviality and irrelevance arising from skewed set memberships and for contradictory statements.

The explanation and presentation of results will require the researcher to bear in mind the likely familiarity or unfamiliarity of the audience with configurational analysis, linking the results back to theory and, ideally, comparing these results with other forms of data analysis. In many ways the application of configurational analyses to business strategy or public policy is much clearer than for variable-based analyses, since causally sufficient configurations can be seen as alternative 'recipes' for bringing about a desired outcome.

The results of a fuzzy set analysis of the alcohol marketing dataset show results that are very different from those that would have been obtained from a variable-based analysis. Configurational analysis has its strengths and its weaknesses. These are explained in detail and compared with the strengths and weaknesses of variable-based approaches in Chapter 9. The researcher, however, does not have to choose between these approaches; they can be mixed in various ways, which are explored also in Chapter 9.

Exercises and questions for discussion

1. Is the 'fuzzification' of ordered categories just another way of scaling a variable?
2. Why is it sometimes better to treat a nominal variable with two categories (a 'dichotomy') as two separate binary sets?
3. Try minimizing the following five expressions for sufficiency for the outcome Y:

$$\sim A \sim B \sim C + \sim A \sim BC + \sim AB \sim C + \sim ABC + A \sim BC \longrightarrow Y$$

4. Why is it necessary to avoid, as far as possible, fuzzy membership values of [0.5]?
5. How does counterfactual analysis help to focus the results produced by fsQCA?
6. In what ways is the output from fsQCA analyses closer to 'reality' than the output from variable-based analyses?
7. Download the alcohol marketing fuzzy set dataset `Fuzzysetalcohol marketing.dat`, from www.https://study.sagepub.com/kent and save it onto your system. Download fsQCA. Just Google fsQCA and select `fs/QCA software`. Download `fsQCA 2.0`. Select `File|Open|Data` and browse to your saved .dat file and click on `Open`. Rerun the alcohol marketing data (a) using 0.8 as the consistency cut-off, then (b) the negation of the outcome intention to drink alcohol in the next year, then (c) trying both higher and lower frequency thresholds, and then (d) indicating in the `intermediate solution` all the conditions as `absent or present`. Repeat (a)–(d) but include `drink status` in the conditions.
8. Go to the website www.surveyresearch.weebly.com. Here you will find lots of interesting information about social surveys created by John Hall, previously Senior Research Fellow at the UK Social Science Research Council (1970–6) and Principal Lecturer in Sociology and Unit Director at the Survey Research Unit, Polytechnic of North London (1976–92). Download the 'Quality of Life: 2nd Main Survey'. Select `Subjective Social Indicators (Quality of Life)|SSRC Survey Unit Quality of Life Surveys in Britain 1971-1975|Abstracts, data and documentation`. For the 2nd National Survey 1975, download the data from `SPSS saved file` and save onto your system. Download and print off the `Abstract` and the `Questionnaire`.

   This survey covers many variables, but several of the questions like QA13 ask respondents to give a 0–10 rating. These can be readily converted into fuzzy set values so that 0=[0], 1=[0.1], 2=[0.2] up to 10=[1]. Copy all 13 items in QA13 (these are labelled `var136-var148`), all the items in QA18 and QA19 into a separate SPSS file and convert to fuzzy sets using SPSS `Recode`. Give the file a name like FuzzysetQOL. At this stage it is better to rename the variables so that you can recognize them in fsQCA, so `var136` could become `kitchen` and so on (this can be done in the `Variable View`). This SPSS file now needs to be saved as a .dat file. Select `File|Save As`. Under `Save as type` select `Tab delimited (*.dat)`. Check that the box `Write variable names to spreadsheet` is ticked. Click on `Save`. Now download fsQCA. Just Google fsQCA and select `fs/QCA software`. Download `fsQCA 2.0`. Select `File|Open|Data` and browse to your saved .dat file.

   QA19 can be taken as the 'outcome' to be studied. This is an overall measure of satisfaction. The items in QA13 can be taken as potentially sufficient causal conditions. A truth table analysis will show which combinations of conditions may be sufficient for overall satisfaction. Follow the instructions in Boxes 7.3 and 7.4. See if you can interpret the results. You can try the same for QA18.

Further reading

Ragin, C. (2008) *Redesigning Social Inquiry: Fuzzy Sets and Beyond*. Chicago: University of Chicago Press.

This is Ragin's latest comprehensive offering, focusing on set-theoretic thinking and fuzzy set analysis. It is not a guide to the use of his own software, fsQCA, although there are selected instructions at the end of some of the chapters.

Ragin, C. (2008) *User's Guide to Fuzzy-Set/Qualitative Comparative Analysis*.

This is the guide that comes with the software if you download it. It tells you what all the buttons do, but there is little help on how to interpret the results.

Ragin, C. (2009) 'Qualitative comparative analysis using fuzzy sets', in B. Rihoux and C. Ragin (eds), *Configurational Comparative Methods*. Thousand Oaks, CA: Sage.

This focuses on using fsQCA to analyse the survival of democracy data which is referred to in this chapter. Again, it is not a manual on how to use the software.

Schneider, C.Q. and Wagemann, C. (2012) *Set-Theoretic Methods for the Social Sciences: A Guide to Qualitative Comparative Analysis*. Cambridge: Cambridge University Press.

This is a comprehensive guide to set-theoretic methods and to the use of configurational data analysis. It covers both the basics and more advanced aspects. It is not a manual on how to use fsQCA, but explains very carefully what it does and does not do. If you want to pick just one book to follow up on this chapter, then this is the one to go for. There is an online appendix which is a guide to the use of software, including fsQCA, but, in fact, relatively little space is devoted to this software. Most of it focuses on using Stata and R.

Suggested answers to the exercises and questions for discussion can be found at the end of this text, pp. 293-321, and on the companion website, (https://study.sagepub.com/kent), which also give links to relevant free online Sage journal articles, PowerPoint slides, an overview of data analysis packages, an introduction to SPSS and weblinks to alternative datasets.

Notes

1. The data are derived from Berg-Schlosser and Mitchell (2000, 2003). The data from this project are used extensively by Rihoux and Ragin (2009).
2. Strictly speaking, to be binary, the negative of 'survived' should be 'not survived' rather than 'breakdown'. To 'not survive' may be seen as a different concept from 'breakdown', which could be a temporary interruption of democratic elements. However, in keeping with the original research, the term 'breakdown' is continued in this example.

8

CLUSTER AND DISCRIMINANT ANALYSIS

Learning objectives

In this chapter you will learn about:

- approaches to cluster analysis as a case-based method that differs from configurational analysis;
- how cluster analysis may be related to fuzzy set analysis;
- the contrast between cluster analysis and discriminant analysis, which is also essentially case-based;
- how to undertake cluster analysis and discriminant analysis using SPSS;
- how both these techniques may be applied to the alcohol marketing dataset;
- how a cluster analysis of cases on the need for popular brands on a range of products shows the unremarkable result that the main cluster includes cases that emphasize the importance of popular brands for clothes and trainers.

Introduction

It should be clear from Chapter 7 that set-theoretic methods and configurational data analysis are case-based in the sense that they group together cases that have identical configurations of set memberships, counting up the number of cases in each possible configuration. Other techniques use properties of cases, not as set memberships, but as traditional variables to analyse grouping structures among cases, based on measures of 'distance' between them. This is an approach used by cluster analysis and the first part of this chapter explains what cluster analysis is and then compares it with fuzzy set configurational data analysis.

Discriminant analysis, it can be argued, is case-based because it begins with groupings of cases that are already known, and which cases belong to each group is also known, but calculates which specific linear relations of selected explanatory variables can best predict each case's group membership. The groupings may be seen as a dependent categorical variable while the predictors are metric.

Both cluster and discriminant analysis are largely exploratory techniques that focus on classification; they are not usually used to 'test' or evaluate models. They focus on either the generation of taxonomies or figuring out what way variables may be 'responsible' for a pre-specified taxonomy. Cluster analysis, however, may be seen as an interdependence technique, but discriminant analysis does distinguish between dependent and independent variables.

Approaches to cluster analysis

Cluster analysis is a range of descriptive and exploratory (but not inferential) techniques for grouping cases (although it can also be used for grouping variables) into different clusters such that members of any cluster are more similar to each other in some way than they are to members in other clusters. There is no a priori information about group or cluster membership for any of the cases. Each case is assigned to a cluster suggested by the data, not defined beforehand, based on specified properties as variables (rather than being assigned to a given configuration based on properties as set memberships, as in Chapter 7).

Cluster analysis is used for a variety of purposes including the identification of 'natural' clusters in the data, the construction of useful conceptual schemes or taxonomies for classifying entities, data reduction, generating hypotheses or testing hypothesized groupings believed to be present. It has been referred to by a number of names including numerical taxonomy, classification analysis or Q analysis. This arises from its usage in many disciplines including psychology, biology, sociology, economics, engineering, business and marketing.

Cluster analysis involves a number of considerations and decisions that need to be taken by researchers. First, the variables on which the clustering is to be based need to be defined. These should be selected either based on past research and relevant to the research problem at hand, or based on existing theory. The variables used for clustering techniques may be metric or categorical, although Cohen and Markowitz (2002) point out that the inclusion of binary or dummy variables, while quite appropriate for regression models, creates a problem in cluster analysis since no amount of recoding will transform nominal items into metric ones whose differences imply distances.

The second consideration is to select a distance measure since the objective of clustering is to group together cases that are similar in terms of the specified variables. This is usually done in terms of distances between pairs of cases. The default measure in SPSS is what is generally called **Euclidean distance**, which is the square root of the sum of the squared differences in values for each variable. SPSS offers five other types of measure including, for example, Pearson correlation. Using different distance measures may produce different solutions so it may be advisable to use several different measures and to compare the results. If the variables are measured in very different units (e.g. euros and age in years), then it may be advisable to standardize the values by rescaling each variable to have a mean of zero and a standard deviation of one.

A third consideration is to select a clustering procedure. What is called **hierarchical clustering** develops a tree-like structure and may be approached in two ways. Agglomerative clustering is based on taking individual cases and combining them on the basis of some measure of similarity, such as the degree of correlation between the cases on a number of variables. Each case is correlated with each other case in a correlation matrix. The pair of cases with the highest index of similarity is placed into a cluster. The pair with the next highest is formed into another cluster and so on. Each cluster is then averaged in terms of the index being used and combined again on the basis of the average similarities. The process continues until, eventually, all the cases are in one cluster. The index of similarity may be achieved in several different ways including being based on the computation of the distance between pairs of cases, on the minimization of the squared Euclidean distance between cluster means, or the distance between pairs of clusters based on the means of all the variables.

The other approach to hierarchical clustering is divisive clustering, which begins with the total set of cases and divides them into sub-groups on a basis specified by the researcher. Thus the researcher may want a four-cluster solution of 1,200 respondents on 10 variables. An iterative partitioning computer program might begin by setting up four equal-sized groups at random. The centre of each cluster on the 10 variables is then calculated and the distances between each of the 1,200 respondents and the centres of the four groups are measured. On the basis of these distances, respondents are reassigned to the group with the nearest cluster centre. The new cluster centres are recalculated and the distances again measured, with a further reassignment taking place. This process is repeated until no further reassignments are needed.

Non-hierarchical clustering, often called *k*-means clustering, involves predetermining the number of cluster centres and all cases are grouped within a specified threshold from the centre. Again, there are several ways in which this can be achieved: some are 'single pass', some involve iterative partitioning, while others include density searching or fuzzy clustering (which is explained in the next section). Unlike hierarchical clustering where once cases are assigned to a cluster they are never reassigned in the clustering process, in non-hierarchical clustering cases may be reassigned until maximum homogeneity within clusters is achieved.

The output from cluster analysis may be displayed in a number of ways. For agglomerative hierarchical clustering, the most common way is as a **dendrogram** (see Figure 8.1). Each case is listed and the vertical lines represent clusters of cases that are joined together. The position of these lines on the horizontal scale indicates the distances at which the clusters were formed. This information is helpful for deciding on the number of clusters. However, with a large number of cases (over 100 or so) it is very difficult to read any pattern, so it may be a good idea to sample randomly a subset of cases, perhaps even to take several such subsets and see whether similar clusters emerge.

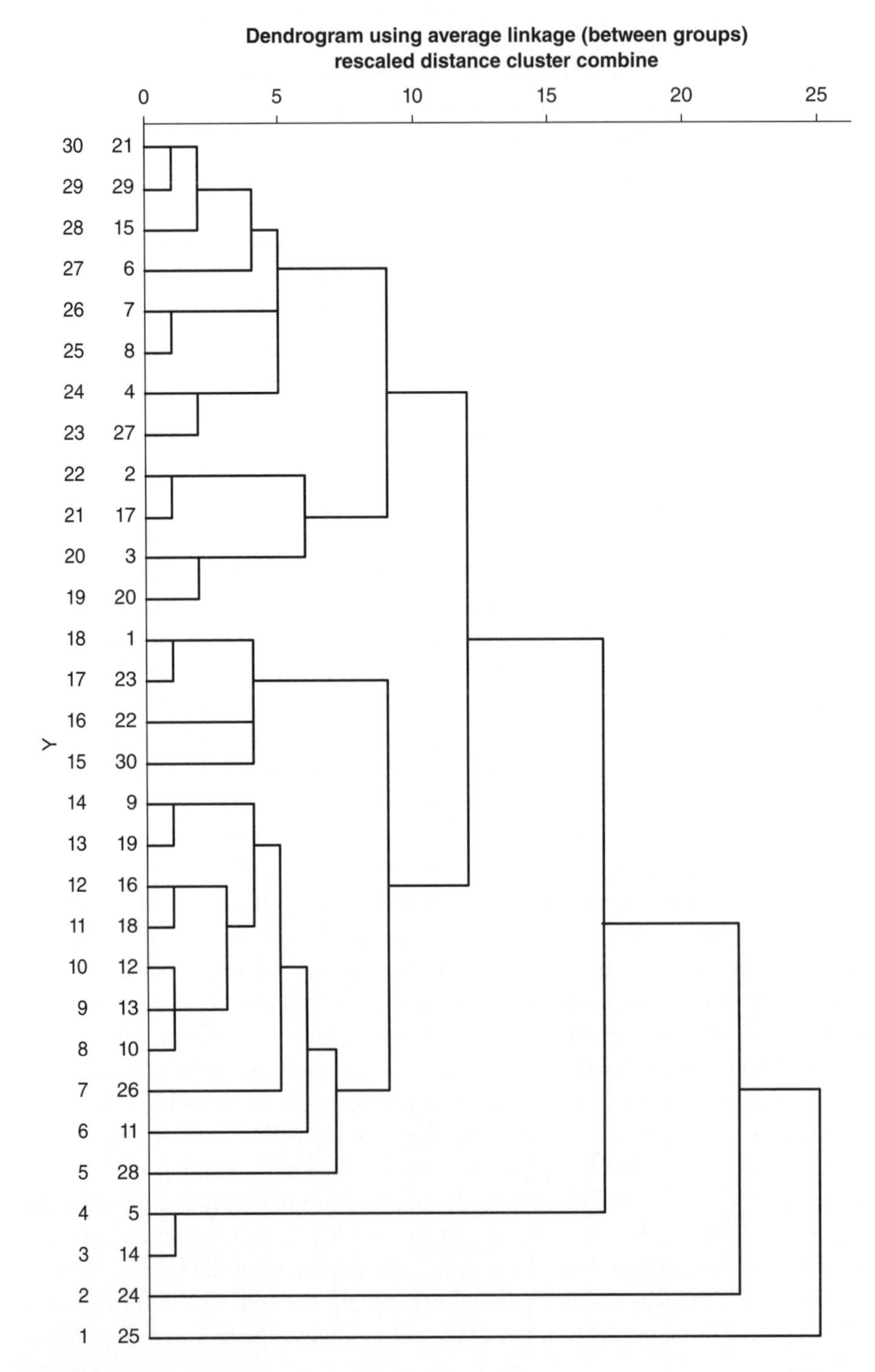

Figure 8.1 A dendrogram

Hierarchical cluster analysis does not produce a single final solution; rather, it identifies a set of preliminary solutions. One approach to determining the appropriate number of clusters is to judge from the agglomeration schedule (see below) when there are large increases in heterogeneity when moving from one stage to the next. There may, alternatively, be practical, conceptual or theoretical reasons for deciding on the desired number of factors. Thus managers may want a particular number, or previous research may have suggested a particular quantity. Alternatively, various criteria can be set, for example specifying the minimum number of cases in a cluster may determine the number of clusters sought.

Interpreting the clusters involves examining the mean values of the cases in a cluster (the **centroids**) on each of the variables. This will enable researchers to describe the clusters and maybe help them to assign labels. This stage can be quite difficult and, if there are many cases, a lot of work. Doubts have also been raised about the validity and reliability of cluster solutions. Bottomly and Nairn (2004), for example, showed that random data can generate apparently interesting clusters of cases.

Cluster analysis limitations

Cluster analysis is an exploratory, descriptive technique. It cannot, furthermore, be used for prediction, as with regression, but only to profile the clusters that are generated. There are dozens of different methods of clustering and a variety of different distance measures may be used, each combination is likely to produce a different result. Researchers need to make many choices and although there may be clues in outputs that suggest what to try or explore next, there is always a large element of subjectivity present. To validate cluster solutions, some researchers rely on seeing whether or to what extent the different methods produce similar clusters. Others may analyse separate subsamples (randomly splitting the sample) and compare the results.

Each case has its place in the output; accordingly, when the number of cases being clustered is over 100 or so, the interpretation of the results can become complex. Furthermore, cluster analysis, like factor analysis, always produces clusters, even when there are, in fact, no natural groupings in the data. The various techniques work by imposing a cluster structure on the data rather than allowing the structure to emerge from the analysis. Accordingly, researchers should always have a strong conceptual basis for why such groups should exist in the first place and the results of cluster analysis should always be seen as tentative rather than final.

Uprichard (2009) points out that measures of similarity (or dissimilarity) are always problematic and tend to be treated as symmetrical, that is, if case X is deemed to be similar to case Y, then it is assumed that Y is similar to X, but this may not be the situation. The relationship may be asymmetrical, as in a subset relationship seen in the previous chapter. Furthermore, because similarity and dissimilarity measures ultimately dictate the results of any cluster analysis, the question of 'What is a cluster?' becomes problematic as well. There is no way of knowing whether the emergent clusters are an artefact of the method or an intrinsic part of the data. Uprichard goes on to consider how a researcher's assumptions about the existence or 'reality' of the cases themselves may help to explain some of the issues and controversies about cluster analysis as a method.

Cluster analysis and the alcohol marketing dataset

As part of the interview survey, young people are asked how important they think it is for people to choose a popular 'well-known' brand when deciding to

Agglomeration Schedule

Stage	Cluster Combined		Coefficients	Stage Cluster First Appears		Next Stage
	Cluster 1	Cluster 2		Cluster 1	Cluster 2	
1	21	29	3.000	0	0	11
2	12	13	3.000	0	0	9
3	7	8	3.000	0	0	18
4	1	23	4.000	0	0	16
5	9	19	4.000	0	0	17
6	16	18	5.000	0	0	13
7	2	17	5.000	0	0	21
8	5	14	5.000	0	0	27
9	10	12	5.500	0	2	13
10	3	20	8.000	0	0	21
11	15	21	8.500	0	1	15
12	4	27	9.000	0	0	18
13	10	16	11.167	9	6	17
14	22	30	13.000	0	0	16
15	6	15	13.333	0	11	19
16	1	22	14.500	4	14	25
17	9	10	14.800	5	13	20
18	4	7	16.000	12	3	19
19	4	6	18.750	18	15	24
20	9	26	19.000	17	0	22
21	2	3	19.500	7	10	24
22	9	11	21.625	20	0	23
23	9	28	23.667	22	0	25
24	2	4	30.688	21	19	26
25	1	9	31.200	16	23	26
26	1	2	41.250	25	24	27
27	1	5	55.269	26	8	28
28	1	24	71.786	27	0	29
29	1	25	83.207	28	0	0

Figure 8.2 The agglomeration schedule

buy each a of range of nine different products including alcohol, along with cigarettes, perfume/aftershave, magazines, clothes, trainers, crisps, fizzy drinks and chocolates or sweets. There are issues of wording here since it is not clear whether they are being asked about 'people' in general or about themselves. However, it is possible that there are groups or clusters of individuals who combine similar perceptions about the importance of brands – groups who are different from other groups. A good first stage is to begin with hierarchical clustering to help determine how many clusters would be appropriate. Since there are many cases, the output from hierarchical cluster analysis will be huge, so for Figures 8.1 and 8.2, only the first 30 cases are used. Figure 8.2 shows the `Agglomeration Schedule`. There are $n - 1$ or 29 stages of agglomeration for 30 cases. In the first stage, cases 21 and 29 have been combined, in stage 2, cases 12 and 13, and so on. These have the shortest 'distance' between them and reappear in the dendrogram (Figure 8.1) with the distance indicated along the horizontal scale. At stage 11, case 21 is combined with case 15, but case 21 has already appeared at stage 1, and this is indicated in the `Stage Cluster First Appears` column. Values of 0 indicate that the case is still a single-member cluster; the case has not been combined before that stage. `Next Stage` indicates that case 15, for example, is next combined in a new cluster at stage 15. Again, this can be seen in the dendrogram. The agglomeration `Coefficient` measures the increase in heterogeneity (reduction in within-cluster similarity) that occurs when two clusters are combined. There is no sudden jump in the coefficient (Figure 8.2), but the biggest jump occurs at stage 19 when cases 4 and 6 are combined. From the dendrogram, we can see that there are about eight clusters at this stage (working from right to left), so an eight-cluster solution may be appropriate. Box 8.1 explains how to undertake cluster analysis using SPSS.

Box 8.1 Cluster analysis using SPSS

To select the first 30 cases in the alcohol marketing dataset, go to `Data|Select Cases`. In the `Select Cases` dialog box choose `Based on time or case range` and enter range 1–30. Any analyses will now be carried out on these 30 cases. To return to the original 920 cases, go back to `Select Cases` and select the `All cases` radio button.

Cluster analysis is found under `Analyze|Classify` and offers a choice of `Hierarchical Cluster, K-Means Cluster and TwoStep Cluster.` Hierarchical clustering is probably the most flexible. It can handle both metric and categorical data, it has a variety of methods of clustering available (these are algorithms for defining how similarity between multi-member clusters is measured) and it can standardize metric variables if different units have been used. Click on `Hierarchical Cluster`. In the `Hierarchical Cluster`

(Continued)

(Continued)

Analysis box click on Method. The default is Between-groups linkage, which iteratively combines cases in a way that maximizes the between-group variance between clusters. Other options include Within-groups linkage, Nearest neighbor, Furthest neighbor, Centroid clustering, Median clustering and Ward's method. There is also the option of standardizing variables in various ways if the clustering variables have used different metrics. The default distance measure for metric variables (SPSS refers to metric variables in this procedure as Interval) is Squared Euclidean distance. This is the measure normally used, but Pearson correlation may be an alternative. For categorical variables (SPSS calls them Counts) Chi-square is normally used, but Phi-square is an alternative. Note that for Binary variables, it is still possible to use squared Euclidean distance and a range of other measures. To obtain a dendrogram, click on Plots and then tick the Dendrogram box. Orientation is normally Vertical but Horizontal is an alternative.

K-Means Cluster requires that all the variables selected are metric and that the researcher specifies the number of clusters required. The algorithm uses Euclidean distance measures and results are sensitive to the order in which the cases are placed into the analysis, so it is advisable to obtain several different solutions with cases sorted in different random orders to verify the stability of a given solution. The procedure assumes that the researcher has selected the appropriate number of clusters and has included all relevant variables. Having decided on the number of clusters, it is now possible to use K-Means Cluster to analyse the full 920 cases and to 'fine-tune' the allocation of

Number of Cases in each Cluster

Cluster	1	88.000
	2	48.000
	3	87.000
	4	119.000
	5	34.000
	6	108.000
	7	112.000
	8	147.000
Valid		743.000
Missing		177.000

Figure 8.3 SPSS number of cases output

cases to those clusters. Under Options it is possible to check the box Cluster information for each case. The output shows the number of cases in each cluster (Figure 8.3). Note that there are 177 missing cases. The eight clusters are all well populated with cases. The output also shows the cluster membership for each case and its distance from the cluster centre. Alternatively, click on Save instead of Options and SPSS will create a new variable in the data matrix.

TwoStep Cluster was developed to cope with large samples. In the first step the hierarchical method is used to create a number of intermediate clusters and the second step uses these to cluster further. The algorithm employed by this procedure enables it to handle both categorical and metric variables and can automatically determine the optimal number of clusters by comparing the values of a model-choice criterion across different clustering solutions. Furthermore, by constructing a cluster features tree that summarizes the records, it allows researchers to analyse large data files. Two distance measures are available, one based on log likelihood, allowing a probability distribution to be placed on categorical variables and a Euclidean measure which is only for metric variables. There is also an option for the treatment of **outliers**. Cases that are in sparse leaves of the cluster features tree will be placed in a 'noise leaf'. In the TwoStep Cluster dialog box, put any clustering variables that are binary, nominal or ordered category into the Categorical Variables box and any metric variables into the Continuous Variables box. If any of the variables selected are categorical, SPSS will have the Log-likelihood radio button under Distance Measure selected. If they are all continuous, then the researcher can choose between this and Euclidean distance. SPSS will assume that continuous variables are to be standardized. If they are all measured in the same units, then select Options and transfer such variables to the Assumed Standardized box. Under Number of Clusters, the default is to Determine automatically with a maximum of 15 clusters, but it is usually better to specify a fixed number of clusters. If you wish to know which cases are in which clusters then select Options and click on the Create cluster membership variable box and SPSS will put a new variable into the data matrix indicating cluster membership for each case. Click on Continue and OK and SPSS will give you a model summary and an indication of cluster quality based on measures of cohesion and separation. If it is Fair or Good, double-click anywhere in the area to activate the solution. SPSS then gives you a number of options for viewing cluster sizes and the predictor importance of the selected cluster variables.

For the alcohol marketing dataset, it is probably easier and better to use TwoStep Cluster. Cluster quality is only just short of Fair. Double-click on the area and on Clusters. Under Display select Sort inputs by within cell importance. Over 70 per cent of cases are in cluster 3 that emphasizes the importance of brand for clothes and trainers. This finding is hardly surprising. It does confirm, however, that the procedure 'works', but also hints at some of the limitations of using cluster analysis that were considered in the previous section. Again, it is possible to get SPSS to create a new cluster membership variable.

Key points and wider issues

Although cluster analysis is widely used in the social sciences, it is, curiously, often missing from textbooks, even 'advanced' ones, on statistics and is seldom mentioned in handbooks and manuals on SPSS. An example is Field's (2009) 800-page book *Discovering Statistics Using SPSS*, which makes no mention of cluster analysis, although there are a few pages on discriminant analysis. Perhaps one reason is that although there are clear considerations involved in doing such analyses, at each stage there are choices to be made by researchers about applications of the technique, all of which are likely to give different results. Accordingly, there are considerable subjective elements in its procedures. Another reason might be that outputs from cluster analysis are daunting to interpret.

Cluster analysis is essentially an exploratory rather than a verificational method and it is descriptive rather than inferential. Cluster solutions, furthermore, are not generalizable because they are totally dependent on the variables used as the basis for similarity measures. Selecting optimal cluster solutions is a subjective process so researchers should take care in validating them, for example by cluster analysing separate samples or splitting the sample if other samples are not available.

It is also an approach that is probably best when used in combination with other approaches rather than giving a final 'stand-alone' result. Thus the allocation of cases to clusters may be used as an input to discriminant analysis, which is explained later in this chapter, or as an input to fuzzy set analysis; how is discussed in the next section.

Cluster analysis and fuzzy set analysis

Both fuzzy set analysis and cluster analysis work with the idea of positioning cases in multidimensional space. However, fsQCA (see Chapter 7) uses set membership values to justify its partitioning of a dataset (into 2^k of all possible configurations), while cluster analysis relies on geometric distance measures of similarity or dissimilarity and concepts of variance minimization (Cooper and Glaesser, 2011). In conventional cluster analysis, at each stage of the agglomeration or division process, each case is allocated to just one cluster. There is however, a less well-known procedure of fuzzy cluster analysis in which cases have fractional degrees of membership in several clusters in a manner analogous to fuzzy set memberships, but membership fractions are set to add to unity. Cooper and Glaesser (2011) show that if few cases are near the [0.5] partitioning point in an fsQCA analysis using the minimization rule for intersection along with its rule of allocating cases to the configuration in which they have a membership greater than [0.5], then a fuzzy cluster analysis, set to produce the same number of clusters as 2^k configurations, will produce very similar results. Results can, however, be very different if there are many cases near the [0.5] fuzzy set partition. Cooper and Glaesser (2011) warn that, while in fsQCA the partitioning point of [0.5] is, or should be, determined on theoretical grounds and not inductively from the data, with cluster analysis it is the reverse, with clusters created in an inductive, iterative process from the data and are thus sample dependent.

Cluster analysis can fruitfully be used in combination with fuzzy set analysis. One of the limitations of fuzzy set analysis is that, because it is combinatorial, the number of combinations expands exponentially with the number of conditions. Cluster analysis can help by taking cluster membership as a summary or typification of a number of conditions that can then be fed into an fsQCA analysis.

Discriminant analysis

While in cluster analysis neither the clusters nor each case's membership of them are known in advance, in discriminant analysis the reverse is true. The groupings and each case's membership are known. The objective of discriminant analysis is to know which specific linear relationships of selected metric variables best predicts each case's group membership.

Discriminant analysis is useful for situations where researchers want to build a predictive model of group membership based on observed characteristics of each case. The functions are generated from a sample of cases for which group membership is known; the functions can then be applied to new cases with measurements for the predictor variables but unknown group membership, for example product success or failure from a range of metric predictor factors, or a blood condition from smoking frequency and alcohol consumption. The procedure is very similar to logistic regression (see Chapter 6), but it is essentially case-based rather than variable-based in the sense that the focus is on making a prediction for each case rather than calculating the overall odds or probability of group membership for cases in general. Both, however, like multiple regression (see Chapter 6), are dependence techniques used for prediction and explanation of relationships, but the dependent variable is categorical, that is, the 'groupings' are defined by a researcher-selected nominal variable.

Discriminant analysis determines which weightings of metric independent variables best discriminate between two or more groups of cases (a categorical dependent variable) and do so better than chance. The weightings form a new composite variable that is known as a **discriminant function** and is a linear combination of weightings and metric values on the variables in the composite. This has the same form as a classical **regression** model, but instead of a dependent metric variable, there is a score that is used to classify the cases into the groupings using cut-off points. The weights are set to maximize the between-group variance relative to the within-group variance. Discriminant analysis can thus be seen as the equivalent of a multiple analysis of variance (MANOVA, see Chapter 6), but where the status of the dependent and independent variables is reversed; it determines which of the metric independent variables account for most of the differences in the average metric score profiles of the two or more groups.

Along with other parametric methods, discriminant analysis assumes that the independent variables are normally distributed with more or less equal variances. It also assumes that there is no **multicollinearity** – that the independent variables are not themselves highly correlated. Extreme values or **outliers** can create problems so these are best removed before analysis begins. If the categorical dependent variable is binary, then Hair et al. (2010) suggest that **binary logistic regression** (see Chapter 6) might be better since its assumptions are somewhat less stringent.

Discriminant analysis and the alcohol marketing dataset

A key dependent variable that the researcher may wish to explain is whether or not, or the extent to which, youngsters intend to drink alcohol in the next year. It would be useful to be able to predict from metric variables whether they are, for example, in the 'definitely' or 'probably yes' category, in the category that contains 'not sure', 'don't know' or 'no answer', or the 'definitely' or 'probably no' category. `Recode` can be used to regroup the existing six categories into these three categories.

The SPSS output includes a case processing summary that shows that there are 518 valid cases (402 had at least one discriminating variable missing). With three groupings, there are two discriminating functions, one for discriminating group 1 from groups 2 and 3, and one for discriminating group 2 from group 3. Figure 8.4(a) shows that the first function accounts for nearly all the variance. The overall statistical significance is evaluated using Wilks's lambda, which can be transformed into a chi-square whose significance level can be determined. Figure 8.4(b) shows that both functions (1 through 2) are statistically significant, but if the first function is removed (2 in the table) then it is not significant. Figure 8.4(c) tells us the relative contribution of each variable to the function. The importance of brands and involvement in alcohol advertising make the biggest contribution to function 1, but for function 2 the number of channels seen and involvement make a very strong but opposite contribution (but remember that this function was not statistically significant, so is not really important for this analysis). From Figure 8.4(d) we can see that function 1 discriminates between definitely or probably yes and definitely or probably not since the signs are opposite. Box 8.2 shows how to do this analysis in SPSS.

a

Eigenvalues

Function	Eigenvalue	% of Variance	Cumulative %	Canonical Correlation
1	.053[a]	99.6	99.6	.224
2	.000[a]	.4	100.0	.015

a. First 2 canonical discriminant functions were used in the analysis.

b

Wilks' Lambda

Test of Function(s)	Wilks' Lambda	Chi-square	df	Sig.
1 through 2	.949	26.658	6	.000
2	1.000	.114	2	.945

c

Standardized Canonical Discriminant Function Coefficients

	Function	
	1	2
Total importance of brands	.615	.202
Total number of channels seen	.237	.928
Total involvement	.564	-.904

d

Functions at Group Centroids

	Function	
Intention to drink alcohol	1	2
Definitely or probably not	-.225	-.005
Not sure, DK or no answer	.069	.032
Definitely or probably yes	.283	-.009

Unstandardized canonical discriminant functions evaluated at group means

Figure 8.4 SPSS discriminant analysis output

Box 8.2 Discriminant analysis using SPSS

Discriminant analysis is found under `Analyze|Classify| Discriminant`. Transfer into the `Grouping Variable` box the key nominal variable that contains the groupings you wish to be able to predict (e.g. the three groupings of intention to drink alcohol in the next year). You will need to define the range of codes; click on the button and enter the highest and lowest code values. Into `Independents`, transfer the metric variables to be used to predict the grouping, for example `Total importance of brands`, `Number of channels seen` and `Total involvement`.

SPSS offers two methods of discriminant analysis: direct, in which the independent variables are entered into the discriminant function together; and stepwise, in which statistical criteria are used to determine the order of entry. The default is direct. To obtain an actual prediction for each case, click on `Save` and then the box `Predicted group membership`. The predicted group membership will appear as a new column in the data matrix labelled `Dis_1`. Click on `OK`.

Key points and wider issues

The objective of discriminant analysis is to know which specific linear relationships of selected metric variables best predicts each case's group

(Continued)

(Continued)

membership. This can be achieved for a sample of cases that sets up the appropriate number of discriminant functions that can then be used on further cases for which group membership is unknown and which can be predicted better than doing so by chance. Discriminant analysis is affected by the number of cases, in particular its ratio to the number of predictor variables. According to Hair et al. (2010), many studies suggest a ratio of at least 20 cases for each predictor variable. In addition, there should be at least 20 cases in each category being predicted and, ideally, no wide variations in the number in each category.

For validating a discriminant analysis, researchers will usually divide the sample into two subsamples, one for estimation of the discriminant functions and the other for validation, both of which need to meet the requirements on sample size mentioned above. Validity is commonly measured by comparing the number of cases correctly classified by the discriminant function with the predicted accuracy expected by chance.

Implications of this chapter for the alcohol marketing dataset

The scope for cluster analysis on the alcohol marketing dataset is rather limited because there are so few metric variables and even these are derived from summated ratings rather than from genuinely continuous variables. The clusters derived in this chapter are fairly 'obvious' and lead to no great insight, but at least the example illustrates the application. There should be more scope for discriminant analysis since it would be helpful to use a limited number of metrics to predict future drinking behaviour. However, in this example, the categories being predicted are respondents' own assessment of whether or not they are likely to have an alcoholic drink in the next year. There is a degree of circularity here since it would be simpler just to ask them for their own predictions. Discriminant analysis is best used, then, when the categories being predicted are actual behaviour in the future (like prisoner reoffence) or the case is an organization or an object (like product or organizational success or failure) that cannot be questioned.

Chapter summary

Cluster analysis uses traditional variables to analyse potential grouping structures among cases, based on measures of 'distance' between them. While it is perhaps not as essentially case-based as configurational analysis, it is more case-based than variable-based in the sense that the focus is on predicting groupings for individual cases, as evidenced by the fact that SPSS will assign – if asked – each case to a grouping, creating a new variable in the process. The same is true for discriminant analysis; the difference is that cluster analysis generates the groupings from the data, while in discriminant analysis the researcher selects a nominal or ordered category variable as the basis for groupings, and a sample or subsample is used to estimate how good a selection of metric variables is in predicting those groupings. In terms of the distinctions made in Chapter 6, cluster analysis is an interdependence technique in that no distinction is drawn between dependent and independent variables, while in discriminant analysis the grouping variable is always dependent on the metric predictor variables and so is a dependence technique.

Both methods are complex in the variety of algorithms that may be used and the SPSS outputs of both can be quite daunting to interpret. Both use the equivalent of crisp set (binary) category membership of the groupings, except for fuzzy cluster analysis that allows for degrees of membership. However, the procedures for doing this are not available in SPSS. The advantage of fuzzy cluster analysis over fsQCA, furthermore, is unclear, particularly since interpretations in terms of asymmetrical relationships and necessary and sufficient conditions cannot be made.

Exercises and questions for discussion

1. Select the first 50 cases from the SPSS alcohol marketing dataset and undertake a hierarchical cluster analysis. Try crosstabulating cluster memberships against gender.
2. Undertake a discriminant analysis, taking `Drink status` as the grouping variable and `Total importance of brands`, `Number of channels seen` and `Total involvement` as the independents.
3. In what circumstances would you employ discriminant analysis rather than cluster analysis?
4. What assumptions are being made by cluster analysis and discriminant analysis?
5. Why is one approach a dependence technique and the other an interdependence technique?

Further reading

Cooper, B. and Glaesser, J. (2011) 'Using case-based approaches to analyse large datasets: a comparison of Ragin's fsQCA and fuzzy cluster analysis', *International Journal of Social Research Methodology*, 14 (1): 31–48.

This article very much expands on the comparison of cluster analysis and the software fsQCA, which is explained in Chapter 7 of this book. There is also a clear explanation of fuzzy cluster analysis.

Hair, J., Black, W., Babin, B. and Anderson, R. (2010) *Multivariate Data Analysis: A Global Perspective*. Upper Saddle River, NJ: Pearson Education.

Chapter 7 explains discriminant analysis (and logistic regression, which is covered in Chapter 6 in this book). Chapter 9 is a very detailed account of cluster analysis with lots of examples.

Uprichard, E. (2009) 'Introducing cluster analysis: what can it teach us about the case?', in D. Byrne and C. Ragin (eds), *The Sage Handbook of Case-Based Methods*. London: Sage.

A succinct and very clear introduction to the uses and types of cluster analysis.

Suggested answers to the exercises and questions for discussion can be found at the end of this text, pp. 293–321, and on the companion website, (https://study.sagepub.com/kent), which also give links to relevant free online Sage journal articles, PowerPoint slides, an overview of data analysis packages, an introduction to SPSS and weblinks to alternative datasets.

IV

COMPARING AND COMMUNICATING RESULTS

9

COMPARING AND MIXING METHODS

Learning objectives

In this chapter you will learn about:

- what variable-based analyses are good at;
- the limitations of variable-based analyses;
- what quantitative case-based analyses are good at and their limitations;
- how the two methods and the different types of data they generate can be mixed;
- the implications for the alcohol marketing study, showing that all the various approaches to data analysis have something to add, but none is sufficient on its own.

Introduction

Part Two of this book considered univariate, bivariate and multivariate approaches to variable-based data analyses, while Part Three covered quantitative case-based approaches including set-theoretic methods, configurational data analysis, cluster and discriminant analysis. All approaches to data analysis have their strengths and their weaknesses. This chapter compares variable-based and quantitative case-based analyses by reviewing what each is good at and what are the limitations of both. The aim is not to come to some overall evaluation of which is 'better', but to help researchers to come to some conclusion about which approach may be more appropriate to their particular research objectives. Nor is it necessarily a case of having to choose between the two. The second part of this chapter outlines various ways in which the approaches may be combined or mixed so that advantage may be taken of the strengths of both while compensating for some of their weaknesses.

What variable-based analyses are good at Data reduction

Variable-based analyses are essentially data reduction tools. With a few graphs, tables, summary statistics or equations, variable-based analyses can show the size and shape of the phenomena under investigation and reveal patterns and trends in the way variables relate together, even in very large datasets with thousands of cases and hundreds of variables. A review of all the variables one at a time can sometimes produce some unexpected results, for example, in the alcohol marketing dataset very nearly two-thirds claim that they have never had a proper alcoholic drink, and over half (53.6 per cent) said they definitely or probably would not have an alcoholic drink in the next year. Only 30 per cent said they definitely or probably would. There could, of course, be problems here in reluctance to admit having drunk or intending to drink alcohol. Reviewing frequency distributions for categorical variables can reveal unevenness between categories or, for metric variables, whether there is some semblance of a normal distribution as, for example, for the frequencies in the number of channels on which adverts for alcohol have been seen (see Figure 9.1).

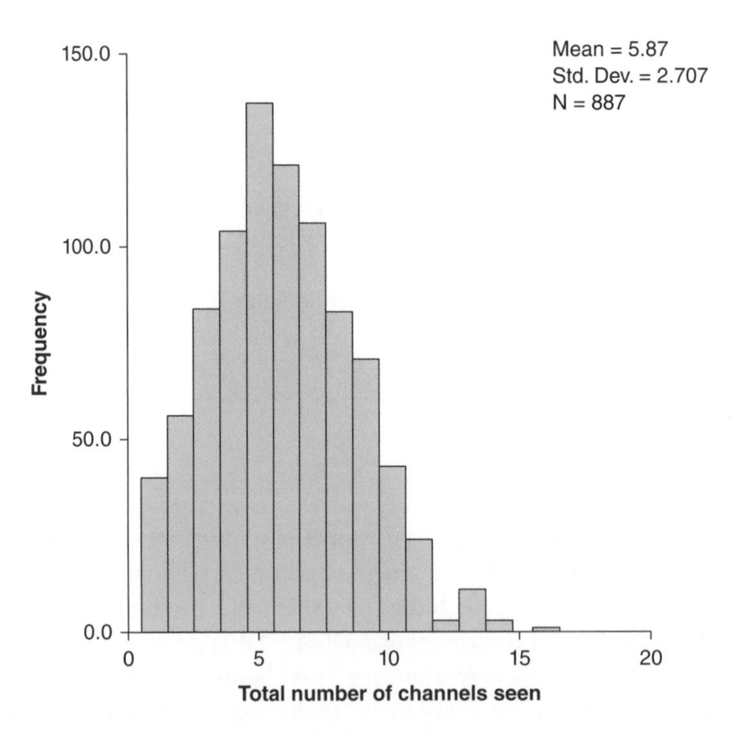

Figure 9.1 The distribution of total number of channels seen

Making comparisons

One of the specialities of variable-based analyses is to be able to compare cases, or groups of cases, in very precise quantitative terms so that we know exactly how and how far they differ.

Thus the median number of channels on which young people said they saw adverts for alcohol is six for males and five for females. This, in turn, can raise other issues, for example in what other ways do males and females differ in terms of the other variables?

Where there are differences between more than two groups or categories, then either these can be taken two at a time, or the differences between the observations for each category and some theoretical expectation (e.g. that there are no differences) can be used to generate an overall index of difference such as offered by the statistic **chi-square**.

Showing covariation

One of the major uses and strengths of variable-based analyses is the ability to measure exactly the tendency for changes either of category or metric value on one variable to be exactly mirrored by changes in category or metric value on another variable. Thus if there is a tendency for those who have seen adverts for alcohol on many channels to be highly involved in alcohol marketing, there is a corresponding tendency for those who have seen adverts on few channels to be not highly involved.

Such bivariate analyses can be seen as a form of data reduction since the degree of covariation can be expressed in terms of a single coefficient. However, they can also act as a basis for exploring patterns of relationships between variables, or they can constitute a way of evaluating bivariate hypotheses. Covariation is something that can take place between variables only two at a time, although it is possible to check or control for the effect that different values of a third, fourth or more variables have on that bivariate relationship using three-way or ***n*-way crosstabulation** for categorical variables and **partial correlation** for metric ones. For metric variables, it is also possible to produce a correlation matrix of all the bivariate correlations between each variable and each other variable in a researcher-defined set of metric variables. Covariation, however, only makes sense if both variables are binary, or when both are ordered category, ranked or metric.

Establishing the relative importance of independent variables

Where, in a piece of research, there is a clear outcome being investigated, then, if the variables are categorical, it is possible to compare the degree of clustering or covariation between an outcome and each of a number of variables separately, one at a time, that are potentially causal factors, and to say which ones are the most strongly associated. Each can be subjected, furthermore, to an elaboration analysis to check for relationships that may be mediating, conditional or spurious. Where the variables are metric, it is no longer necessary to undertake analyses between variables two at a time, but to include them all in a single multiple regression equation that evaluates the contribution of each variable to the outcome, the values of the other variables in the equation being held constant.

Exploration and verification

Variable-based analyses are very good both for exploring patterns in a dataset and for evaluating or testing hypotheses that have been suggested before the analysis begins. An extreme form of exploration is **data dredging** in which the degree of category clustering or covariation is established between each variable and each other variable in a dataset. This is very easily accomplished using the variable-based coefficients available in SPSS. Interesting or unexpected patterns may be detected in this way, but it is very much a 'shot in the dark'. Coefficients will tend to be low or very low; if they are high, the researcher should suspect that the two variables are, partly at least, measuring the same characteristic.

Variable-based analyses are often focused on the idea of generating a model that approximates the data to a greater or lesser extent. Thus an arithmetic mean can be seen as a model of a particular variable which may or may not be a good approximation, depending upon the extent and type of variation around the mean. We can then use the model to make predictions, so if, for example, we know the average income of a group of people, we can use this to make a better guess of the income of each individual. A regression line can also be seen as a model of the way the values of two metric variables approximate a straight line. That line can then be used to predict the values of one variable from the other. If a researcher hypothesizes that the greater the value on variable *X*, the greater the value on variable *Y* (and, being a symmetrical relationship, the lower the value on *X*, the lower the value on *Y*), then a linear model may be a good approximation and, if so, the hypotheses can be said to correspond with the data.

The limitations of variable-based analyses

Many cases are needed

Variable-based analyses are not good when the number of cases available for analysis is small. What counts as 'small' or 'too small' is, of course, an open question. If simple percentages are being calculated, then there should be at least 100 cases, but if these are to be broken down into sub-groups, then each sub-group should, ideally, have at least 100 cases. In short, several hundred cases may be needed, otherwise if, for example, crosstabulation is to be used, there may be too many cells with very few or no cases.

Many of the multivariate procedures discussed in Chapter 6 have requirements on the minimum number of cases and these were considered along with each technique. Researchers need to take an overall view of the minimum number of cases they need, given the statistical procedures they are proposing to use on the dataset. More cases are nearly always better than fewer, so as many as possible is usually the best strategy. That said, there may, for example, only be, say, 50 or 60 cases that are the focus of the researcher's attention or for which the researcher has data. Here, variable-based analyses may not be appropriate, particularly if, as is generally the situation, most of the variables are categorical rather than metric. Configurational data analysis is, however, possible with these sorts of numbers and researchers might be wise to consider whether or not this would be sensible.

Cases become invisible

Although many cases are needed, variable-based analyses quickly lose sight of them by focusing on variables instead of cases. Variables are researcher-selected attributes of cases, not the cases themselves. It tends to get assumed that variables are forces and can 'do' things to other variables, as in 'Variable *X* is the cause of variable *Y*'. However, variables are often abstract and unobservable. We cannot, for example, 'see' a person's attitude or opinion or their social class. We cannot observe the mechanisms by which such attitudes might affect behaviour. Variables, if they exist at all, have a kind of virtual existence that lies outside the entities treated as cases. The 'relationships' between variables, furthermore, are a further abstraction. Differences, clusterings or covariations are researcher constructs that are represented by the models they create. They are attempts to say that the distributions of researcher-coded variable values are patterned in such a way that approximates such models.

It might be argued that only cases can 'do' things and behave in particular ways; but this might be true only for individuals as cases, not for entities like organizations or 'cases' that are themselves researcher creations like small to medium enterprises. The world is constructed by the actions of people, not variables. We can only understand cases by treating them holistically – as complete entities in their own right – and in great detail. They are complex systems nested inside other systems and themselves consisting of sub-subsystems within. A hospital ward, for example, is a complex system that exists within the hospital itself, which in turn may be part of a local national health trust, itself part of the National Health Service. The ward itself may consist of a number of subsystems, each complex, for example a nursing subsystem, a cleaning subsystem and a food delivery system. In a data matrix, the cases selected by researchers are along the rows, but variable-based analyses ignore the rows and focus on the columns. As soon as a table or a graph is produced or a statistic is calculated, the identity of cases is lost. Any understandings that researchers may have about individual cases are not included in the analysis. Individual cases are seldom mentioned in the analysis of the results.

Asymmetrical patterns are ignored

Variable-based analyses focus mainly on discovering covariations between variables. Covariation implies symmetrical relationships so that if, among the cases, high values on both variables are found (or tend to be found) together, then the same is true for low values (or, for a negative covariation, if high values on one variable are found with low values on the other, then low values are found with high). However, it is often found that while high values on one variable may be found with high values on the other, low values on the first variable may be accompanied by high or low values on the second variable. In terms of binary variables, if, for example, all or nearly all the men say 'yes' in answer to a question, it may not be the case that all or nearly all the females say 'no' – they may be equally split. Variable-based statistics cannot – or at least cannot without great difficulty – handle **asymmetrical relationships** of this kind.

Regression is linear and additive

In the social sciences, regression-based techniques predominate and, according to Berk (2004: 1), are ubiquitous in the development and evaluation of public policy and in the production of the research on which it is sometimes based. These techniques include bivariate regression, multiple regression, logistic regression, analysis of variance, factor analysis, structural equation modelling and many extensions for longitudinal data and multilevel data.

Regression focuses on linear relationships that approximate a straight line and typically centres on estimating the assumed fixed contribution that each 'independent' variable makes to an outcome or dependent variable. The task focuses around assessing the relative importance of the relevant variables and picking out the 'best' predictors, for example by looking at the standardized beta coefficients in a multiple regression equation. It is assumed that this contribution is irrespective of what other variables it is combined with and irrespective of the values of those variables. Each variable is assumed to have its own autonomous capacity, but even this view gets confounded by **multicollinearity** – independent variables that are themselves correlated. These reduce the 'net' impact of variables on the outcome.

The problem here is that many relationships between metric variables are not best modelled with a straight line – or even a curved one. Sometimes the pattern between two variables is triangular and no line, of whatever shape, will be a good approximation. Sometimes the pattern is one of phase shifts with jumps in the values of one of the variables. Sometimes a pattern suddenly become unstable and descends into a state of chaos.

Limited diversity gets overlooked

In the context of variable-based analyses, limited diversity arises when there are empty, or nearly empty, cells in a crosstabulation. This means that certain combinations of characteristics have no or very limited empirical existence. This is fine if such cells are 'on the diagonal' and a high coefficient of covariation will arise. If this is not the situation, empty cells will cause different coefficients to do wildly different things, so that, for the coefficient **lambda**, empty cells may cause it to revert to zero, but for **gamma** this may cause it to revert to one. For metric variables, limited diversity tends to get ignored as a **regression** line gets drawn through empty spaces in a **scattergram**. Ragin (2008: Chapter 10) argues that limited diversity often gets understood as the result of correlated independent variables and thereby overlooked.

Too much focus on statistical inference

Some variable-based techniques like **analysis of variance (ANOVA)** depend on statistical inference – the end result is always a ***p*-value**. This in turns means that they assume that the cases in the dataset are a random sample of some wider population of cases and that sources of error other than those from random variation can be ignored. In reality, many sets of cases in a dataset are attempts at a census of all the cases that meet researcher specification, attempts that are often not totally successful; however, an incomplete census cannot be considered to be a random sample. Sometimes non-random techniques of selection are used or at least there are elements of non-randomness in the sample design. Even attempts at random samples are seldom totally successful; there will nearly always be non-response from some randomly selected individuals or cases that are for some reason unavailable. There may be response errors, recording errors or data entry errors. All this makes any research findings that depend on the calculation of a *p*-value suspect and to be treated with caution.

Researchers are apt to treat *p*-values as a 'test' of a hypothesis. They will pick an arbitrary, conventional value like $p = 0.05$ and conclude that any result with a *p*-value less than this means that there is a 'real' result that cannot be explained by random sampling error and so the result is 'statistically significant'. In reality, the actual *p*-value is simply the probability of obtaining the data for those variables in the dataset provided the researcher makes the assumption that the null hypothesis is true for the population of cases from which the sample was drawn. It says nothing about the probability of the research hypothesis or the null hypothesis being true. Statistical inference is only needed because even proper random samples with no other sources of error can still produce errors. It is generally better for researchers to focus on the size of differences, the degree of category clustering or the amount of covariation to be found in the dataset under investigation. Some researchers call this the '**effect size**', perhaps an unfortunate term since it says nothing about the effect of one variable upon another

Not good at establishing causality

Despite researchers' often expressed interest in establishing causal relationships between variables, variable-based analysis in fact is not very good at doing just that. At most it can establish that a covariation exists (to a greater or lesser degree) between variables that have been identified by the researcher as dependent or independent, and that that covariation is unaffected by the values of other variables. While it is true that if neither of these features hold, then even considering the possibility of causality is not appropriate; if they do hold, this is by no means sufficient to establish causality. The statistics themselves cannot establish which variables are dependent nor which ones are independent – that is for the researcher to decide or theorize about. Nor can they establish which variables come before or after the others, although evidence of temporal order can come from certain kinds of research design.

Perhaps the main failing of variable-based analysis, however, is the failure to consider or distinguish between causal necessity and causal sufficiency. Relationships that may be necessary but not sufficient or sufficient but not necessary are asymmetrical and, as argued above, such relationships are not addressed by traditional variable-based analyses. If relationships are necessary but not sufficient, then necessary causes or conditions may need to be combined with other potential causes or conditions for an outcome to happen. Alternatively, if they are sufficient but not necessary, then outcomes may be produced in other ways. Issues surrounding the establishment of causal relationships, particularly the more complex ones, are taken up in more detail in the next chapter.

What case-based analyses are good at

Keeping the focus on cases

In configurational data analyses, the focus of the analysis is on the rows of cases in the data matrix and the frequencies with which certain combinations of case characteristics happen or do not happen. In fsQCA, there is an option on the `Select Variables` screen to show which individual cases are identified with which solutions that have met the criteria for causal sufficiency. This is fine when the number of cases is limited and it is possible for researchers to add their own understandings of each case to add to the plausibility of each solution. Where the number of cases is large, this, of course, becomes increasingly difficult.

For researchers undertaking cluster analysis it is still possible to get SPSS to put each individual case into clusters that have been identified. The focus can remain on the cases and any researcher understandings about cases can be added to the analysis.

Handling small-*n* research

Configurational data analysis (CDA) was originally developed in the context of comparative social research where the cases are often nation-states. So, by definition, there is a limited number of such cases. While variable-based analyses require large numbers of cases, CDA is possible with as few as 10–15 cases, although 30 or more is preferable. There are, of course, severe constraints on the number of conditions that can be included in an analysis run when the number of cases is this small. Three conditions will produce 8 configurations and four conditions will produce 16. This means there will be a severe lack of diversity with many **logical remainders**. As the number of cases increases, so it becomes possible to add more conditions, but at the same time it becomes more difficult to keep track of individual cases.

It would be wrong of researchers to choose to undertake CDA simply because the number of cases is limited, but at least it becomes a possibility if, in addition, theory is sufficiently well developed such that a clear outcome to be studied exists and a limited number of potential causal conditions can be identified. It should also make sense that the outcome might come about as a result of different combinations of conditions.

Facilitating causal analysis

CDAs facilitate the analysis of causality by focusing on necessary and sufficient conditions. Perhaps one of the greatest strengths of CDA is that it can handle relationships between set memberships that are asymmetrical. This in turn can be interpreted in terms of subsets such that, for example, all those cases with membership on condition X are also members of condition Y, but not vice versa. Where it makes sense to do so in terms of researcher understanding of cases, or in terms of accepted theory or hypothesized relationships, the established subset relationships can be separately analysed for causal necessity and causal sufficiency. More complex causal relationships can be addressed so that conditions may be conjunctural – an outcome is possible only when certain conditions are combined. Furthermore, an outcome may come about in several different ways, while individual conditions may play very different roles depending on what other conditions with which they are combined.

Controlling counterfactual assumptions

A counterfactual is an assumption that is made about what would be the consequences of an unobserved event. All causal statements make assumptions about what would have occurred if the causal condition or set of conditions had not manifested themselves. Thus, if we say that A causes B, we are making the assumption that, in the absence of A, B will not happen. Causal statements, in short, are based on comparing an outcome when a causal event or condition did occur (the 'factual') with an outcome it is assumed would have happened if a causal event or condition did not occur (the 'counterfactual').

In the `Standard Analysis` in fsQCA, researchers are asked, for each condition, whether they feel its presence or absence contributes to an outcome. This means that for any remainder (a configuration that has no empirical existence in the dataset) a counterfactual assumption is made only if it is simplifying and only if it does not contradict the directional expectation indicated by the researcher. However, while constraints can be placed on particular types of counterfactual assumption, the actual assumptions being made in the process of Boolean minimization are not identified by the fsQCA software.

Producing recipes for achieving a desired outcome

The solutions offered by CDA and fsQCA in particular are good for producing recipes for action to achieve desired outcomes like reducing blood pressure, increasing the consumption of fruit and vegetables, improving school performance in league tables or increasing profits for commercial organizations. There may indeed be different recipes that can then take account of the contexts in which desired outcomes may be sought.

The limitations of case-based analyses

CDA can be used only when a number of conditions are met

In order to be able to use fsQCA it is necessary that there is a clear outcome that the research is designed to investigate and, furthermore, that the outcome can be assessed in terms of fuzzy set memberships. There need also to be hypotheses or at least justifiable hunches about the underlying causal structure for that outcome. Each potential causal factor needs to be thought of in terms of a category or 'quality' to which cases may or may not belong, or belong to a certain extent, and the assessment of which is related to theoretical and empirical knowledge external to the data. In addition, fsQCA and other forms of CDA are appropriate only if the researcher wishes to interpret the results in terms of necessary and sufficient conditions, and it is expected that the causal structure is complex, that is equifinal (there are different pathways to an outcome) and conjunctural (conditions are often sufficient only in combination). Finally, there needs to be a minimum of about 15 cases or so, preferably over 30.

Solutions are very sensitive to the decisions and assumptions made by the researcher

The results from fsQCA are very sensitive to the decisions and assumptions the researchers will have made in the process of data construction and data analysis. These will include the ways in which set memberships have been assessed, the number of conditions that are included in a particular run of fsQCA, the thresholds chosen both for defining frequencies of configurations that will count as remainders and for the acceptable level of raw consistency, and the assumptions made about the inclusion of remainders in the analysis.

It is true that variable-based analyses are also sensitive to decisions made by researchers, but whereas in these analyses the variables included in a given analysis are retained in the solution, in fsQCA, some conditions may be eliminated as not relevant to the outcome. Furthermore, which ones are eliminated will often depend crucially on the chosen minimum level of consistency.

CDA is not well suited to either exploration or verification

CDA is not an exploratory tool that can generate hypotheses – it operates on hunches or theories that are sufficiently developed to be able to determine a clear outcome that is being studied and a limited number of potentially causal factors. At the same time there is no 'test' that allows ideas to be accepted or rejected; rather, there are varying degrees of consistency with proposals that some conditions may be necessary for an outcome to happen and some combinations of conditions may be sufficient to bring it about. There is no element of statistical inference in CDA, so even if the cases are a random sample, there is no measure of the extent to which random sampling might have produced the solutions offered.

CDA sits somewhere in between exploration and verification in an area that we might call interpretation, refinement or elaboration. From solutions offered by fsQCA it is often possible to interpret from detailed knowledge of particular cases why it may be understandable that particular configurations of characteristics may bring about certain outcomes. Vague hunches or tentative hypotheses may be refined so that it might be concluded that some relationships between case characteristics only hold when certain conditions are present. Sometimes, when analyses are carried out on the negative, looking at characteristics that may be necessary or sufficient for an outcome not to happen, patterns of relationships become further elaborated.

There are dangers of triviality, irrelevance or contradiction

Where set memberships are heavily skewed such that all or nearly all cases hold very high or very low membership in a set, then solutions may tend towards triviality when all or nearly all the cases in the dataset manifest the outcome, or towards irrelevance when a condition never or very seldom occurs. Contradiction consists of two statements that say the opposite of one another, for example, claiming that a configuration is sufficient for an outcome to occur and also sufficient for its negation – for it not to occur. For fuzzy set analysis where a degree of inconsistency is generally acceptable, and particularly when there is a severe lack of diversity, the software may produce such contradictions.

Key points and wider issues

Variable-based statistics, which focus on univariate, bivariate and multivariate techniques for summarizing and relating the distributions of frequencies across a set of cases, were developed in the nineteenth century for purposes entirely other than for the analysis of survey data. They have withstood the test of time and in circumstances appropriate for their use will produce transparent and parsimonious solutions. In practice, however, they are commonly used without recognition of the parametric assumptions that they make or the limitations on their use and interpretation. The 'health warnings' that should,

(Continued)

(Continued)

ideally, accompany the reports of findings from their use are often missing. Some critics, like Schrodt (2006), have gone further to suggest that the 'linear frequentist orthodoxy' seeks to impose as the sole legitimate form of social science a set of 'rather idiosyncratic and at times downright counterintuitive methodologies that come together to solve problems quite distinct from those encountered by most political scientists'. What is true for political scientists can be equally true for social scientists in general. Frequentist statistics create a kind of virtual reality which is logical and consistent in its own terms, but the equations and the causal models are reifications. They are materializations of the ephemeral; momentary traces or reflections of a reality.

In terms of what variable-based and case-based analyses are good at, the former facilitate data reduction, they can make precise comparisons of differences between groups of cases, they can precisely measure covariation, they can establish the relative importance of independent variables, and they are good for exploration and verification. In contrast, case-based analyses keep the focus on cases, they can handle small-*n* research where the number of cases available is limited, they facilitate causal analysis, counterfactual assumptions can be controlled, and they are good at producing recipes for achieving a desired outcome.

On the downside, variable-based analyses need many cases, but those cases become invisible, asymmetrical patterns are ignored, regression-based techniques are mostly linear and additive, limited diversity gets overlooked, there is too much focus on statistical inference, and they are not good at establishing causality. On the other hand, case-based analyses can be used only when a number of conditions are met, solutions are very sensitive to the decisions and assumptions made by the researcher, they are not well suited to either exploration or verification, and there are dangers in the solutions of triviality, irrelevance or contradiction.

Both approaches, then, have their strengths and their weaknesses. It is not necessary, however, for the researcher to have to choose between them. It is possible to mix them in various ways that are explored in the next section.

Mixed methods

Researchers have long recognized that all methods have their strengths and their limitations and that the limitations of one approach can to some extent be compensated for by mixing it with another or even several others. What is relatively recent is a small but growing literature that is devoted specifically to what has become known as 'mixed methods'. Several authors have, for example, made attempts to create taxonomies of mixed methods designs including Greene et al. (1989), Morse (1991), Tashakkori and Teddlie (1998), Creswell (2009) and Creswell and Plano Clark (2011). The literature is, however, largely about mixing traditional quantitative with traditional qualitative procedures; indeed, the term 'mixed methods' is commonly defined in this way. However, mixed methods can also be seen as mixing different styles of qualitative research (e.g. mixing depth interviews with focus group discussions) and different methods of quantitative research (e.g. mixing randomized control experimental designs and questionnaire surveys). This section considers ways in which variable-based, quantitative case-based and qualitative methods can be mixed. The following section looks at mixing different kinds of data.

Developing taxonomies of different kinds of mixed design may be a helpful way of summarizing what tends to happen in practice, but, as argued in Chapter 1, each research design tends to be a unique combination of different elements. Researchers need to consider three key dimensions when thinking about mixed methods: which methods, their sequence and their relative emphasis.

The methods may be grouped into three broad categories: qualitative, variable-based and quantitative case-based, each with its own sub-types. Thus qualitative methods may be based on interviews of various kinds, on focus groups of different kinds, on unsystematic observational methods, on creating images or on some combination of any of these. Variable-based methods may be experimental or non-experimental as in social surveys or systematic observation. Quantitative case-based methods may be configurational or cluster-based.

Research methods may be mixed in terms of different sequences, for example phased, concurrent or overlapping. In phased designs, there are two or more stages, each one completed before the next and contributing its results to the next stage. An example might be a research design that begins with qualitative exploration of the topic and which may be used to generate typologies that will form either the basis for the measurement of quantitative variables or the basis for forming groups, clusters or segments of cases, followed up by exploratory variable-based analyses that look for differences, clusterings or covariational patterns that might have been suggested in the first phase, followed by configurational analyses on some tentative hypotheses that pursue asymmetrical relations of causal necessity or causal sufficiency, and finally evaluating formally specified, refined and elaborated hypotheses. Alternatively, qualitative research may be used to follow up in more detail findings from variable-based and case-based analyses.

Concurrent mixed designs entail undertaking two or more methods of research at the same or overlapping times, or even at separate times, but as independent enterprises and considered as a single phase of research. This type of mixing may be undertaken for a number of purposes including triangulation, comparison or expansion. **Triangulation** is the use of two or more approaches to research to see if they come to similar conclusions. Such an approach is generally considered to have emerged from the work of Campbell and Fiske (1959) who used several quantitative methods to measure a psychological trait. They called their method the 'multimethod–multitrait matrix', although the term 'triangulation', coined by Denzin (1978), has now become standard. Denzin, in fact, described four different methods of triangulation: data triangulation, investigator triangulation, theory triangulation and methodological triangulation.

Comparison might involve looking for differences, paradoxes or contradictions rather than convergence. Thus a qualitative study of teenage drinkers using focus groups may be used to see if there is any discrepancy between the stated amounts consumed and the amounts coming from a questionnaire survey. Mixing research methods for expansion might be for reasons of supplementation or incorporation. Methods may be mixed in order to add breadth, depth and scope to a project. Thus a separate project involving depth interviews may be carried out to explore patient reaction to a new medical treatment to add to a questionnaire survey aimed at analysing different patient backgrounds.

Whether the mixing is sequential or concurrent, however, there can also be variation in the weight or priority given to the various elements. These may be equal or weighted towards one or the other. Morse (1991) distinguishes between situations where one procedure is used to supplement another 'base' method and situations where different methods are used in projects that themselves are relatively complete, but as part of a wider research programme. The latter she calls 'multi-method' as opposed to 'mixed method' designs. In multi-method approaches, it is the *results* of each method that inform the emerging analysis or interpretation.

Barker et al. (2001) suggest a new paradigm for the twenty-first century that is not quantitative, not qualitative nor even a hybrid, but a new approach that they called 'informed eclecticism' in which the research is itself positioned within a wider knowledge mix drawing on disciplines and worlds outside research. Being eclectic means utilizing and combining any approaches that might be useful. Also mentioned in this context is the idea of 'bricolage'. A bricoleur is a maker of quilts, an assembler of images, a 'jack of all trades' deploying and adapting whatever strategies, methods or empirical materials that are at hand. Gordon (1999) talks about 'pro-search' rather than research. This focuses on the possibilities of the future – a pluralistic vision rather than an analysis of historical records and their simplistic projection into the future. It is looking at the 'big picture' and anticipating future scenarios. Uncertainty and ambiguity are celebrated rather than avoided.

Mixed data

Besides mixing methods, researchers may choose also to mix their data. Mixing methods is likely to result in different datasets, each analysed separately with results either feeding into another method and another dataset or being compared with the results of another dataset. Mixing data means transforming data in such a way that they can be incorporated into another data matrix. This might mean combining the results of two or more surveys that have been independently carried out. **Data fusion**, for example, combines data from two or more surveys so that they can be analysed as if they came from one complete sample. A single 'virtual' source is created by attaching respondents from a 'donor' survey to respondents from a 'recipient' survey by matching them on variables common to both surveys. In effect, virtual respondents have been created who have given answers to questions from both surveys.

Data mixing might, alternatively, involve mixing data of different kinds. In in-depth interviews or focus group discussions that generate largely qualitative data in words, phrases or commentary, researchers may also record for each participant their gender, age, level or type of education, hospital admissions, and so on. Incorporation here can take two forms. In one form the properties as variables or properties as set memberships can be incorporated into computer-assisted qualitative data analysis software like NVivo or MaxQDA so that, for example, the comments made by males about their experiences of hospital admissions can be compared with those of females. Alternatively, the software may be used to generate codings of types of verbal commentary. These can then be transformed either into degrees of set membership, for example the fuzzy set 'negative hospital experience', or into binary, nominal, ordered category, ranked or metric variables that can be added to an SPSS dataset. In another example, surveys often include both open-ended and fixed choice questions in the same questionnaire. Text in the open-ended questions might be coded into categories of answer and treated as a nominal variable.

Key points and wider issues

Research designs may include mixing not only methods, but also data. With phased mixed methods the research programme is divided into two or more phases, each phase feeding into the next. Normally there is a 'base' method to which previous phases contribute and which subsequent phases might supplement. With concurrent designs, each component is a relatively complete project and the mixing is likely to occur only at the interpretation stage of the overall research programme. This interpretation may involve triangulation, comparison, supplementation or incorporation. Eclectic designs take the idea of mixture a stage further by mixing anything that might be useful, including incorporating ideas, theories, models or other data from outside the research itself.

Mixing data involves transforming data in one dataset so that they can be incorporated into another. Qualitative data as text may be classified into categories of answer, and coded and treated as nominal variables provided the categories are reasonably mutually exclusive and exhaustive. These may readily be incorporated into SPSS. However, to convert these into fuzzy sets for use in fsQCA requires that the categories be seen in some kind of order. If this is not possible or sensible, then using multi-value software like Tosmana might be considered.

Data from SPSS can be transformed into crisp or fuzzy set memberships and used in fsQCA. While fsQCA will recognize only values that lie between 0 and 1 and so all properties must be in this format, SPSS will recognize any values, so fuzzy set values could be treated as variables in SPSS without transformation.

Implications of this chapter for the alcohol marketing dataset

Part Two of this book has shown many of the advantages of variable-based analyses for the alcohol marketing dataset. Univariate analysis is particularly good at data reduction and making comparisons between groups. It can give a kind of bird's-eye view of the entire dataset, producing single-figure data summaries for each variable, so that, for example, among those who had had a proper alcoholic drink, the mean number of units consumed on the last occasion was 4.48 units, which, interestingly, was slightly higher for females (4.69) than for males (4.45). Bivariate analysis is very good for measuring the degree of covariation between two variables and is thus particularly suited to evaluating simple bivariate hypotheses that are symmetrical; for example, the more aware young people are of alcohol marketing, the more likely they are to (say they will) drink alcohol in the next year. Multivariate analysis is good at establishing the relative importance of variables that have been designated as independent for the outcome variables being analysed. Thus a multiple regression was able to show that only total involvement in alcohol marketing is a moderate predictor of total units consumed.

While there is no problem with the number of cases in the dataset (920), which is fully adequate for any variable-based analyses that may be required, identity of cases is soon lost so that it may be difficult to see why females may be (or may be more willing to admit to) drinking less than males. Variable-based analyses are not good at establishing **asymmetrical** relationships and are unable to pick out relationships that may be necessary but not sufficient or sufficient but not necessary. By contrast, an fsQCA analysis was able to show, for example, that *not* having siblings who drink, *not* being involved in alcohol advertising, and *not* liking ads is sufficient in nearly all cases to ensure that the person thinks that they will *not* drink alcohol in the next year.

Chapter summary

A key message of this chapter is that no single approach to data analysis is suitable for all occasions or for all research objectives. There are some contexts in which variable-based analyses may be better and others where quantitative case-based approaches are more suitable. Researchers have to make their own decisions, preferably based on a firm understanding of the advantages and limitations of the two main approaches. Thus variable-based analyses facilitate data reduction, they can make precise comparisons of differences between groups of cases, they can precisely measure covariation, they can establish the relative importance of independent variables, and they are good for exploration and verification. In contrast, case-based analyses keep the focus on cases, they can handle small-*n* research where the number of cases available is limited, they facilitate causal analysis, counterfactual assumptions can be controlled, and they are good at producing recipes for achieving a desired outcome.

At the same time, researchers need to bear in mind that variable-based analyses need many cases, that those cases tend to become invisible once the analysis begins, that asymmetrical patterns are unseen, that regression-based techniques are restricted to linear patterns, that limited diversity gets overlooked, that there is too much focus on statistical inference, and that they are not good at establishing causality. On the other hand, quantitative case-based analyses can be used only when a number of conditions are met, solutions are, furthermore, very sensitive to the decisions and assumptions made by the researcher, they are not well suited to either exploration or verification, and there are dangers in the solutions of triviality, irrelevance or contradiction.

Researchers can, however, mix these methods and mix their data in various ways. Methods may be phased in different sequences, or they may be concurrent. There may be differences in emphasis. Mixing data involves transforming data in one dataset in various ways so that they can be incorporated into another.

Exercises and questions for discussion

1. Download the alcohol marketing fuzzy set dataset `Fuzzysetalcohol marketing.dat`, from https://study.sagepub.com/kent and save it onto your system. Download fsQCA. Just Google fsQCA and select `fs/QCA software`. Download `fsQCA 2.0`. Select `File|Open|Data` and browse to your saved .dat file and click on `Open`. Select `Analyze|Statistics| Frequencies` and transfer all the `Variables` to the `Frequencies` box and click on `OK`. Now open SPSS, select `File|Open|Data`. In `Open Data`, under `Files of type` select `Text` and `Fuzzysetalcohol marketing.dat` and `Open`. Follow the `Text Import Wizard`. Now go to `Analyze| Descriptive Statistics|Frequencies` and select all the variables. Are the results identical to fsQCA?
2. Using your new SPSS from Exercise 1, try regressing `fsintention` against `cssibsdrink`, `fslikeschool`, `fslikeads`, `fsinvolve` and `fsaware`. Compare the results with Figure 7.23 in Chapter 7, which uses the same model, but on fsQCA.
3. Given the stated hypotheses in the alcohol marketing study, which approach to data analysis would you favour or recommend?

Further reading

Creswell, J. and Plano Clark, V. (2011) *Designing and Conducting Mixed Methods Research*, 2nd edn. London: Sage.

Chapter 1 will probably tell you as much as you want to read about mixed methods. There is a long discussion of all the different definitions of the subject.

Yang, K. (2010) *Making Sense of Statistical Methods in Social Research*. London: Sage.

Read Chapter 2, which considers what (variable-based) statistics are good/not good at. Also have a look at his 10 rules of using statistics.

Suggested answers to the exercises and questions for discussion can be found at the end of this text, pp. 293–321, and on the companion website, (https://study.sagepub.com/kent), which also give links to relevant free online Sage journal articles, PowerPoint slides, an overview of data analysis packages, an introduction to SPSS and weblinks to alternative datasets.

10

EVALUATING HYPOTHESES, EXPLAINING AND COMMUNICATING RESULTS

Learning objectives

In this chapter you will learn about:

- how hypotheses might be evaluated;
- how causal relationships might be established;
- what is meant by 'explanation';
- the different formats in which results might be communicated;
- the implications for the alcohol marketing data, in particular that patterns discovered or evaluated may, to varying degrees, be consistent with various kinds of causal interpretation.

Introduction

The main theme of this chapter is on interpreting and communicating results once the researcher has them. The chapter begins by considering how research hypotheses might be evaluated. This goes far beyond what is normally considered under the heading of 'hypothesis testing'. The chapter then moves on to the very complex issue of how causality might be established in the process of data analysis. It draws a distinction between using variable-based or set-theoretic methods to review patterns of relationships that might be consistent with interpretations of causality and the warranting of that interpretation from outside the data themselves. To many social scientists, establishing causality is what is meant by 'explanation'. However, this chapter goes on to argue that explanation is a process of communication that attempts to persuade audiences that understanding is being offered. This is accomplished through various rhetorics which, to audiences of other social scientists, does usually entail some form of causal analysis. The chapter concludes by reviewing the various formats in which the results of data analysis might be presented.

Evaluating hypotheses

The notion of hypotheses was introduced in Chapter 4 where it was suggested that a hypothesis is a carefully worded statement, as yet untested, about one or more properties of a set of cases. Hypotheses need to be carefully worded since the manner of their evaluation will depend on how they are phrased. They are as yet untested since, once evaluated and found to be in some manner valid or at least partially substantiated, they become 'findings' or 'results'. Even hypotheses found to be inconsistent with the data can still be considered as findings. There are different types of hypothesis, and in Part Two of this book a distinction has already been drawn between univariate, bivariate and multivariate hypotheses and, in the sections on statistical inference, a further distinction was made between null hypotheses and research hypotheses.

Bivariate and multivariate hypotheses may, furthermore, be non-directional or directional. For example, the bivariate hypothesis 'Customer satisfaction and brand loyalty are associated together' is non-directional in that it does not state or even imply any direction of influence, impact or causality. Directional hypotheses, by contrast, distinguish, implicitly or explicitly, between the variables that happen or exist first – the causes, conditions or so-called 'independent' variables – and those that come afterwards – the effects, outcomes or 'dependent' variables. Directional hypotheses may in addition imply or state different degrees of influence or impact, for example, high awareness of alcohol advertising may be 'an influence', 'a major influence', 'a cause', 'the cause' or 'a sufficient but not necessary cause' of high consumption of alcohol.

Hypotheses, whether directional or not, may be symmetrical or non-symmetrical. If a researcher hypothesizes that 'Men will say "yes" to this question', then a symmetrical relationship will imply that, by the same token, women will say 'no'. In an asymmetrical relationship, even if all the men say 'yes', the women may say 'yes' or 'no'. In this situation, we can say that all As are Bs, but not all Bs are As. All men are 'yeses', but not all 'yeses' are men. Being male can be seen as a subset of all those who say 'yes'. In terms of causal analysis, we could say that being male is a sufficient, but not necessary, condition for saying 'yes'. The analysis of causality is taken up in more detail later in this chapter.

So, where do hypotheses came from? Some are deduced from general **theory** in the social sciences or by looking at the literature and adopting or adapting hypotheses from earlier research. Other hypotheses may be induced from the data by looking for patterns. In practice, these are often mixed so that deduced hypotheses may be refined or elaborated after studying the data, or tentative hunches from a vague theory may be explored in the data.

There are, in short, many different kinds of hypotheses and the manner of their evaluation will depend on the type and on the way in which they are worded. In variable-based analyses it is traditional to consider only what is usually called 'hypothesis testing'. This means checking the results for statistical significance – calculating the probability of obtaining a finding, assuming the null hypothesis is true, as a result of random sampling fluctuations. 'Test' implies accepting or rejecting a conclusion, so statisticians have tended to use an arbitrary level of probability as the dividing line between acceptance or rejection of a null hypothesis – traditionally a *p*-value of 0.05 or 0.01. These values, however, are a hangover from the days before SPSS (which can calculate exact probabilities) when researchers had to look up critical values in tables which, originally, Fisher had produced in 1925 only for 0.05, 0.02 and 0.01.

The idea of hypothesis evaluation suggested here is a much wider concept that takes on board many different ways in which hypotheses can be measured against a dataset. Nor does it imply any kind of 'test', but rather some assessment of how far a particular hypothesis stacks up against data in a given dataset. For variable-based analyses, the size of difference, the extent of clustering or the degree of covariation between the variables contained in a hypothesis give some indication of the nature of the pattern involved. If a researcher is crosstabulating a series of categorical variables, he or she may find, for example, that a Cramer's V of 0.4 is the highest value discovered; thus, in terms of that particular piece of research, such a value might be considered to be (relatively) 'high' and lends some support to a bivariate hypothesis. In another piece of research, such a value might be relatively low. In quantitative case-based analyses, such as those carried out by fsQCA, researchers will be looking for minimum levels of **consistency** with a proposition of causal sufficiency for selected conditions and, at the same time, for maximum **coverage**. Again, this is no 'test'; hypotheses are neither accepted nor rejected, but evaluated in the wider context of the concepts, theories or models being proposed in that research.

Some researchers have proposed effect size as an alternative to significance testing. This is an attempt to standardize measurements of the size of an observed effect across different pieces of research. Thus the differences between means of groups of cases may be stated in terms of **standard deviations**. The correlation coefficient **Pearson's r** is usually considered to be a standardized measure of correlation for two metric variables. The same could be argued for a statistic like **Cramer's V** for two categorical variables. It is important to be clear, however, that effect size is measuring only the magnitude of an effect once it has been established that it makes sense to talk about cause and effect.

In principle, hypotheses that are to be evaluated or tested should always be set up and formally stated in advance of data analysis. Any generalizations that emerge from the data should be tested or evaluated against data derived from further or other research. In practice, researchers may modify or refine hypotheses after they have undertaken some preliminary data analysis. In a more extreme form, researchers, in writing up their results, may pretend that hypotheses were set up in advance, even though they had in fact generated them from the data. Whether or not this amounts to academic malpractice is an issue on which researchers may take different views. Some may see such procedures as simply styles of writing up.

Key points and wider issues

The generation, evaluation and testing of hypotheses is the stuff of academic research. Curiously, there is little discussion in textbooks on research methods of what hypotheses are, what different kinds of hypotheses may be generated, where they come from and how they might be evaluated against data in a dataset. Most commentary under the heading of 'hypothesis testing' relates to the calculation of a probability that a random sample has generated a difference between the observed result and a specified value. Where this probability is below a given threshold, the result is said to be 'statistically significant'. Testing hypotheses for statistical significance is, in fact, only one way in which hypotheses can be evaluated – a way, furthermore, that is appropriate only when a random sample of cases has been taken. Evaluation is a matter of interpretation of the results in the wider context of the objectives for which the research that generated the dataset was designed. Gorard (2013: Chapter 4) puts all this in the context of warranting research claims – a step in the research logic that connects research findings to conclusions drawn from them. If, for example, in the alcohol marketing research, it is shown that there is a tendency for those who are more aware of alcohol marketing to be more likely to have had a proper alcoholic drink and more likely to indicate that they are likely to do so in the next year, then if the researcher concludes from this that there should be a legally imposed minimum price for alcohol, this claim would be unwarranted by the research findings.

Establishing causality

What researchers mean when they say that 'X causes Y' is open to an enormous range of possibilities. Historically, there has been fierce debate about what causality is and how it is established. The classical analyses are those of Aristotle and Hume. Aristotle recognized that people meant a variety of things by 'cause'. Causes could be material, formal, final or efficient. To Hume (1911) a cause is 'an object, followed by another, and where all the objects similar to the first are followed by objects similar to the second'. Causality is thus a matter of invariable sequence or constant conjunction. Since Hume, the literature on the philosophy of causality has become vast, controversial and inconclusive (Abbott, 1998).

The modern philosophy of causality – and here it gets tied up with an equally complex literature on the nature of explanation – was set forth in Hempel's (1942) 'covering law'. Where event A is followed by event B this is 'explained' by a covering law of the form that B always follows from A, that is A causes B. This formulation, however, has been criticized as either being 'trivial' or failing to explain the covering law. Real explanation, so its critics argue, has to make sense of what social actors do.

According to Yang (2010: 176), the ideal, but impractical, condition for establishing a causal relationship is that the expected outcome happens with the cause present, and does not happen without the cause, but at the same time and on the same set of cases – otherwise there may be interfering effects from events between times of observations or from characteristics of different cases. However, since cases cannot both have experienced and not experienced the cause, there is a **counterfactual** involved. Research needs to be designed in such a way as to offer an alternative. An experimental design with cases randomly assigned to treatment and control groups is probably best, but this is often not possible. In non-experimental research it is necessary to select cases so that some have experienced the cause or causes and some have not, and to compare the outcomes. Evidence of causality in variable-based analyses then comes from a series of statistical propositions that show (a) there is covariation between cause and effect, (b) there is a temporal sequence of cause first, followed by the effects, and (c) the manner in which that relationship is affected by the values of other variables.

Covariation and temporal sequence are usually seen by variable-based analysts as necessary, but not sufficient, to establish causality. Unless they are present, then causality can be ruled out. However, covariation refers only to symmetrical patterns. If the pattern is triangular, then statistics like **Pearson's r** or **Cramer's V** will show no or little covariation, but there will be a distinct possibility of interpretation in terms of causal sufficiency or causal necessity. There are, furthermore, difficulties in establishing the presence of covariation or temporal sequence. Covariation is a matter of degree; some researchers will argue that if it is 'statistically significant' then it exists. Others may pick a 'satisfactory' level of covariation based on a chosen coefficient. Establishing temporal sequence may be possible from **longitudinal data** or by asking respondents when they did something. Even here there will be exceptions or perhaps just a general tendency for events to be in a particular sequence. If 'causes' are conditions that exist for periods of time, then establishing 'when' they happened will be problematic.

The manner in which potentially causal relationships are affected by the values of other variables can be investigated by looking at the impact of holding the values of other variables constant, a process often referred to as **elaboration analysis**. If, for example, it is found that female nurses earn less than male nurses of the same grade, this may be because female nurses work fewer hours. Holding hours worked constant would make the original relationship 'disappear' – there would be no difference between males and females who work the same hours. Hours worked, in this example, may be interpreted as an intervening or **mediating relationship** such that females work fewer hours which in turn lessens their income. The original relationships might, alternatively, disappear for other reasons. Suppose, for example, it is found that pupils attending private fee-paying schools perform better than those attending state schools. This relationship is often assumed to be causal – that fee-paying schools produce higher performance. However, suppose IQ is controlled and the relationship disappears. This would suggest that the fee-paying schools select pupils with higher IQs who also perform better. The original relationship is the outcome of a common cause. The original bivariate relationship is said to be a **spurious relationship**. Sometimes the original bivariate relationship does not disappear but changes according to the values of another variable, for example the effect of a new curriculum on academic performance turns out to be greater for the females than for the males. The original relationship is thus a **conditional relationship** – conditional upon or at least moderated by another variable. The only difference between intervening and spurious relationships is a matter of sequencing. In the former, the control variable comes after the original cause and before the original effect. In the spurious relationship, the control variable is prior to both the original cause and the original effect. These relationships can be studied to some extent using **three-way analyses** for categorical variables and **conditional correlation** or **partial correlation** for metric ones (see Chapter 6 for details and examples). However, temporal sequence would still need to be understood to be able to distinguish between spurious and intervening or mediating relationships.

In set-theoretic analyses such as offered by fsQCA (see Chapter 7 for details), the focus is on establishing those conditions, viewed as set memberships, that singly or, more likely, in combination are sufficient, or mostly sufficient, for an outcome to occur, and those conditions that may be necessary, or mostly necessary. Evidence of causality comes from establishing the existence of subset relationships that are consistent with propositions of causal necessity or causal sufficiency. Once again, this is a matter of degree and researchers may conclude that a given configuration of conditions is 'almost always' sufficient for an outcome to occur based on a chosen level of **consistency**. These analyses allow for the possibility that there may be more than one set of conditions that appear to be sufficient – that an outcome may come about in more than one way such that each pathway, while sufficient, may not be necessary. There may, in short, be alternative explanations. The equivalent of elaboration can take the form of entering more conditions into the analysis run (although there are severe limitations on the extent to which this can be done) or taking extra conditions one at a time and comparing the solutions with and without that condition. From the alcohol marketing study it was possible, for example, to conclude from a fuzzy set analysis carried out by fsQCA that the combination of liking alcohol adverts, being involved in alcohol marketing and having siblings who drink is a combination that is sufficient to an acceptable degree for respondents to admit an intention to drink alcohol in the next year.

These models, whether variable-based or case-based, are not, however, direct representations of the causal process; rather they reflect what we would observe if a causal process exists. They are quantitative frameworks to assist our understanding; they are not reality itself. The notion of cause and effect still needs to 'makes sense' from a theoretical perspective. Unfortunately, this caveat is frequently forgotten and researchers have come to believe that social, business, educational or health and welfare activities look the way they do because certain forces and properties, as measured by constructed variables or calibrated set memberships, 'do' things to other forces and properties. In variable-centred statistical methods it is the variables that do the acting. Cases in effect do very little; they have lost their identity, their complexity and their narrative order that describes each as a story (Abbott, 1992). Within a set of cases, there may, indeed, be patterns of various kinds, but while individuals can act as agents, 'causing' things to happen, variables are sets of values with distributions that might covary with other distributions or show some other kind of pattern; they are not and cannot, strictly speaking, act as causal agents. Social status cannot 'do' anything to repeat purchase; both are distributions of values accessed or constructed by the researcher. Variable-centred analysis is, then, incompatible with any demonstration of causality, unless 'causality' is defined in terms of a demonstration of covariation, temporal sequence and supporting evidence from elaboration analysis, in which case causality is a convenient shorthand for describing certain kinds of pattern.

Neither variable-based statistics nor case-based set-theoretic methods can by themselves establish or verify which variables or set memberships ought to be treated as 'independent', 'causes' or 'conditions' and which ones are 'dependent', 'effects' or 'outcomes'. It is up to researchers to warrant such claims. Often, however, they do not do so and it is merely presumed. It is common, for example, even in texts on multivariate analysis such as by Hair et al. (2010), to define a dependent variable as the 'presumed effect' and the independent variable as the 'presumed cause'. By contrast, Tabachnick and Fidell (2001) see independent variables as either the different conditions to which cases are exposed (e.g. shown an advertisement or not shown it), or characteristics that the cases themselves bring into the research situation (e.g. marital status or level of bureaucracy). Dependent variables are only 'outcome' variables in the sense that they are the ones in which the researcher is mainly interested. The distinction, the authors emphasize, is for the convenience of the researcher and to identify variables that belong on one side of the equation or the other, without causal implication.

The warranting of causal claims from quantitative models is always going to be a challenging issue. The hypotheses on which such claims might be based are themselves unlikely to be 100 per cent supported by the data. There will always be exceptions and often there will be only a degree of support for the existence of differences, clusterings, covariations or subsets. The warrant, furthermore, needs to come from outside the data themselves. This may be in the form of some degree of consensus in previous literature or research. In their article on 'Assessing the cumulative impact of alcohol marketing on young people's drinking', Gordon et al. (2010a), after documenting the increase in alcohol consumption, particularly among young people and the health and social problems that arise from it, suggest that in the proliferation of research into factors influencing adolescent alcohol drinking behaviour, one factor that has been identified is alcohol marketing, the spend on which has grown considerably. They argue that systematic reviews of consumer research into how knowledge, attitudes and behaviour are influenced by exposure to alcohol marketing suggest that this is indeed the case (e.g. Booth et al., 2008; Smith and Foxcroft, 2009). This literature is used, implicitly, to warrant their hypotheses that it is awareness of and involvement in alcohol marketing that affects – is a cause of – current and projected alcohol consumption behaviour (and not, for example, vice versa).

If there is no consensus in the literature about causal processes or there are gaps in such literature, then the warranting of causal claims may come from researcher understanding of the cases in the dataset and having a coherent and plausible account of the mechanisms connecting cause and effect. There may be evidence of such mechanisms from an elaboration analysis of the kind described earlier.

Much of the discussion about causality is about situations where there is one cause and one effect. It tends, furthermore, to be without reference to asymmetrical relationships and the possibility of necessary and sufficient conditions. Causality, however, is more, much more, than studying the nature of the relationships between properties of cases two at a time. Causes tend to be contingent, that is, the effect may happen only under certain circumstances – a fire will break out only if there is oxygen present, flammable material and a means of ignition. They tend also to be conjunctural, that is, an outcome may happen only if certain conditions are combined together. They tend to be multi-pathway – an outcome may come about in several different ways. They are often asymmetric so that a combination of conditions may be sufficient to bring about an outcome, but not necessary if there are other pathways. They are often non-linear – causes do not always make a fixed or regular contribution to an outcome. There may be sudden phase shifts, tipping points or periods of instability or chaos.

Causes or conditions may be part of complex systems that have emergent properties and open boundaries. The impact of causes and configurations of conditions on outcomes may change over time and depend on what sorts of cases are deemed to be part or not part of the 'system'. Causes may be seen as deterministic or probabilistic. The former involve a degree of 'forcing' of impact of one event upon another. In the strongest sense causes should be both necessary and sufficient; in a weaker sense they should at least be one or the other. In probabilistic causality, events may tend towards or be 'mostly' necessary or sufficient or are neither necessary nor sufficient for an outcome to happen – they are just one influence among many.

Causes are usually seen as events or interventions that have some impact on other events. If potential causes, however, are individual attributes like gender or socio-economic status, or conditions of other kinds of cases that are relatively unchanging (like bureaucratic organizations), then there may be some debate about whether or not these can be 'causal' or whether they simply provide the context in which events may or may not happen. Finally, cases may be historical – what events caused other events in the past – or they may be predictive of an outcome that might or will happen if actions are undertaken or events occur in the future. Berk (2004: Chapter 5) argues that in the econometric tradition, causal effect is defined as the ability to predict the outcome of an intervention, one at a time and with other inputs fixed at particular levels.

Key points and wider issues

Establishing causality is a complex issue. To begin with, many different meanings have been applied to the concept. Causes may be seen as deterministic or probabilistic; they may be seen as events or as conditions; they may be historical or they may be predictive. Causality may be simple or complex. Simple causality arises when there is one outcome, one set of potential causal factors (usually each lined up in linear fashion) and no distinction is being made between causal necessity and causal sufficiency. Causal complexity arises when some relationships may be asymmetrical (i.e. necessary but not sufficient or sufficient but not necessary); they may be contingent, non-linear, conjunctural or multi-pathway. They may be part of complex systems that are dynamic (so relationships are always changing) and with open boundaries, so it is not always clear what cases or types of cases are relevant to the study.

Researchers in the variable-based tradition are apt to define cause in terms of a pattern of constant conjunction, temporal sequence and lack of spuriousness. While this may be easier to warrant in terms of statistical analyses, it does not help us to see or understand the mechanisms that may be linking cause and effect. These can be studied to some extent through elaboration analysis, both for variable-based and for case-based analyses. Causality may, alternatively, be seen as more than just an established pattern. It may be seen as needing some kind of influence or forcing of one or more factors upon an outcome. The status of some properties of cases as independent variables or set membership conditions and other properties as dependent variables or membership of outcomes cannot be established by statistical or set-theoretic analyses themselves. The status of each property is determined by the researcher based either on consensuses in other research or in the literature, or on the researcher's detailed understanding of cases. Analyses of datasets can establish that certain patterns exist or tend to exist. It is for researchers to determine, argue or warrant that these patterns are consistent with whatever notions of 'causality' are being understood.

Explaining findings

Whatever patterns researchers locate in their data, whether in exploratory or verificational mode, they may still not understand *why* those patterns exist. To understand is to grasp with the mind. It is a process of following the working, logic or meaning of a situation that is internal to the researcher. Such understandings still need to be communicated to an audience – to colleagues, to clients, to students, to the academic community or to the wider general public. In other words, findings need to be explained.

According to the dictionary, to explain is to make plain or intelligible, to unfold and illustrate the meaning of, to expound, or to account for. While the dictionary thus makes it clear that to explain is a process of communication, it does not say what the process involves or how explanation is achieved. The suggestion in this chapter is that to explain is to provide understanding to an audience so that it understands something that it did not understand before; in particular it is the persuasion of others that intelligibility is being offered. *How* things are explained will thus depend on the audience. We would not explain things to children in the same way we might explain them to our professors. The 'success' of an explanation may be a matter of degree, but it entails acknowledgement from members of an audience that at least some puzzlement has been removed or resolved.

The techniques used in everyday life to explain might include the use of analogy, extrapolation or metaphor. It might mean establishing connections between beliefs, motives, purposes, reasons and actions. It might involve the telling of a story or the unfolding of events that can be judged in terms of their believability or plausibility. In the physical and social sciences, the persuasion of colleagues, clients, students or academic peers that intelligibility is being offered is achieved through the use of particular kinds of rhetoric. Rhetoric is the theory and practice of eloquence; the art of using language to persuade others. It is not necessarily false and showy; it is more about anticipating grounds for acceptance or criticism. There are several different rhetorics to which scientific nostrils may be attuned, although all of them assume that to explain is to offer evidence of causal relationships, despite all the complexities involved that were hinted at in the previous section.

Experimental rhetoric

An experiment entails the manipulation of one or more variables that the researcher wishes to test the effects of, a comparison of at least two measures of a dependent variable, and the control of extraneous factors. Experimental rhetoric means persuading the audience that an appropriate or correct experimental design has been selected, for example a randomized control trial; that discovered differences, covariations, or pattern clusters have been tested for statistical significance and maybe **effect size**; and that, therefore, the treatments or interventions are efficacious in bringing about the outcome. This is the only way, according to experimentalists, that causality can be properly established.

Statistical rhetoric

According to this rhetoric, used by variable-based analysts on non-experimental datasets, explanation is achieved by the persuasion of other variable-based analysts that there is evidence of causal relationships from clever statistical modelling. It may be recognized that the model is not reality, but if social scientists operate as if such models represent reality, then causal mechanisms can be adduced and predictions can be made.

Some philosophers argue that causal analysis is itself inadequate as a form of explanation. Some would argue that the notion of cause is itself an abstract concept – it is a 'black box', a mystical concept. Establishing association, temporal sequence and the impact of other variables on the relationship may be evidence of causality – but it is not *proof*. That can never be established. Some philosophers have doubted that cause inheres in the nature of things. Scientists can often observe that variable *A* is always associated with variable *B* – but they cannot always observe what binds them together.

If we accept that explanation is the provision of understanding to an audience, then whatever an audience accepts as providing such understanding counts as an 'explanation'. From this perspective, cause is explanatory, but only to those who accept a largely positivistic natural science view of the world.

Quantitative case-based rhetoric

Quantitative case-based methods were outlined earlier in Part Three. The explanatory rhetoric that emerges from these methods entails making comparisons across cases, typically in small numbers, in a systematic manner to establish which combinations of characteristics are sufficient, or largely sufficient, for an outcome to arise and which individual characteristics may be necessary. Different combinations may be associated with an outcome. Combinatorial logic is then applied to either binary or fuzzy set memberships to eliminate characteristics whose absence or presence appears not to be relevant to the outcome.

While, for any science, explanation involves the elucidation of causes that extend beyond a single specific case, quantitative case-based rhetoric considers that no laws or propositions are universal, so any understanding of causation must specify the limits of applicability of any account of causal mechanisms. Furthermore, an outcome or similar outcomes may be produced by different mechanisms – the antithesis of experimental and statistical rhetoric, which seeks the one universal model that fits the data. This means that there may be several 'explanations' for an outcome.

Qualitative case-based rhetoric

Experimental, statistical and combinatorial rhetorics all depend on the analysis of quantitative data using systematic methods. The audience must be persuaded that the data are of high, or at least sufficiently high, quality and that the methods and techniques deployed demonstrate that certain kinds of causality are at work. Audiences who are, however, less attuned to quantitative rhetorics are more likely to be persuaded by methods that focus on the totality and uniqueness of each individual case. Cases need to be understood in their own terms as complex systems wherein various causal mechanisms are at work that need to be explained in terms of narratives that make sense to an audience.

There are different forms of qualitative case-based rhetoric. Some of these focus on cases as individual people. The researcher tries to get 'inside the head' of social actors to understand their choices, decisions, behaviour, attitudes, opinions or beliefs. Researchers study the meanings that social actors attach to their own behaviour and the behaviour of others and try to reconstruct their subjective experiences. Each case has a story that can be traced and offered as an account that is persuasive that intelligibility is being provided. Such persuasion, it is often argued, is achieved through producing 'thick descriptions'. These focus closely on the details of cases to create in-depth knowledge; they are detailed accounts of complex social processes in specific environments and sites for social action. They are sometimes presented as an alternative to causal analysis, even of the complex variety; however, alternatively, they may be seen as another view of causality that entails developing an understanding of complex relationships and interconnections between individuals. Explanation, then, takes the form of transmitting this detailed understanding to particular audiences and persuading them that other forms of explanation or causal understanding can be ruled out through cross-case comparisons.

Other forms of qualitative case-based rhetoric take social systems or organizations as cases. Explanatory rhetoric may use analogy or extrapolation, for example seeing social systems as analogous to living organisms that function and develop in certain ways. Thus the way in which the various 'parts' of society interrelate may be 'explained' by saying that it is 'like' a human body with arms, legs, and so on. Certain structures may be seen to exist in order to achieve particular outcomes; the focus is on various means–end mechanisms. Max Weber, one of the founding fathers of sociology, tried to establish the method of constructing 'ideal-types' so that we could understand more easily the relationship between means and ends. Functional analysis looks at the relationship of the parts of a system to the whole – the parts are there 'because' they perform some function for the system or the system's purpose.

Yet other forms of qualitative case-based rhetoric focus on system dynamics, particularly finding emergent pathways or trajectories that are inevitable unless some outside event intervenes. Karl Marx used the notion of dialectical materialism in which any form of economic development contains the seeds of its own destruction. Thus in capitalism there are inherent contradictions that eventually will result in the overthrow of the capitalist system. Dialectical analysis, however, is not necessarily a deterministic approach to knowledge. It is possible to intervene to change the course of history once the internal, underlying dynamics have been revealed and understood. For Marx a dialectic entailed a final point of arrival. For contemporary writers, the approach signifies more a trajectory, an underlying process in which things will progress, develop or change in particular ways if left to themselves. The product life cycle thesis used by marketers can be considered in this light.

Key points and wider issues

To explain is to attempt to provide understanding to an audience so that it understands something that it did not understand before – it is the persuasion of others that intelligibility is being offered. *How* things are explained and the 'success' of that explanation will thus depend on the audience. Social systems and the individuals within them may be explained in different ways – there is no such thing as a final or ultimate explanation. For every 'explanation' being offered, a listener may ask a further 'Why?' question.

Researchers as scientists use a variety of rhetorics in their attempts at explanation. These may be based on experimental design, statistical control or set-theoretic methods. Alternatively, rhetorics may home in on using qualitative data to persuade audiences by way of deploying thick descriptions of market or social behaviour of individuals or exposing system dynamics. The problem with all these forms of qualitative case-based explanation is that they are ultimately untestable in the way that quantitative rhetorics may be subjected to empirical validation.

It can be argued that all forms of scientific explanation involve notions of causality at some point, but there are very different ideas about what 'causality' means and how it is established. In the final analysis, it is what the audience or the listener will accept as an explanation.

Presenting results

The results of data analyses are likely to be presented to an audience in one or more of four main contexts: in staff seminars, at conferences, in academic journals or in reports to management, clients or research sponsors. Seminars and conferences involve face-to-face presentation. They are an opportunity for the researchers to give preliminary results and conclusions, and for the audience to ask about the research before an academic article is submitted or the final report is written. One issue that frequently arises in discussions is that the research is often based on samples, so the findings are estimates. In variable-based analyses, which are the ones audiences are most likely to be familiar with, estimates have **confidence intervals** and hypotheses are normally 'tested' only at a given **level of confidence**. However, if all the results are presented with all the appropriate qualifications, the audience may begin to feel that the results are suspect. There is a fine line to tread between openness and honesty on the one hand, and clarity and simplicity on the other. When data are analysed and few differences, clusterings or covariations are found (which, in fact, is quite often the case), researchers may be tempted to 'see' findings that are not fully supported by the data. For example, a researcher may present a **regression** equation and endow it with meaning when in fact few of the beta coefficients have **statistical significance**. To arrive at a rational, logical and convincing conclusion is more satisfying than to admit that the findings are inconclusive or inconsistent. Where samples are small, then results are less likely to be statistically significant, but non-significant results may nevertheless be important as negative findings. By the same token, where samples are large then even very small effects may be statistically significant, but not important from the point of view of the research.

The key to effective presentations is preparation. Use of the `Notes Page` facility in PowerPoint can be very helpful. The audience should be given a copy of the PowerPoint slides using the `Handouts` facility. Presenters should learn as much as possible about who will be in the audience and take this into account, for example in deciding how much technical detail to include. The presentation should be carefully timed, allowing ample, or the agreed amount of, time for questions at the end. A presentation, however, it is not about using slick technology, but about convincing the audience that the research was well done, that the results are interesting and will have an impact on or implications for current understanding of the topic, issue or problem being investigated. The presentation should pick out key points and findings; it should not attempt a complete summary of the results. Too many slides with too much detail on each will quickly lose an audience. Maintaining eye contact, the use of humour and some interaction with the audience are all important. Presenters need to be aware of their use of gesture and body language. The appropriate use of gesture can keep the presentation lively and animated. Closure is very important and a summary of the key points made can be quite helpful.

The presentation of results of set-theoretic methods such as from fsQCA to an audience often causes problems and misunderstandings, particularly to scholars unfamiliar with this approach (Schneider and Grofman, 2006). A truth table is an intermediate result of fsQCA analysis and should always be presented wherever possible. This can:

- reveal the most frequent configurations along with the similarities and differences between them;
- evaluate the limits to diversity, for example combinations that are theoretically feasible but not empirically observed or fewer in number than expected;
- indicate combinations with inconsistent outcomes.

Another form of presentation from fsQCA is the **XY plot**. This produces a two-dimensional plot of membership values for each case for the outcome, and for either a single condition or for a causal configuration. By trying all the causal conditions in turn, the researcher can see visually any triangular patterns and one or two of these can be shown to the audience. These will show any conditions that may tend towards causal sufficiency or necessity. In presenting the fsQCA solution terms, it is a good idea to give an accompanying interpretation of the results, including a reference back to actual cases wherever possible. The treatment of **logical remainders** to arrive at these solutions needs to be transparent and justified.

Academics, however, are likely to be judged not by their staff or conference presentations, but by the number of scholarly articles they publish in academic journals. The length of such articles tends to be severely restricted, often to about 16 pages, so researchers need to be fairly selective about what they write for publication. There are some components, particularly for quantitative research, that any reviewer of academic articles will expect to see, so the article is unlikely to be accepted for publication unless they are there. Normally these will include:

- a title;
- an abstract;
- an introduction;
- a literature review;
- a research methods section;
- a results section;
- a discussion of the results and/or conclusions;
- references.

The titles of academic articles are very important since they are often used for the purpose of making electronic searches for articles on specified topics or using particular methods. The title, then, must match up with or convey very clearly what the article is about. The abstract plays a similar role to an executive summary in a report, but it will tend to be shorter and will focus on the contribution made by the research to understanding the phenomena being investigated. A number of key words may be listed to help further with the electronic search for articles on a specified topic.

The introduction will explain the rationale for undertaking the research. This will have, or should have, a number of components that typically will include:

- the background of what the research is about and why it was undertaken;
- the theories or models that are to be used in the research;
- the purposes and objectives for which the research is designed.

A statement of the research objectives should let the reader know exactly what the researchers intend to accomplish as a result of undertaking their research. There should be an overall research purpose statement in which key words like 'explore', 'measure', 'investigate', 'examine' or 'explain' are likely to be used, for example 'The purpose of this research is to measure the extent to which the application of social marketing techniques can affect fruit and vegetable consumption and physical activity'. Overall the purpose is likely to fall into one of three main groups: exploratory, descriptive or verificational. In addition to the overall purpose of the research, there should be a further elaboration of more specific research objectives that will be sub-issues of this general-purpose statement and may appear either as research questions or, for quantitative research, as **hypotheses** to be tested. Nearly all researchers do provide a research purpose statement, but in one or two cases it may need to be deduced from the way the article is written. Most do list the hypotheses being tested or make clear what the research questions are.

The literature review in an academic article needs, of necessity, to be brief, so instead of providing a comprehensive review of all past publications that may be relevant to the topic, it must be a highly selected review of the articles from which the research questions or hypotheses have emerged or against which the results of the research are going to be compared.

The research methods section should contain the following elements:

- how the key variables or set memberships were measured;
- the population of cases covered;
- any sampling procedures used, the response rate and evidence of characteristics of non-responders;
- the data capture instruments;
- the data collection methods used, including a justification of the design choices that were selected;
- any specific techniques that were used;
- how the data were analysed.

The results section will be a major part of the article. The results may well be presented in a manner that is more technical than in a management report and will assume a degree of statistical competence on the part of the reader. There are also conventions that should be followed, for example Field (2009: 193) suggests that in reporting correlation coefficients, there should be no zero before the decimal point for the correlation coefficient or the probability value, for example $r = .87$, $p < .05$. There will often be a discussion of the results, relating them back, for example, to previous research, while the conclusions will consider their contribution to theory, any implications for public policy or management, the limitations of the research that was undertaken and suggestions for how research might be carried out in the future. The references will include details of any authors that are mentioned in the article.

Academics may be required to write reports when the research is for a client, perhaps on a commissioned or sponsored basis or to communicate the results to a grant-giving body. Reports are tangible products of the research activity, serving as a historical record of the project and quite possibly the only documentary evidence that remains once letters or emails have been deleted or archived. The report may be a specified part of the research contract, which may spell out the format of the report and the number of copies to be made available.

The formal report, like the face-to-face presentation, is above all a method of communication, so the author needs to bear in mind the kinds of people who are likely to read it and what their needs are. Reports will normally be written in management report style. This means clear, concise, grammatical English, free of jargon or complex sentences, and organized in a way that allows the reader quickly to assemble and digest the content of the report. A fairly standard approach is to use plenty of headings and subheadings, and arranged into a format that includes a number of elements. A title page should state the title of the report, who commissioned it, who prepared it, and the date submitted.

A contents page will systematically number sections and subsections and list tables and figures. An executive summary is a one-page abstract of the entire report, and may be all that a busy person reads. It should explain the terms of reference, the purpose and scope of the report, state the key methods and approach used, list the main conclusions and list the key recommendations. The main body of the report will consist of a series of sections arranged under headings and subheadings that typically would include:

1. Background (an analysis of the current situation or problem)
2. Research methodology

   2.1 Objectives of the research
   2.2 Measurement of the key variables or set memberships
   2.3 The population of cases covered
   2.4 Sampling procedures
   2.5 Data capture instruments
   2.6 Data collection methods
   2.7 Specific techniques and applications
   2.8 Data analysis

3. The key results
4. Conclusions and recommendations
5. Limitations, caveats and suggestions for future research
6. Appendices, which might include:

   - any explanatory notes that would clutter up the main report;
   - tabulations and calculations not included in the text;
   - references;
   - copies of questionnaires or visuals used.

The results section will often be the longest and may be organized in a number of different ways. If the research has used different forms of data collection, for example significant amounts of secondary data, qualitative research and quantitative research, the results may be best presented by taking each of these in turn. Alternatively, the research might be organized by objectives, taking each key objective in turn, showing how the different kinds of data collection or research methods used relate to each objective. Wherever possible, graphs, tables, charts and diagrams should be used to illustrate and clarify arguments (see Chapter 4). Ideally, these should be incorporated into the text, each numbered according to the appropriate section, for example 'Table 4.1.2', and referred to in the text, which gives an interpretation or extracts the key points or lessons to be derived from it.

A good report is persuasive and convinces the reader that the conclusions and recommendations make good sense. It is important that the findings are related back to the objectives of the study and that the recommendations are based on the data, not on speculations that could have been made without the research. In a sense, the whole purpose of the research is to come to a conclusion and perhaps to make recommendations, so this section of the report is vital.

Key points and wider issues

The results of data analyses are likely to be either presented face to face to an audience in staff seminars or at conferences, or incorporated in articles for academic journals and in reports to management, clients or research sponsors. These formats may be seen as opportunities to persuade audiences that the research focuses on an important issue, that it is well designed and carried out, and that it results in conclusions that have implications for contexts outside of the research itself. The actual analysis and presentation of the dataset is thus seen as only part of the research enterprise.

Implications of this chapter for the alcohol marketing dataset

The form of the hypotheses implied in the alcohol marketing study were discussed in some detail at the end of Chapter 5. It was clear from that discussion that (a) the hypotheses as stated in the original research article could be interpreted in several different ways and (b) the degree of empirical support for them in the data was somewhat limited. Interpreted as a series of bivariate hypotheses, one example was 'The more aware young people are of alcohol marketing the more likely they are to have consumed alcohol'. Although the researchers have phrased the hypothesis as a form of covariation, the form of the subsequent analysis implies that drink status is the dependent variable and that awareness and involvement must have happened before the drink took place. There is, in fact, no evidence of temporal sequence in the dataset and it is equally possible that young people are more likely to be aware of alcohol marketing if they have already taken an alcoholic drink. It should be clear from this chapter that it is the researchers who have chosen the status of variables as dependent or independent. The warranting of the implied causal claim comes from outside the data – in this research, in the form of some kind of consensus from earlier research that alcohol marketing has some impact on the drinking behaviour of young people.

Also implied in this hypothesis is that the relationship between awareness and drink status is symmetrical – those who are aware of alcohol advertising are likely to have had an alcoholic drink and those who are not aware (or less aware) are unlikely to have had an alcoholic drink. Set-theoretic methods such as fuzzy set analysis would, however, look at the possibility that the relationship might be asymmetrical – while those who are aware of alcohol advertising are likely to have had an alcoholic drink, those who are not aware may or may not have done so. Awareness, then, might be interpreted as a sufficient but not necessary condition for taking an alcoholic drink.

All these arguments could be applied to understanding the causality as in the other direction; for example, could taking an alcoholic drink be a sufficient but not necessary condition for becoming aware of alcohol advertising? All these arguments can – and should – be applied to all the other suggested bivariate relationships. However, even bivariate relationships should be subjected to **elaboration analysis** by holding constant the values of other potentially confounding variables. Thus, for example, could the relationship between drink status and awareness of alcohol marketing be the same for males and females? For those who have or do not have parents who drink alcohol? For any other variables that might be important? All bivariate relationships should be checked in this way because they might turn out to be mediating, spurious or conditional.

None of the variable-based analyses can help the researcher to sort out which conditions or combinations of conditions may be sufficient for an outcome to occur and which ones might be necessary. A set-theoretic analysis carried out using fsQCA, however, showed, for example, that although no individual conditions were necessary, a combination of being aware of alcohol advertising, being involved in alcohol marketing, liking alcohol adverts and disliking school met an acceptable level of sufficiency for young people to say they intend to drink alcohol in the next year.

Tackling the alcohol marketing dataset for evidence of causal relationships is, in short, a matter of some complexity – complexity that in practice few researchers address. Patterns discovered or evaluated in a dataset may, to varying degrees, be consistent with various kinds of causal interpretation. However, even if such relationships can be warranted, it still may not be clear why. Can we understand why it makes sense that alcohol marketing will have an impact on the drinking behaviour of young people? This may require an exploration of causal mechanisms linking these two ideas and then presenting our understanding of these mechanisms to an audience in a manner that is persuasive, coherent and acceptable.

Chapter summary

This chapter has shown that while the generation, evaluation and testing of hypotheses is at the heart of academic research, testing them for statistical significance is, in fact, only one way in which hypotheses can be evaluated – a way, furthermore, that is appropriate only when a random sample of cases has been taken. The idea of hypothesis evaluation suggested here is a much wider concept that takes on board many different ways in which hypotheses can be measured against a dataset. Nor does it imply any kind of 'test', but rather some assessment of how far a particular hypothesis stacks up against data in a given dataset.

Hypotheses frequently imply or explicitly state relationships in terms of causality, but establishing causality is a complex issue and many different meanings have been applied to the concept. Causes may be seen as deterministic or probabilistic; they may be seen as events or as conditions; they may be historical or they may be predictive. Causality may be simple or complex. Causal complexity arises when some relationships may be asymmetrical (i.e. necessary but not sufficient or sufficient but not necessary), they may be contingent, non-linear, conjunctural or multi-pathway. They may, in addition, be part of complex systems that are dynamic (so relationships are always changing) and with open boundaries so it is not always clear what cases or types of cases are relevant to the study.

Researchers may define causality in terms of a pattern of constant conjunction and a temporal sequence. While these may be easier to warrant in terms of statistical or set-theoretic analyses, it does not help us to see or understand the mechanisms that may be linking cause and effect. These can be studied to some extent through elaboration analysis, both for variable-based and for case-based analyses. The status of some properties of cases as causes and other properties as effects cannot be established by statistical or set-theoretic analyses themselves. The status of each property is determined by the researcher and needs to be warranted from evidence outside the dataset based either on consensuses in other research or in the literature, or on researchers' detailed understanding of cases. Analyses of datasets can establish that certain patterns exist or tend to exist. It is for researchers to determine, argue or warrant that these patterns are consistent with whatever notions of 'causality' are being understood.

To establish causality is not necessarily to explain how or why those relationships hold. To explain is to attempt to provide understanding to an audience so that it understands something that it did not understand before. How things are explained and the 'success' of that explanation will thus depend on the audience. Researchers as scientists use a variety of rhetorics in their attempts at explanation. These may be based on experimental design, statistical control or set-theoretic methods. Alternatively, rhetorics may home in on using qualitative data to persuade audiences by way of deploying thick descriptions of market or social behaviour of individuals or exposing system dynamics. It can be argued that all forms of scientific explanation involve notions of causality at some point, but there are very different ideas about what 'causality' means and how it is established. In the final analysis, it is what the audience or the listener will accept as an explanation.

The presentation and communication of the results of data analysis either face to face to an audience in staff seminars or at conferences, or their incorporation into articles for academic journals and in reports to management, clients or research sponsors, is an important part of the research process and has to be seen in the context of the overall research design and the implications of those results for the wider community.

Exercises and questions for discussion

1. Make a list of the different types of hypotheses and suggest how each might be evaluated.
2. In seeking to explain their findings, do social scientists have any alternative than to attempt to establish some form of causality?
3. To what extent are the manner and style of presentation of research findings part of the 'rhetoric' of explanation?

Further reading

Abbott, A. (1998) 'The causal devolution', *Sociological Methods and Research*, 27 (2): 148–81.

There are not many clear discussions of the nature of causality and how causal relationships are established, but this article is well worth reading.

Gorard, S. (2013) *Research Design: Creating Robust Approaches for the Social Sciences*. London: Sage.

A critical focus on the warranting of research claims, all related to the idea of design in research. If you ever read a book on research design, then this is the one to read. Chapter 4 on warranting research claims is a must.

Yang, K. (2010) *Making Sense of Statistical Methods in Social Research*. London: Sage.

Chapter 11 is on causal analysis. It covers basic concepts and issues in causal analysis, examines the extent to which regression analysis can establish causes and discusses the role of counterfactuals.

Suggested answers to the exercises and questions for discussion can be found at the end of this text, pp. 293–321, and on the companion website, (https://study.sagepub.com/kent), which also give links to relevant free online Sage journal articles, PowerPoint slides, an overview of data analysis packages, an introduction to SPSS and weblinks to alternative datasets.

ANSWERS TO EXERCISES AND QUESTIONS FOR DISCUSSION

Chapter 1

Exercise 1

Are all data 'manufactured' in some way or are there some data that we can accept as 'given'?

If we accept that all data arise from systematic record-keeping, then they are all constructed by somebody at some point in time within a social, economic, political and moral matrix of possibilities and constraints. The records created are not reality itself; rather they are a result of researchers' or other individuals' attempts to observe or measure traces or evidence of phenomena situated within complex systems. The extent to which the records are 'manufactured', however, is a matter of degree. Much demographic data – age, sex, educational background, area of residence, and so on – we would probably accept as 'given' or factual, since there is a 'true' value that exists independently of the researchers' attempts to measure it. However, some demographics like social class are, by definition, researcher creations. Attempts to measure cognitive properties of individuals as cases are always going to require a manufacturing process.

Exercise 2

If a social researcher wanted to measure the extent to which individuals are 'religious', suggest how this could be achieved in a way that is (a) direct, (b) indirect, (c) derived or (d) multidimensional.

(a) A direct measure of religiousness (or religiosity) would involve asking respondents to put themselves on a scale of degrees of religiousness, for example 'How religious would you say you are?':

- ❑ Very religious
- ❑ Fairly religious
- ❑ Not religious

(b) An indirect measure would mean taking an indicator like asking people when they last went to church or when they last said a prayer.

(c) Religiousness almost certainly has several dimensions like church attendance, frequency of praying, belief in life after death, beliefs about God, and so on. Each could be a separate question with a five-point scale for responses like Never, Seldom, 1–3 Times a Month, Weekly, Daily. These could be given a score of 1–5 and scores totalled for each item.

(d) It could be argued that church attendance, praying and various beliefs about God are such disparate dimensions that adding them together is meaningless. An alternative would be to treat each as a separate dimension so that in explaining, for example, the various social factors that affect people's religion, church attendance, praying and beliefs could be crosstabulated separately against these other factors. An alternative would be to profile each respondent on each separate aspect.

For a commentary on the dimensions of religiosity, see Jong et al. (1976).

Exercise 3

Make a list of variables that (a) are naturally binary, (b) can sensibly be made binary and (c) would be unwise to convert into binary.

(a) We tend to think in binary terms, but often the distinction is not very clear-cut as between a 'democratic' and a 'non-democratic' state or organization. A true binary variable is a record of the presence or absence of a property, so keeping a hospital appointment and not keeping it would be a natural binary variable. Some binary variables that are clear-cut may in fact be administrative creations, for example living at a distance from school that entitles a free bus pass or does not entitle.

(b) Nominal variables with few categories can often sensibly be made binary by taking one of the categories as the property possessed and the others as not possessing that property. Thus an assortment of different types of housing may be classified into 'local authority rented' and 'not local authority rented'.

(c) To convert any metric variable into a binary one will entail a decision about a value that is a cut-off point between possessing and not possessing a property like 'high income'. Simply taking the average will often not be sensible. Choosing other values might be quite arbitrary and different values are likely to affect crucially the outcome of many data analysis procedures. It will usually be wiser to keep the variable as metric (or convert into a fuzzy set) and to choose a method of data analysis that takes into account the distribution of different values.

Exercise 4

What type of measure would you use for each of the following?

(i) Degree of satisfaction or dissatisfaction with the services offered by the local social services department.
(ii) Attitudes towards the BBC's Radio 2.
(iii) The degree of local support for the creation of a 'free' school in an area of urban deprivation.

(i) This would need to begin by asking respondents which local social services, if any, they had used within a defined period of time, for example in the last year. Satisfaction can be measured directly by asking for an overall evaluation on a five- or seven-point rating scale. The resulting variables would, strictly speaking, be ordered category, although many researchers would treat them as if they were metric and will calculate averages, standard deviations or use the results in factor analysis. Satisfaction may be thought of as multidimensional, so several indicators may need to be combined, for example satisfaction with the speed of response, helpfulness of the staff, the outcome and the follow-up. The codes allocated, provided the highest code is given to the highest level of satisfaction, might be summed to give a summated rating scale. Alternatively, they may be seen as separate dimensions that cannot be summed, in which case some kind of profile will need to be given.
(ii) The problem with 'attitudes' is that it is a blanket term used usually to mean any form of positive or negative evaluation of some phenomenon, situation or person. We need to specify attitudes to what exactly, for example, in this case to the quality of the programming as a whole, to particular programmes or time segments, or to the quality of the sound. Attitude measurement is usually based on derived techniques, particularly summated rating scales, for example Likert scales.
(iii) This is a complex issue since it would involve not only attitudes of support or hostility, but also what actions respondents might be prepared to take, for example, leafleting or canvassing in support of various forms of protest, writing to newspapers, going on demonstrations, organizing or attending protest meetings. These behavioural properties could be used to construct some kind of index of hostility which could then be compared across time to measure trends or across different types of local resident.

Exercise 5

Examine Table 1.1 and consider which variables are demographic, which ones are behavioural and which ones are cognitive. Also consider which ones have been measured directly, which ones indirectly and which ones are derived.

Demographic properties relate to features that researchers have chosen to characterize the nature or condition of a case. They are not behaviours or cognitive. In Table 1.1 the last three variables, namely gender, social class and religion, may be seen as demographics, so might age at which respondents had their first proper alcoholic drink. **Behavioural properties** relate to what cases did in the recent past, to what they usually or currently do, or to what they might do in the future. The first five questions, for example 'Watched television in the last 7 days', clearly fall within this category, so do the items relating to the channels on which adverts for alcohol were seen, involvement in alcohol marketing, for example 'Received free samples of alcohol products', drink status ('Have you ever had a proper alcoholic drink?'), likelihood of drinking alcohol in the next year, how often they drink alcohol, total units last consumed, whether sibling or parents drink alcohol and smoking behaviour. **Cognitive properties** relate to mental processes that go on within individuals and include their attitudes, opinions, beliefs and images. In Table 1.1 this will include brand importance, liking or disliking of alcohol adverts or school.

Most of the variables are measured directly, but total importance of brands, total number of channels seen, and total involvement are derived. Social class is measured indirectly taking occupation of the household chief income earner as an indicator of social class.

Exercise 6

Explain the type of measure indicated in Table 1.1 for each of the variables in the alcohol marketing study.

Binary variables consist of a record of the presence or absence of a property. The items that relate to whether or not respondents have seen the promotion of alcohol in a number of channels are indicated in Table 1.1 as binary because either they say 'Yes' or some other answer is given. In the dataset, these have been coded as 1 for 'Yes' and 0 for 'No' and 'Don't know'. The same is true for involvement in marketing of alcohol, whether or not they have ever had a proper alcoholic drink and whether or not siblings or parents drink alcohol. Gender, strictly speaking, is not binary, and in the dataset has been coded as 1 = male and 2 = female rather than 1 and 0. However, where there really are only two categories it may be treated as if binary and is indicated as binary in Table 1.1.

Nominal variables consist of contrasting groups. Usually there are three or more categories, as in the first five items in Table 1.1, where 'Don't know' is treated as a separate category. This could, of course, be rapidly made into a binary variable by recoding 'DK' into 'No'. What it makes sense to do is up to the researcher. If there are many 'DK' answers then it may be better to treat this as a separate category, or the researcher may, for theoretical reasons, be particularly interested in analysing 'DK' answers. The only other nominal variable in Table 1.1 is religion, which has four categories.

Ordered category variables consist of two or more categories that are arranged in relationships of greater than or less than, although there is no metric that will indicate by how much. Brand importance in Table 1.1 is a good example, provided 'DK' is left out. This has been coded as 6 in the dataset. It is crucial, if the nine items are to be totalled, that 'DK' is treated as a missing value, otherwise a

'DK' answer has a higher score than 'Very important'! The degrees of liking or disliking alcohol ads or liking or disliking school are single-item questions so will not be totalled. However, beware that, for alcohol ads, liking a lot has been given a low code in the dataset while liking school a lot has been given a high code. There is no rationale for this, but researchers are not always consistent, so data analysts need to be careful. Note that 'Neither' is a middle category that is being treated as different from 'DK'. For intention to drink alcohol, 'Not sure' has not, in the dataset, been coded in the middle between 'Probably not' and 'Probably yes'. This is not a problem unless an ordinal statistic like gamma is used, in which case the ordering is important. 'Not sure' can be either recoded as the middle category or left out as a missing value. The process of recoding is explained in Chapter 2.

How often respondents have an alcoholic drink and how often they smoke are best considered as ordered category, although if a calculation about number of times or number of cigarettes over a period of time could be calculated, then it would be discrete metric. Again DK and not stated answers would need to be excluded for it to be considered ordered category. The same applies to social class.

Total importance of brands is discrete metric because it is treating codes as numeric. These are whole numbers or integers, so is not continuous. The total number of channels seen and total involvement are more clearly only discrete metric. Only age at which respondents had their first proper alcoholic drink and total number of units last consumed are continuous metric. Even though age, for example, is usually reported as rounded down to age last birthday, age itself is still a continuous variable and could, potentially, take any value. Similarly, fractions of units of alcohol are possible.

Chapter 2

Exercise 1

To what extent can treating codes allocated to ordered category measures as if they are numeric values be justified in the data analysis process?

Ordered category variables define the relationships between values in terms of categories that not only are exhaustive and mutually exclusive, but are also arranged in relationships of greater than or less than, although there is no metric that will indicate by how much. For data entry purposes, these categories will usually be coded such that the numbers allocated preserve the order of the categories with the highest code allocated to the highest or most positive category. If, for example, there are five categories, these will normally be coded 5 down to 1. If, and this is a big 'if', the categories can be considered as more or less equally spaced, then researchers will often treat the codes as numeric values and, accordingly, the variable as metric. Researchers may then calculate (or, rather, get SPSS to calculate, see Chapter 3) an arithmetic mean by totalling the values and dividing by the number of cases used in that calculation. This may be done separately for different groups so that means may be compared. Whether or not this can be justified really depends on the legitimacy of the assumptions made about the 'distances' between the categories. Likert categories from 'strongly agree' to 'strongly disagree' are generally accepted as legitimate for this purpose. The categories used for degrees of satisfaction or dissatisfaction are often more problematic and may be best treated as ordered category variables. Where the categories are to be a component of a summated rating scale, then the totalling is done across the items (using the `Compute` procedure on SPSS, see Box 2.4). The totals may then be averaged across cases. The legitimacy of this data analysis process tends to be generally accepted on the basis that treating ordered categories as if they are metric results in relatively little error. Researchers should – and sometimes do – check out the **reliability** and **validity** of their summated rating scales by, for example, taking repeat measures, checking for internal consistency by getting SPSS to calculate Cronbach's coefficient alpha (see Box 1.1) or reviewing the content and adequately satisfying themselves that the items included in the scale adequately sample the domain of features that should be included.

Exercise 2

When transforming variables, researchers make many decisions for which there are no 'rules' or even rough guidelines. What impact might these decisions have on the validity of the data?

There are many ways in which variables might be transformed before analysis begins, or even after it begins, for example regrouping values on a nominal or ordered category measure to create fewer categories, creating class intervals from metric measures, computing totals or other scores from combinations of several values of variables, treating groups of variables as a single multiple response question, upgrading or downgrading measures, handling missing values and 'Don't know' responses, coding open-ended questions, or creating crisp or fuzzy set memberships from nominal, ordered category, ranked or metric measures.

Data transformation is an important part of the data analysis process. There are no 'right' or 'wrong' ways of engaging in data transformation and there are usually several different ways in which it can be done. Perhaps the best strategy is what is sometimes called 'sensitivity analysis' whereby transformations may be tried in different ways to see how sensitive the results are to such processes. There is still the difficult question, however, of the degree of sensitivity that is considered to undermine the validity of the data.

Exercise 3

What are the key circumstances in which missing values might be a severe problem for the data analyst?

In any survey, not all respondents will, for a variety of reasons, answer all the questions. The result is that some values will always be missing from some of the cells in the data matrix. Where this is a result of questionnaire design whereby not all the questions are relevant to all the respondents, then this is not so much a problem unless it leaves too few cases to analyse.

Where a question *would* be appropriate to a given respondent, but an answer is not recorded, then such 'item non-response' may be a more serious issue. Most researchers are inclined just to accept that there will be item non-response for some of the variables and will simply exclude them from the analysis. This is fine when the number of cases entered into the data matrix is large or at least sufficient for the kinds of analyses that are required. However, there is always the danger that this approach may reduce the number of cases used in a particular analysis to such an extent that meaningful analysis is not possible. Many techniques have been suggested in the literature for ways of dealing with this situation, most of which involve filling the gaps caused by missing values by finding a replacement value.

Most of the techniques assume, however, that question items not responded to are done so at random, but it is quite possible that certain types or categories of people are not responding. Furthermore, when the amount of item non-response is small – less than about 5 per cent – then applying any of the methods is unlikely to make any significant difference to the interpretation of the data. Ideally, of course, researchers should, in reporting their findings, communicate the nature and amount of item non-response in the dataset and describe the procedures used to remedy or cope with it.

Exercise 4

Open IBM SPSS on whatever system you are using and enter the nine key variables for the first 12 cases for the alcohol marketing dataset that are illustrated in the next chapter in Figure 3.1. The procedures for doing so are explained in Box 2.1.

This is just an exercise in entering data into SPSS. It is best to begin by naming your variables and entering labels as appropriate. Select `Variable View` and follow the instructions in Box 2.1. Notice that there are several missing values for `Initiation`. This is because those who say they have never had a proper alcoholic drink will not have an age at which they first had such a drink. They should have been coded as 0 under `Drinkstatus`.

Exercise 5

Figure 2.10 shows the total scores for the importance of well-known brands in choosing products. Try creating class intervals in various different ways using SPSS. The procedures for doing so are explained in Boxes 2.2 and 2.3.

First, you need to decide how many intervals you want. To create two intervals, for example, then, from the `Cumulative Percent` column in Figure 2.10, you can see that nearly half had total scores of up to 26 and the rest 27 or more. To create more intervals it is usually preferable to make them as equal in size as possible, for example 0–9, 10–19, 20–29, 30–39, 40–45. These are not exactly equal, but you could have intervals of 9 rather than 10.

Access the full alcohol marketing dataset (available at https://study.sagepub.com/kent). Total importance of brands has already been created under `Totbrand`. Select `Transform` then `Recode into Different Variables`. Scroll down to `Total importance of brands` and move across to the `Input Variable` box. Now follow the instructions in Boxes 2.2 and 2.3.

Exercise 6

Go to the website www.surveyresearch.weebly.com. Here you will find lots of interesting information about social surveys created by John Hall, previously Senior Research Fellow at the UK Social Science Research Council (1970–6) and Principal Lecturer in Sociology and Unit Director at the Survey Research Unit, Polytechnic of North London (1976–92). Download the `Trinians` dataset. To do this, select `Survey Unit`, `Social Science Research Council`, then `Surveys by SSRC Survey Unit` and then the `'Trinians'` survey. Read the background to the survey, download the article in *Folio* and the questionnaire. Finally download and save the dataset from trinians.sav. Not all the questions in the questionnaire appear as variables and they are not all in the same order as in the questionnaire, but the question numbers are clearly marked. Check out the values being used from the `Values` column. Under `Measure`, they are all indicated as `Scale`. This is the default if researchers do not change any of these. Go down the variables and change to `Ordinal` or `Nominal` as appropriate (left click on `Scale` and the other two options will appear).

The variables that should be changed to `Nominal` are:

The items in Q14
Q15
Q16
Q18
Q19
Q20
Q21
Q24
Q25
Q26
Q32
Q2
Q28

The variables that should be changed to `Ordinal` are:

Form
The items in Q11

Q17
Q27
Q9
Political bias
Approval of political protest methods

The rest are `Scale` (discrete or continuous metric in the terminology used in this text). Month of birth is not entirely equal interval, but can still be measured in terms of number of days. Q33 is a semantic differential and has been treated as metric.

Chapter 3

Exercise 1

Is data analysis any more than choosing the right kinds of statistics to apply to a dataset?

The answer to this question, if you follow the argument in this text, is an emphatic 'Yes'. Data analysis is much more than just applying statistics to a dataset. Data analysis is not just about performing statistical calculations on numerical variables; it is about making sense of a dataset as a whole and thinking about a range of alternative ways of approaching its analysis, taking a well-rounded view of what all the evidence is saying. Analysis becomes a dialogue between ideas and evidence. Data analysis is the process whereby researchers take the raw data that have been entered into the data matrix and create information that can be used to address the objectives for which the research was undertaken. The processes that that analysis entails will include at least preparing and describing the data. Some 'descriptive' analyses may stop at that, but to create information that is useful to policy makers or to clients or will help academics to understand social phenomena, it is usually necessary to go beyond just giving an account of a dataset so that data are, in addition, interpreted, related, evaluated, explained, applied and presented. This is true whatever the kind of data, but which particular activities are involved within each of these tends to differ according to whether the data are quantitative or qualitative and whether the approach is variable-based or case-based.

Exercise 2

Access the alcohol marketing dataset which is available at https://study.sagepub.com/kent and check out the values that have been entered for the codes for the nine variables in Figure 3.1 for the non-metric measures. From Figure 3.1, try to summarize each variable by looking down each column. Try to summarize each case by looking across the variables.

The values of the nine variables are:

Drinkstatus 1 = Yes and 0 = No. This is clearly binary and has been coded as such. However, under Measure, this has been entered as Nominal in SPSS.

Intentions Codes 1–4 from 'Definitely not' to 'Definitely yes' are ordered category. 'I'm not sure' (code 5) could be seen as in the middle between 'Probably yes' and 'Probably not', but here is treated as outside this scale and separate from 'Don't know/not stated' (code 6). It has, accordingly, been entered as Nominal in SPSS. It is not clear what the intended difference is between 'Not sure' and 'Don't know'. Perhaps codes 5 and 6 are best treated as missing values. It could then be entered as Ordinal.

Initiation This is a continuous metric variable, so should be left as Scale in SPSS. It is continuous because it is a result of a calibration process in which the value could, in principle, be any fraction of a measure (like Adrian Mole aged 13¾). Under Values, None has been recorded.

Totalseen and Totalinvolve These are discrete metric variables because they are a result of counting and can only be whole numbers. They, too, should be left as Scale.

Likeads This is an ordered category variable so has been entered as Ordinal in SPSS. However, note that this has been reverse coded so that 'I like alcohol adverts a lot' has been given the lowest code and so on. This is the kind of thing researchers often do, so you need to watch out for inconsistencies like this.

Gender This has been treated as nominal with 1 = Male and 2 = Female.

Socialclass This is clearly ordered category and had been entered as Ordinal. However, 'Don't know/not stated' would need to be treated as a missing value if this variable is to be used in an ordinal capacity, for example in the statistic **gamma**.

Religion The categories are 1 = Christian, 2 = Other religion, 3 = None and 4 = DK/unstated. This, clearly, is a nominal variable.

A matrix like this can be summarized variable by variable by looking down the columns so that, for example, we can see that 5 out of the 12 have had a proper alcoholic drink, 3 would 'definitely not' take an alcoholic drink in the next year, and so on. Note that all three also claim never to have had a proper alcoholic drink. The metric variables like Totalseen could be summarized by calculating an average score.

An alternative, and one that is developed in Chapter 7 of this text, is to summarize each case across the variables, so that case 1 has never had a proper alcoholic drink, has no intention of taking one in the next year, has an awareness score of 14, and so on. This case can now be compared with other cases for similarities and differences. Chapter 7 explains how this can be done in a systematic manner and with the ability of showing, for example, which combinations of characteristics may be sufficient to explain the outcome of intention to drink alcohol in the next year.

Exercise 3

What, do you think, are the key ethical issues raised by the alcohol marketing study?

Ethics are moral principles or standards that guide the ways in which individuals treat their fellow human beings in situations where they might cause actual or potential harm whether economic, physical or mental. Ethics in social research are concerned with professional standards of conduct and with the use of techniques in ways that avoid harm to respondents, to clients or to other parties. The main ethical issues that arise in the conduct of social research concern privacy, confidentiality, deception, imposition, integrity and misrepresentation.

The fact that the alcohol marketing study involves interviewing young people aged between 12 and 14 raises ethical issues in particular of transparency and consent. Certainly the consent of parents was sought with a consent form for them to sign, while an information sheet attempted to make the objectives of the study transparent. What degree of 'consent' the young people thought they had in agreeing to be interviewed may be an issue. In the publications of the results, no individuals and no schools are identified. However, there may be an issue of selectivity in the choice of statistical techniques so that they can be used to show that alcohol advertising has an adverse impact on young people's drinking behaviour.

Chapter 4

Exercise 1

Is there a danger that the procedures used to analyse a dataset become largely a function of the procedures that happen to be available on a particular computer package like SPSS?

To a degree this must be true so that, for example, the particular forms of bar chart that a researcher may use will be conditioned by the fact that SPSS offers simple, clustered and stacked bar charts. Having said that, most of the statistics offered by SPSS are pretty standard, so the amount of 'conditioning' in that sense is probably limited. I would also add that some procedures have been available on SPSS for a long while, but appear to be little used. I am thinking of all the statistics that are available on the SPSS `Crosstabs|Statistics` procedure like lambda, gamma, Cramer's V, and so on. These are not commonly used and are often not even mentioned or explained in introductions to SPSS. In short, I think the amount of 'driving' done by SPSS is limited.

Exercise 2

Do pie charts have any advantages over bar charts?

Pie charts certainly have more visual impact when the proportions of various segments are the key point of interest. Otherwise bar charts have most of the advantages. They focus more on the actual frequencies than on relative proportions. They also preserve the order of the categories, and the use of stacked and clustered bar charts means that other variables can be introduced.

Exercise 3

You can get SPSS to produce any kind of nonsense. The trick is to know what counts as 'nonsense'. Suggest some of the main ways in which the researcher might produce nonsensical tables and charts.

The most common way in which nonsense gets produced is to treat categorical variables as if they were interval. Try `Descriptives` on `Gender`. You obtain an 'average' sex of 1.53! Even treating a discrete metric variable as continuous metric can also produce results that do not make a lot of sense, for example the average number of channels on which adverts for alcohol have been seen is 5.87. Some researchers might use this figure, but not a single respondent can have watched 0.87 of a channel. The other commonly met way is to take a continuous metric variable and use it in a crosstabulation or using the `Frequencies` procedure without first putting the data into **class intervals**. This should not be too much of a problem for discrete variables provided there are not too many categories.

Exercise 4

Get SPSS to produce a one-way table for each of the variables either in the alcohol marketing dataset (available at https://study.sagepub.com/kent) or in the `Trinians` dataset (see Chapter 2, Exercise 6 for instructions). Look at the frequency distribution of each and think about which ones might require some data transformations.

In the alcohol marketing dataset, the most likely transformation needed is to convert some of the 'Don't know/not stated' answers into missing values so that the remaining measure is properly ordered category. In addition, where some of the distributions are very uneven, it may be sensible to add together some of the categories, for example there are only 12 cases where the social class of the chief income earner in the household is social class A, so these could be added into social class B and the new category becomes social class AB.

In the `Trinians` dataset, Q11 asks respondents to pick out the three most important and the three least important things that the school should try to achieve for them from a list of nine items. This is not a rating scale, so do not try to get SPSS to add up the scores. Nor is it a fully ranked measure. If you run a `Frequencies` procedure on each item, you can pick out which ones have the highest number of 'Most important' evaluations and which ones have the highest number of 'Least important' evaluations. Similar considerations apply to Q14 and Q16. Q25 can be treated as a multiple response question (indicating the code 1 as the counted value). SPSS will then give you how many 'Yeses' each item has received. For Q27 it might be tempting to add up the codes allocated (remembering that 'Often' is given the lowest code), but you would be adding up very different forms of protest so that writing to a newspaper will have the same evaluation as assassination. These items really need to be treated separately. Q33 is a set of semantic differential items, so, again, do not try to add them up. The accompanying article from the *Folio* school magazine has picked out one of the items – left wing and right wing – and looked at the factors that appear to be associated with this perception of themselves.

Exercise 5

Try out the `Explore` and `Descriptives` functions in SPSS on some of the variables.

These are very similar. `Descriptives` will give you minimum, maximum, mean and standard deviation for a list of metric variables. `Explore` is also only for metric variables, but will in addition give you confidence intervals, interquartile range, skewness and kurtosis for the selected variables. These can also be generated separately for other factors, for example by gender. Oddly, the resulting table is headed `Descriptives`.

Chapter 5

Exercise 1

Bivariate data summary using SPSS is so quick and simple that the temptation must be to crosstabulate everything in sight. Is this a good idea?

On the whole, no; it amounts to **data dredging**. This is why it is a good idea that hypotheses are formulated before the analysis begins. This is not to say that researchers should not try out vaguely formulated 'hunches' about what variables might be associated, but just producing hundreds of crosstabulations to see if anything emerges can be a waste of time and will probably focus on the 5 per cent of 'accidental' results that sampling theory says will emerge anyway! If students – or any researchers – do this it probably means that they have not thought sufficiently about the objectives of the research, and may, indeed, have chosen the wrong style of research. If they really are that vague about the variables then perhaps qualitative research should have been undertaken instead.

Exercise 2

What result from Cramer's V do you think would count as a 'high' degree of association?

What counts as 'high' in one piece of research may not be in another. What most researchers are looking for is what variables might be *relatively* most highly associated with the factors they are trying to explain or study. Having said that, a Cramer's V of 0.1 or 0.2 would in any case be quite low, and one of 0.7 or more would be high, although I would be suspicious of such results. The chances are that the two variables are in fact measuring the same thing, so there is a degree of circularity.

Exercise 3

Access the alcohol marketing dataset (available at https://study.sagepub.com/kent).

(i) Crosstabulate brand importance for alcohol by `likeads` and request the coefficients gamma and Cramer's V. Is there any discernible pattern? Try collapsing the table to a three by three. Have the coefficients changed?
(ii) Regroup total importance of brands (`totbrand`) into three categories and crosstabulate against gender, requesting the coefficient Cramer's V. Interpret the results.
(iii) Recreate Figure 5.15, requesting both gamma and Cramer's V. Can you explain why the two statistics differ?
(iv) Redo the analysis for (ii) above, but using total involvement.

(i) The result is a 5×5 crosstabulation with 25 cells and some very low frequencies in many cells. It is difficult to interpret. Note that phi and Cramer's V give different coefficients. **Cramer's V** is a measure of departure from independence adjusted for size of table, so this is the appropriate one to review. The value of 0.127 is very low, but would still be statistically significant if the 774 cases were a random sample – which, in this research, they are not. Since both variables are ordered categories, **gamma** is also an appropriate statistic that takes account of the frequencies in pairs of categories that are on the diagonal. It has a higher value than Cramer's V at 0.231. Notice that it is negative because brand importance is arranged from low to high rather than the reverse. There is a small tendency, as one would expect, for those who emphasize the importance of branding to like alcohol adverts. Using `Recode` to create three categories for both variables (treating DK as missing, although this makes very little difference since there are very few) results in slightly higher coefficient for both Cramer's V and gamma. The table is, however, easier to interpret and it is possible to see the relatively high number (245) who see the branding of alcohol as unimportant and who also dislike alcohol ads.
(ii) Have a look at the frequency count for `Totbrand` and check out the cumulative frequency. To create three groups of more or less equal size, just under a third have total scores of 23 or less, another third between 24 and 29, and a third 30 or more. Use `Recode`, putting in these ranges and labelling the new variable into `Low`, `Medium` and `High`. If you now crosstabulate against sex of respondent, you will find that there is very little difference between males and females.
(iii) Cramer's V comes out at 0.097 and gamma at 0.187. Both variables are ordered category and gamma is looking only at patterns of diagonality; there seems to be a slight tendency for those who have seen alcohol ads on up to four channels to say they do not intend to drink alcohol in the next year and for those who have seen seven or more channels to say they do intend to do

so. Cramer's V is looking at departure from independence for each cell. If you calculate expected frequencies for some of the cells you will find that they differ little from the observed frequencies. Chi-square is quite low (17.3) and with this many cases to divide by gives a very low value for Cramer's V.

(iv) The distribution here is very different with 396 claiming no involvement and most of the rest having only one involvement, so could be split into zero, one and more than one. Crosstabulating by gender gives a Cramer's V of 0.251. From inspection of the table it is the females who tend to claim no involvement.

Exercise 4

Access the `Trinians` dataset and reread the *Folio* article (see Exercise 6, Chapter 2 for instructions). The researchers have taken one item from the very last question (a semantic differential on the sixth item, left wing … right wing) and made it the 'outcome' variable to study. However, they have just compared percentages on an item-by-item basis. Try turning the items in Q27 (how often certain protest actions are justified) into a summated rating scale and correlate with the left-wing/right-wing scores. Interpret your answer and compare with the comments in the *Folio* article. Would a test of statistical significance be appropriate?

If you just added up the allocated codes for each item, you would be adding together very different degrees of 'radicalism' – writing to a newspaper is not equivalent to sabotaging factories or assassination. The items would, at the very least, require some form of weighting. However, if you run a frequency count on each item, very few indicated 'often' or 'sometimes' for the more radical forms of protest. You might be justified in adding together just the non-violent or non-damaging items. I used `Compute` to do this, remembering that 'often' is given the low value so the more radical have the lowest total scores. A correlation of Pearson's r against left wing/right wing on the last semantic differential question gives $r = 0.35$. Unsurprisingly, those who describe themselves as left wing are more likely to see as justified a number of non-violent forms of protest. However, it is r^2 that indicates the degree of correlation, and this is only 0.12. Certainly a test of significance would not be appropriate – the cases are not a representative sample.

In the original *Folio* article, left- and right-wing self-definition are distinguished for each form of protest. It is possible to pick out which forms make the biggest distinction between left and right, for example unofficial strikes. It can also be seen that for each form of protest, left wingers have a higher justification rating than right wingers. In many ways this can be seen as more helpful and more informative than just reporting $r^2 = 0.12$. Statistical summaries can be very helpful and indeed are very parsimonious, but they can also hide more than they reveal.

Chapter 6

Exercise 1

This chapter checked the bivariate association between seeing ads for alcohol on more channels (and therefore having higher awareness) and stress on the importance of well-known brands for alcohol against gender, and found in a three-way crosstabulation that it was a little greater for males than for females. From the alcohol marketing dataset, using three-way crosstabulation on SPSS, check whether the same is true of the four hypotheses listed in Chapter 5 under implications of the chapter for the alcohol marketing dataset. Pick out one or two categorical variables other than gender that you think might affect these relationships and try a three-way crosstabulation with all four hypotheses.

The four hypotheses listed in Chapter 5 are:

H_1 The more aware young people are of alcohol marketing, the more likely they are to have consumed alcohol.
H_2 The more young people are involved in alcohol marketing, the more likely they are to have consumed alcohol.
H_3 The more aware young people are of alcohol marketing, the more likely they are to drink alcohol in the next year.
H_4 The more young people are involved in alcohol marketing, the more likely they are to drink alcohol in the next year.

These are bivariate hypotheses. However, if controlled for a third variable like gender, they become multivariate. In fact, for all four hypotheses, the bivariate association for females is slightly stronger than for males. This is the opposite of the relationship between awareness and brand importance for alcohol. Other variables that might be used as controls are whether or not their brothers, sisters or parents also drink alcohol and whether or not they smoke.

Exercise 2

This chapter ran a multiple regression of total number of alcohol units last consumed against total importance of brands, total number of channels seen, total involvement and age at which the first alcoholic drink was taken. The resulting multiple R^2 was very low. Using SPSS, check out the bivariate correlations between total units consumed and the other variables.

Figure 6.6 shows the SPSS results of a multiple regression of total units of alcohol last consumed against total importance of brands, total number of channels seen, total involvement and age at first alcoholic drink. The adjusted multiple R^2 is very low at 0.095. The bivariate correlations between total units consumed and the other variables are:

Total importance of brands, $r = 0.11$
Total number of channels seen, $r = 0.16$

Total involvement, $r = 0.29$
Age at first alcoholic drink, $r = 0.03$

Only total involvement has any notable degree of correlation, although total number of channels seen would also be statistically significant if the cases were a random sample.

Exercise 3

How can the status of any variable as 'dependent' or 'independent' be established?

For dependence techniques, it needs to be emphasized that it is the researcher who decides on the dependent or independent status of variables. The statistics themselves are blind to such allocations. Researchers will often conclude that one variable 'accounts for' a certain percentage of the variability on another variable, but the statistics themselves would also allow the 'accounting for' in the other direction or indeed that both share their variability. The dependent or independent status of the variables comes (or should come) from the research context of the researcher's theoretical ideas, not from the statistics.

Exercise 4

The appropriate use of regression-based techniques depends on a number of assumptions being met. Given that these are seldom met in their entirety, or not at all, to what extent has the use of regression been, in the words of Berk (2004: 203), a 'disaster'?

For multiple regression to be legitimately performed, a number of conditions need to be met. First, there must be an adequate number of cases. Second, regression analysis assumes that the dependent variable is metric. Third, regression assumes that all metric variables are normally distributed. Fourth, multiple regression assumes linearity – that the data are best summarized with a straight line rather than a curved or oscillating one. The final assumption is that the independent variables are not themselves highly inter-correlated. If these assumptions are unexamined, then statistical analysis can easily become a misleading ritual. Readers of the results of regression analyses need to be warned if there are any issues with these assumptions. Quite apart from these statistical assumptions, there are often assumptions or decisions about the interpretation of the results. Thus it is often assumed that the variable selected as the 'dependent' variable in a regression equation is indeed an 'outcome' that is being studied; in reality, other interpretations are possible. Thus in Exercise 2 above it is being assumed that the total number of alcohol units last consumed is somehow a consequence of total importance of brands, total number of channels seen, total involvement and age at which the first alcoholic drink was taken. On another interpretation, however, the total number of alcohol units last consumed may be one of several independent factors that affect awareness of alcohol advertising, which is being taken as the dependent variable.

Apart from the selection of variables as dependent or independent, there is the issue of the meaningfulness of the final multiple R^2. A result of $R^2 = 0.1$ or even 0.2 may be interpreted as not important, not worthwhile or as a negative result. Researchers may be tempted, however, to say that the result is, nevertheless 'statistically significant' if the ***p*-value** is less than 0.05, which it is for $R^2 = 0.095$ in Exercise 2. To argue that the relationship, therefore, 'exists' is overstating the case. The correlation is very, very small, even if it cannot be explained away as an outcome of random sampling variation.

Chapter 7

Exercise 1

Is the 'fuzzification' of ordered categories just another way of scaling a variable?

Ordered category variables that are a result of self-reported ratings (like Likert items from strongly agree to strongly disagree, or degrees of satisfaction) are direct measurements, so converting these mechanically into fuzzy values, as illustrated in Figure 7.1 in the text, can be seen as a simple rescaling in order to be able to use the variable in fuzzy set analysis software like fsQCA. However, this is a somewhat mechanical approach that may or may not 'make sense' in terms of set memberships and runs counter to the idea that set memberships should, ideally, be determined in a process of researcher calibration. It is also likely to produce too many values of exactly [0.5]. However, where ordered categories are used in **summated rating scales**, then the summations can be used by researchers to determine totals above which, for conceptual reasons, can be considered as 'full membership' of a set, the values below which constitute full non-membership and a crossover value of maximum ambiguity.

Exercise 2

Why is it sometimes better to treat a nominal variable with two categories (a 'dichotomy') as two separate binary sets?

A dichotomy represents two contrasting groups, like privately owned and publicly owned organizations. If there really are only two possibilities, then being not privately owned means being publicly owned and can be treated as a single binary set. However, if there are other possibilities, like charities or political parties that are neither privately not publicly owned, then being not privately owned does not necessarily imply publicly owned, so we need two binary sets: privately owned/not privately owned and publicly owned/not publicly owned.

Exercise 3

Try minimizing the following five expressions for sufficiency for the outcome Y:

$$\sim A \sim B \sim C + \sim A \sim BC + \sim AB \sim C + \sim ABC + A \sim BC \longrightarrow Y$$

The first two expressions ~A~B~C + ~A~BC reduce to ~A~B since Y happens irrespective of the presence of C. Similarly, ~AB~C + ~ABC reduces to ~AB. So the expression becomes

$$\sim A \sim B + \sim AB + A \sim BC \longrightarrow Y$$

This may be further reduced since ~A~B + ~AB reduces to ~A. So the final minimized expression is

$$\sim A + A \sim BC \longrightarrow Y$$

This is logically equivalent to the original expression, but in more parsimonious form.

Exercise 4

Why is it necessary to avoid, as far as possible, fuzzy membership values of [0.5]?

Any fuzzy set values of exactly [0.5] are excluded from the truth table since it cannot be determined whether they are more 'in' or 'out' of the set. Fuzzy set membership values of exactly [0.5] are best avoided wherever possible because, although they are included in fsQCA calculations of logical sufficiency, cases where any membership value in the configuration is exactly [0.5] are not included in the frequency column in the truth table. Box 7.2 in the text explains how to avoid these values.

Exercise 5

How does counterfactual analysis help to focus the results produced by fsQCA?

All causal statements make assumptions about what would have occurred if the causal condition or set of conditions had not manifested itself. Thus, if we say that A causes B, we are making the assumption that, in the absence of A, B will not happen. A **counterfactual** is an assumption that is made about the impact of an unobserved event. In the context of fuzzy set analysis it is an assumption that is made about what would have been the outcome if there had been any empirical cases for a configuration that is a **logical remainder** – a row in a truth table without enough cases in it. Probing the historical and logical consistency of counterfactuals is the essence of counterfactual analysis.

The fsQCA analysis will make whatever counterfactual assumptions result in minimizations, as in Exercise 3 above. However, the software allows the researcher to restrict the use of counterfactuals by asking researchers to include their directional expectations about each causal condition. This will ask the researcher, for each causal condition, whether its presence or absence is expected to contribute to the outcome. Thus ~A~B~C + ~A~BC will not be reduced to ~A~B if the researcher has indicated that C is expected to contribute to the outcome. When diversity is limited, there will be many remainders and excluding them from the analysis by using the complex solution will result in little or no Boolean simplification. Likewise, a parsimonious solution, which makes all remainders available as simplifying assumptions (and which ignores any absent or present conditions entered in the intermediate solution), can be unrealistic, over-simple and may make assumptions that are untenable. Intermediate solutions strike a balance between parsimony and complexity by incorporating the researcher's substantive and theoretical knowledge. In general, intermediate solutions are preferred since they are the most interpretable and incorporate only those assumptions that can plausibly be made.

Exercise 6

In what ways is the output from fsQCA analyses closer to 'reality' than the output from variable-based analyses?

Reality is often complex and messy. There will often be more than one way in which an outcome may come about. Case characteristics may contribute to outcomes in different ways depending on what other characteristics they are combined with. Some properties may be essential for an outcome – they are necessary conditions; some combinations of properties may be sufficient or largely sufficient to produce an outcome. The output or 'solutions' from fsQCA allow for all these possibilities. There is not a single result as in many variable-based analyses, for example a multiple regression of $R^2 = 0.46$. Even looking at beta coefficients for each independent variable entered into the equation assumes that each makes a fixed contribution to the outcome, all the other variables being held constant.

Exercise 7

Download the alcohol marketing fuzzy set dataset, `Fuzzysetalcoholmarketing.dat`, from https://study.sagepub.com/kent and save it onto your system. Download fsQCA. Just Google fsQCA and select `fs/QCA software`. Download `fsQCA 2.0`. Select `File|Open|Data` and browse to your saved .dat file and click on `Open`. Rerun the alcohol marketing data (a) using 0.8 as the consistency cut-off, then (b) the negation of the outcome intention to drink alcohol in the next year, then (c) trying both higher and lower frequency thresholds, and then (d) indicating in the `intermediate solution` all the conditions as `absent or present`. Repeat (a)–(d) but include `drink status` in the conditions.

(a) To obtain the completed truth table shown in Figure 7.22 in the text, select `Analyze|Fuzzy Sets|Truth Table Algorithm`. Put `fsintention` into the `Outcome` box and `fsaware`, `fsinvolve`, `fslikeads`, `fslikeschool` and `cssibsdrink` into `Causal Conditions`. Click on `Run`. To choose six as the minimum frequency threshold, click on `5` under `number`, then `Edit|Delete current row to last row`. Under `raw consist`, click on the top cell then

Sort|Descending. To make 0.8 the consistency cut-off, enter 1 under fsintention for the first four configurations then 0 for all those below 0.8. (I think that 0.797599 can be considered to round off to 0.8.) Now click on Standard Analysis. For the intermediate solution, indicate that having siblings who drink, liking alcohol ads, involvement in alcohol advertising and being aware of such advertising all contribute to the outcome, but that the absence of liking school is likely to so contribute. Click on OK. The intermediate solution gives only one solution: that fsaware, fsinvolve and cssibsdrink jointly are sufficient for the outcome to an acceptable level of consistency. Compare this with Figure 7.23 in the text, which has used a consistency cut-off of 0.75 and results in two solutions, but the first has low consistency and the second has low unique coverage.

(b) To rerun the analysis, it is, unfortunately, necessary to go right back to Analyze|Fuzzy Sets|Truth Table Algorithm – there is no 'back' button. Put fsintention as Set Negated in the Select Variables screen. Select the same causal conditions as before. Make the frequency threshold 6 and revert to the original consistency threshold of 0.75. You will see from the truth table that levels of consistency are much higher for this negation of the outcome. It is easier to see what configurations might lead to a lack of intention to drink alcohol in the next year.

(c) If you lower the frequency threshold, for example to 1, then many more rows are included in the analysis. The result includes the two solutions for the higher threshold, but adds three more solutions. This result is, probably, less helpful. Raising the threshold, for example to 12, reduces the solutions to only one and with a lower overall consistency. The original frequency threshold was probably about right.

(d) In the Standard Analysis screen indicate all the conditions as absent or present (this is the default setting). The solutions under Intermediate Solution now turn out to be the same as under the Complex Solution; in other words, no counterfactual assumptions are made.

Not surprisingly, if you include drink status among the causal conditions, then this becomes the only condition that meets an acceptable level of consistency. For the most part, having already had an alcoholic drink is sufficient for the outcome intend to drink alcohol in the next year.

Exercise 8

Go to the website www.surveyresearch.weebly.com. Here you will find lots of interesting information about social surveys created by John Hall, previously Senior Research Fellow at the UK Social Science Research Council (1970–6) and Principal Lecturer in Sociology and Unit Director at the Survey Research Unit, Polytechnic of North London (1976–92). Download the 'Quality of Life: 2nd Main Survey'. Select Subjective Social Indicators (Quality of Life)|SSRC Survey Unit Quality of Life Surveys in Britain 1971–1975|Abstracts, data and documentation. For the 2nd National Survey 1975, download the data from SPSS saved file and save onto your system. Download and print off the Abstract and the Questionnaire.

This survey covers many variables, but several of the questions like QA13 ask respondents to give a 0–10 rating. These can be readily converted into fuzzy set values so that 0=[0], 1=[0.1], 2=[0.2] up to 10=[1]. Copy all 13 items in QA13 (these are labelled var136-var148), copy all the items in QA18 and QA19 into a separate SPSS file and convert to fuzzy sets using SPSS Recode. Give the file a name like FuzzysetQOL. At this stage it is better to rename the variables so that you can recognize them in fsQCA, so var136 could become kitchen and so on (this can be done in the Variable View). This SPSS file now needs to be saved as a .dat file. Select File|Save As. Under Save as type select Tab delimited (*.dat). Check that the box Write variable names to spreadsheet is ticked. Click on Save. Now download fsQCA. Just Google fsQCA and select fs/QCA software. Download fsQCA 2.0. Select File|Open|Data and browse to your saved .dat file.

QA19 can be taken as the 'outcome' to be studied. This is an overall measure of satisfaction. The items in QA13 can be taken as potentially sufficient causal conditions. A truth table analysis will show which combinations of conditions may be sufficient for overall satisfaction. Follow the instructions in Boxes 7.3 and 7.4. See if you can interpret the results. You can try the same for QA18.

This is an exercise that gives you the opportunity to think about how you might transform variables into set memberships for fsQCA analysis. There are many different ways in which this can be done, so there are no 'answers' or correct solutions.

Chapter 8

Exercise 1

Select the first 50 cases from the SPSS alcohol marketing dataset and undertake a hierarchical cluster analysis on the brand importance items. Try crosstabulating cluster memberships against gender.

Cluster analysis is a range of descriptive, exploratory techniques usually for grouping cases rather than variables into different clusters such that members of any cluster are more similar to each other in some way than they are to members in other clusters. Each case is assigned to a cluster suggested by the data, not defined beforehand, based on specified properties as variables (rather than being assigned to a given configuration based on properties as set memberships, as in Chapter 7). Hierarchical clustering develops a tree-like structure, usually based on taking individual cases and combining them on the basis of some measure of similarity, such as the degree of correlation between the cases on a number of variables.

To select the first 50 cases in the alcohol marketing dataset, go to Data|Select Cases. In the Select Cases dialog box choose Based on time or case range and enter range 1–50. Any analyses will now be carried out on these 50 cases. Now select Analyze|Classify|Hierarchical Cluster. To obtain a dendrogram, click on Plots and then tick the Dendrogram box. Put the brand importance items into the Variable(s) box and click on OK. Eight of the first 50 cases are missing, leaving 42 included in the analysis. In the Agglomeration Schedule there are $n - 1$ or 41 stages of agglomeration for 42 cases. In the first stage, cases 45 and 49 have been combined; in stage 2, cases 18 and 31; and so on. These have the shortest 'distance' between them and reappear in the dendrogram with the distance indicated along the horizontal scale. The agglomeration Coefficient measures the increase in heterogeneity (reduction in within-cluster similarity) that occurs when two clusters are combined. There is no sudden jump in the coefficient, but the biggest jump occurs at stage 30 when cases 22 and 30 are combined. From the dendrogram, we can see that there are about six clusters at this stage (working from right to left), so a six-cluster solution may be appropriate.

Now return to the Hierarchical Cluster Analysis screen, select Save, then under Cluster Membership select Single solution and enter 6 for number of clusters. Click on Continue and OK. A new variable should appear on your data matrix giving cluster membership for each of the 42 cases in the analysis. SPSS will have given the variable a name like CLU6, which you can change if you wish. This variable can now be crosstabulated against gender if you wish to compare cluster memberships between males and females.

Exercise 2

Undertake a discriminant analysis, taking Drink status as the grouping variable and Total importance of brands, Number of channels seen and Total involvement as the independents.

In discriminant analysis the groupings and each case's membership are already known. The objective of discriminant analysis is to know which specific linear relationships of selected metric variables best predicts each case's group membership. Discriminant analysis is useful for situations where researchers want to build a predictive model of group membership based on observed characteristics of each case. The discriminant functions are generated from a sample of cases for which group membership is known; these can then be applied to new cases with measurements for the predictor variables but unknown group membership.

Select Analyze|Classify|Discriminant. Transfer into the Grouping Variable box the key nominal variable that contains the groupings you wish to be able to predict – in this exercise Drink status. You will need to define the range of codes; click on the button and enter the highest and lowest code values 0 and 1 in this exercise. Into Independents, transfer the metric variables to be used to predict the grouping, for example Total importance of brands, Number of channels seen and Total involvement.

To obtain an actual prediction for each case, click on Save and then the box Predicted group membership. Click on Continue and OK. The predicted group membership will appear as a new column in the data matrix labelled Dis_1.

The SPSS output includes a case processing summary that shows that there are 518 valid cases (402 had at least one discriminating variable missing). Since there are only two groupings for drink status, there is only one function, which explains all the variance. Total involvement makes the biggest contribution, but the degree of discrimination between the two categories of drink status is fairly limited. However, since the categorical dependent variable is binary, then **binary logistic regression** might be better since its assumptions are somewhat less stringent.

Exercise 3

In what circumstances would you employ discriminant analysis rather than cluster analysis?

Both cluster and discriminant analysis are largely exploratory techniques that focus on classification; they are not usually used to 'test' or evaluate models. They focus on either the generation of taxonomies or figuring out what way variables may be 'responsible' for a pre-specified taxonomy. Cluster analysis is used for a variety of purposes including the identification of 'natural' clusters in the data, the construction of useful conceptual schemes or taxonomies for classifying entities, data reduction, generating hypotheses or testing hypothesized groupings believed to be present. Although the variables used for clustering techniques may be metric or categorical, it is best to stick to metric variables since the inclusion of binary or dummy variables, while quite appropriate for regression models, creates a problem in cluster analysis since no amount of recoding will transform nominal items into metric ones whose differences imply distances.

In discriminant analysis the clusters and each case's membership of them are known in advance. The objective of discriminant analysis is to know which specific linear relationships of selected metric variables best predict each case's group membership.

Discriminant analysis is useful for situations where researchers want to build a predictive model of group membership based on observed metric characteristics of each case.

Exercise 4

What assumptions are being made by cluster analysis and discriminant analysis?

Cluster analysis assumes that there are always clusters even when there are, in fact, no natural groupings in the data. The various techniques work by imposing a cluster structure on the data rather than allowing the structure to emerge from the analysis. Accordingly, researchers should always have a strong conceptual basis for why such groups should exist in the first place and the results of cluster analysis should always be seen as tentative rather than final. Cluster analysis is essentially an exploratory rather than a verificational method and it is descriptive rather than inferential. Cluster solutions, furthermore, are not generalizable because they are totally dependent on the variables used as the basis for similarity measures.

Cluster analysis, furthermore, assumes that there is limited **multicollinearity** and that there are no outliers, which can severely distort the results.

Discriminant analysis assumes that the dependent variable is binary or nominal and represents group differences of interest. Furthermore, it assumes that the metric independent variables are normally distributed with more or less equal variances and that, once again, multicollinearity is limited.

Exercise 5

Why is one approach a dependence technique and the other an interdependence technique?

Cluster analysis is an interdependence technique in that no distinction is made between dependent and independent variables. It cannot, furthermore, be used for prediction, but only to profile the clusters that are generated. Discriminant analysis is a dependence technique in that its objective is precisely to make predictions of a dependent categorical variable from specific linear relationships of selected metric variables.

Chapter 9

Exercise 1

Download the alcohol marketing fuzzy set dataset `Fuzzysetalcoholmarketing.dat` from https://study.sagepub.com/kent and save it onto your system. Download fsQCA. Just Google fsQCA and select `fs/QCA software`. Download `fsQCA 2.0`. Select `File|Open|Data` and browse to your saved .dat file and click on `Open`. Select `Analyze|Statistics|Frequencies` and transfer all the `Variables` to the `Frequencies` box and click on `OK`. Now open SPSS, select `File|Open|Data`. In `Open Data`, under `Files of type` select `Text` and `Fuzzysetalcoholmarketing.dat` and `Open`. Follow the `Text Import Wizard`. Now go to `Analyze|Descriptive Statistics|Frequencies` and select all the variables. Are the results identical to fsQCA?

Yes, the frequencies and the percentages are exactly the same, although the tables in fsQCA are much cruder. The exercise makes the point that both fsQCA and SPSS will use the same data matrix, but do very different things with it.

Exercise 2

Using your new SPSS file from Exercise 1, try regressing `fsintention` against `cssibsdrink`, `fslikeschool`, `fslikeads`, `fsinvolve` and `fsaware`. Compare the results with Figure 7.23 in Chapter 7, which uses the same model, but on fsQCA.

In SPSS select `Analyze|Regression|Linear`. Transfer `fsintention` to the `Dependent` box and `cssibsdrink`, `fslikeschool`, `fslikeads`, `fsinvolve` and `fsaware` to the `Independent` box. Click on `OK`. The adjusted R^2 is very low at 0.197 and the standardized beta coefficients are also low and negative, as we would expect, for `fslikeschool`. Having siblings who drink has the highest coefficient. From the fuzzy set analysis, being aware of, involved in and liking alcohol ads plus not liking school in combination are sufficient to an acceptable level of consistency for the outcome having an intention to drink alcohol in the next year. Alternatively, being aware of alcohol ads and having siblings who drink alcohol are sufficient for the same outcome, but to a lower level of consistency (but higher coverage). Interestingly, using a higher consistency cut-off (Figure 7.24) makes the result more similar to the regression analysis. Having siblings who drink alcohol is important for both expressions. However, for those who like school, young people need to like alcohol ads instead. Remember that regression is trying to measure the fixed contribution that each variable makes to the outcome, while fsQCA is looking at combinations of conditions that might be sufficient or largely sufficient for the outcome to occur. The contribution of any individual condition depends on what other conditions it is combined with. FsQCA outputs are in a way more 'messy' than regression, but are closer to reality and make fewer assumptions, for example, about linearity and fixed contribution. It is also possible, for example, to run a separate analysis on the negative outcome, looking at what conditions may be sufficient for young people to have the intention *not* to drink alcohol in the next year.

Exercise 3

Given the stated hypotheses in the alcohol marketing study, which approach to data analysis would you favour or recommend?

The key research hypotheses as stated by the researchers (Gordon et al., 2010a) are that the more aware of and involved in alcohol marketing that young people are, the more likely they are to have consumed alcohol, and the more likely they are to think that they will drink alcohol in the next year. There are two (presumed) dependent variables or outcomes here: whether or not they have already consumed alcohol and how likely they think they are to drink alcohol in the next year. It is usually best to think in terms of one outcome at a time. The first is binary. If the two independent variables can be considered to be metric then variable-based analysis would suggest that binary logistic regression would be appropriate. An SPSS analysis suggests that the R^2 equivalent is tiny (0.044) and only involvement gives a reasonable prediction of drink status.

The other outcome variable is how likely they think they are to drink alcohol in the next year. This is at best ordinal and only then provided `I'm not sure` is recoded as falling between `Probably yes` and `Probably not` and `Don't know/not stated` is treated as a missing value. The allocated codes of 1–6 in the original dataset should certainly not be treated as metric values. One possibility is to recode intentions, awareness (number of channels on which ads for alcohol have been seen) and involvement into three ordered categories and crosstabulate intentions by awareness and intentions by involvement, taking gamma as a measure of association. These produce very low coefficients of 0.187 and 0.234. However, involvement is slightly more strongly associated with intentions than is awareness. Both of these bivariate associations could be controlled by the other independent variable, for example the association between intentions and involvement can be measured for different categories of awareness by using the 'layering' function in SPSS `crosstabs`.

An fsQCA analysis with just two conditions (involvement and awareness) is not particularly helpful. Try it – there are only four configurations and none meets an acceptable level of consistency. One of the advantages of fsQCA, however, is that it can handle all the different combinations of characteristics that are possible and work out which ones are connected with the outcome. These connections may, furthermore, be asymmetrical so that, although all those with a particular configuration manifest the outcome, not all those who manifest the outcome necessarily have that configuration – the outcome may come about in other ways.

In answer to the question about which approach to recommend, I would suggest both. Each has a contribution to make to our understanding of the various relationships between the advertising of alcohol and alcohol drinking behaviour. Each has strengths and weaknesses as explained in Chapter 9. The approaches could be mixed in various ways suggested in the same chapter. They could be phased in two stages, possibly beginning with the variable-based analyses to look at the patterns between the key variables and going on to include a range of conditions in an fsQCA analysis that shows how various configurations relate to the outcome. Alternatively, they could be concurrent or overlapping, showing how the two approaches compare and either support or contradict one another. The data could be mixed with the fuzzy set membership values incorporated into SPSS as in Exercise 1 above. Notice that the very activity and thought going behind creating fuzzy set memberships may help to transform a variable like intention to drink alcohol in the next year from a nominal or at best ordinal variable into a set of fuzzy set membership values that could be treated as metric.

Chapter 10

Exercise 1

Make a list of the different types of hypotheses and suggest how each might be evaluated.

Hypotheses are carefully worded statements, as yet untested, about one or more properties of a set of cases. They need to be carefully worded since the manner of their evaluation will depend on how they are phrased. A basic distinction is between univariate, bivariate and multivariate hypotheses. These may be evaluated using traditional univariate (Chapter 4), bivariate (Chapter 5) or multivariate (Chapter 6) analyses. Which particular procedures are used depends a lot on whether the research objectives are descriptive, interpretive or verificational and whether the variables are binary, nominal, ordered category, ranked or metric. At a descriptive level, univariate hypotheses may be evaluated using a range of summary measures like proportions or modal categories for binary or nominal variables and measures of central tendency or dispersion for metric variables. Thus a hypothesis like 'The average primary school class size in England is less than 30' can be confirmed or denied by looking at official figures of average class sizes. If the data are based on a random sample, then further evaluation might include confidence intervals or by setting up the null hypothesis that classes are 30 or more. Multivariate hypotheses can be evaluated using either multivariate techniques or configurational data analysis like fuzzy set analysis.

Bivariate and multivariate hypotheses may be directional or non-directional. Non-directional hypotheses can be evaluated directly by using appropriate measures of association. Directional hypotheses bring in issues of sequence and timing. These can be evaluated sometimes from a thorough understanding of the cases in the research, or by using longitudinal data in extensions of traditional or configurational data analysis. Directional hypotheses nearly always imply some form of influence or causality. These can never be proven (at least, not in non-experimental research designs), but if the notion of causality is simple and linear, then checks for the impact of intervening variables and spurious relationships can be made by holding values of other variables constant in a multivariate analysis. If notions of causality are complex, for example involving asymmetrical relationships and the possibility of sufficient but not necessary or necessary but not sufficient relationships, then configurational data analysis can provide evidence that the data are at least consistent or not consistent with such propositions.

Exercise 2

In seeking to explain their findings, do social scientists have any alternative than to attempt to establish some form of causality?

It has been argued in Chapter 10 that to explain is to attempt to provide understanding to an audience so that it understands something that it did not understand before – it is the persuasion of others that intelligibility is being offered. How things are explained and the 'success' of that explanation will thus depend on the audience. If the audience is other social scientists who, on the whole, will only accept the establishment of causal connections as explanatory, then probably the researcher has little alternative but to follow suit. However, there are very different ideas about what 'causality' means and how it is established. Within this context, researchers, then, may follow very different pathways.

Exercise 3

To what extent are the manner and style of presentation of research findings part of the 'rhetoric' of explanation?

The short answer is: very much so. The audience needs to be persuaded that understanding and intelligibility are being offered. Presentations that are clear, structured, detailed, interesting, even amusing, are more likely to succeed in this task. The presenter may be limited to rhetorics that social scientists are likely to appreciate, for example experimental, statistical, quantitative or qualitative case-based rhetorics.

GLOSSARY

Adjusted R^2 A deflated estimate of **multiple R^2** that takes into account the number of variables and the size of the sample.

Alpha value The probability of rejecting the null hypothesis when it is in fact true. It is sometimes referred to as the level of confidence. By tradition it is set at either 0.5 or 0.01.

Analysis of covariance (ANCOVA) An **analysis of variance** that removes the effects of **covariates** through the use of **regression**-like procedures.

Analysis of variance (ANOVA) A procedure for testing the statistical significance of differences in scores in a metric variable between categorical groups, samples or treatments.

Asymmetrical relationship Changes either of category or of metric value on one variable are not mirrored by changes in category or metric value in another variable. The pattern is triangular rather than diagonal as with a **symmetrical relationship**.

Backward elimination A process in **log-linear analysis** in which the hierarchy of interactions effects is gradually eliminated while maintaining the ability to predict the frequencies in the table cells.

Bar chart A graphical display in which each category of a categorical variable is depicted by a bar whose height or length represents the frequency or proportion of observations falling into each category.

Behavioural properties What people actually did in the recent past, what they currently or usually do, or what they might do in the future.

Beta coefficient A standardized **partial slope**. The **regression** analysis is conducted using standardized or **z-scores**.

Bias A form of sampling error that tends to be in a particular direction.

Binary logistic regression A form of **logistic regression** in which the dependent variable is a **binary variable**.

Binary variable A record of the presence or absence of a property, usually coded as 0 for the absence and 1 for the presence.

Bivariate analysis The display, summary or drawing of conclusions from the way in which two variables are related together.

Bivariate crosstabulation A table in which the frequencies of cases that combine a value on one categorical variable with a value on another are laid out in combinations of rows and columns.

Bivariate hypothesis A statement about the existence of or degree of relationship between two variables in a population or sample of cases.

Bootstrapping Several smaller samples are taken from the main sample and these are used to estimate the population standard deviation.

Case The entity whose characteristics are being recorded in the process of data construction.

Categorical variable A **variable** that consists of two or more categories that are exhaustive and mutually exclusive, but which may, in addition, possess order.

Category clustering The tendency for some combinations of categories to predominate in a **crosstabulation** of two **nominal variables**.

Causal analysis A study of the way in which some events or circumstances can produce or bring about other events or circumstances.

Causal research Research that analyses the degree of influence of one or more independent variables upon one or more dependent variables.

Cell The combination of a row and a column in a **crosstabulation**.

Census An attempt by a researcher to contact or to study every **case** in the population.

Centroids The mean values of the cases in a **cluster analysis** used in the final stage in the analysis.

Chart Any form of graphical display of numerical information.

Chi-square A statistic that measures the overall departure of a set of observations from some theoretical proposition. It adds up the squared differences between observed values and the values expected from the theoretical proposition taken as a proportion of the expected differences.

Class interval A range of values on a metric variable that are grouped together for presentation and analysis.

Cluster analysis A range of techniques use for grouping cases which have characteristics in common.

Clustered bar chart A chart that shows the frequency or percentage of the categories of one variable separated out by the categories of another and placed side by side.

Code A number that stands for a category of a binary, nominal or ordered category variable.

Codebook Lists all the variable names (which are short, one-word identifiers), the variable labels (which are more extended descriptions of the variables and which appear as table or chart headings), the response categories used and the code numbers assigned.

Coding The transformation of edited questionnaires into machine-readable form.

Coefficient The value of a measure of association or correlation that varies between zero and one or between minus one and plus one.

Coefficient of determination The proportion of the variance on one metric variable accounted for by the variance on another. It is calculated by squaring the **correlation** coefficient (Pearson's r).

Cognitive properties Individual attitudes, opinions, beliefs and images.

Collinearity A degree of **correlation** between two metric predictor variables in a regression equation that poses a threat to the validity of the **regression** analysis.

Column marginal The totals at the foot of each column in a **crosstabulation**.

Concurrent mixed design A research design that entails undertaking two or more styles of research at the same or overlapping times, or even at separate times, but as independent enterprises and considered as a single phase of research.

Condition A property of a case that is used as a potential causal characteristic to **explain** an outcome.

Conditional correlation A correlation between two metric variables broken down by two or more categories of a third, categorical variable.

Conditional relationship A relationship between two variables that is moderated by a third variable.

Confidence interval See **interval estimate**.

Configuration A combination of **conditions** that describes a group of empirically observed or hypothetical cases.

Configurational data analysis (CDA) A form of **set-theoretic method** that has been implemented as either **crisp set** or **fuzzy set** analysis. It is necessarily causal and requires that one property (assessed in terms of set membership) be singled out as the 'outcome' while the other properties as set memberships are all potential causal conditions. Second, CDA makes use of **truth tables** that enable researchers to study all possible configurations from two or more conditions and to assess the extent to which they are subsets of the outcome, that is logically sufficient. Third, CDA makes use of the principles of Boolean minimization by which the results are expressed in a more parsimonious but logically equivalent manner.

Confirmatory factor analysis A method of **factor analysis** used to test the extent to which a hypothesized factor structure is supported by the data.

Confounding relationship The individual effects of two independent variables on a dependent variable are distorted because the independent variables are themselves related.

Consistency An index of the extent to which membership values for a condition or configuration of conditions and membership values on an outcome are consistent with the proposition that one of the sets is a subset of the other.

Construct validity The extent to which a measure relates to other measures to which it should be related.

Contact rate The number of eligible respondents successfully contacted as a proportion of the total number of eligible respondents approached.

Content validity The extent to which the domain of a characteristic is adequately sampled by the measure.

Continuous variable A **metric variable** whose values can take any number or fraction of a number.

Contradiction Two statements that say the opposite of one another.

Correlation A measure of the extent to which the values of two metric variables covary and approximate a rising or a falling straight line in a **scattergram**.

Correlation ratio A measure of association for situations where the independent variable is binary or nominal and the dependent variable is metric. It measures the proportion of the total variability in the dependent variable that can be accounted for by knowing the categories of the independent variable.

Counterfactual An assumption that is made about the impact of an unobserved event. In the context of fuzzy set analysis an assumption about a **logical remainder**.

Covariance An average **covariation** calculated by dividing the covariation by the number of cases.

Covariate Uncontrolled metric independent variable that creates extraneous (nuisance) variation in the dependent variable, and whose effect can be removed through the use of regression-like procedures.

Covariation A tendency for changes either of category or of metric value on one variable to be exactly mirrored by changes in category or metric value on another variable. The pattern is diagonal either in a **crosstabulation** of two **binary variables** or two **ordered category variables** or in a **scattergram** of two **metric variables**. The concept does not apply to **nominal variables** with three or more categories.

Coverage The proportion of cases that are covered by a configuration that has an acceptable level of consistency, that is, it forms a subset relationship. There are two kinds of coverage: **matrix coverage** and **outcome coverage**.

Cramer's V A statistical measure of association for two categorical variables that have been crosstabulated and based on the notion of departure from independence. It is calculated by taking **chi-square**, dividing by the number of cases multiplied by a value which is either the number of rows minus one or columns minus one, whichever is the minimum, and taking the square root.

Crisp set A record of the presence or absence of category membership. Similar to **binary variable**.

Criterion validity The extent to which a measure successfully predicts some other characteristic to which it is related.

Critical values Values that lie exactly on the boundary between accepting and rejecting the null hypothesis.

Cronbach's coefficient alpha A measure of scale reliability. It takes the average correlation among items in a summated rating scale and adjusts for the number of items.

Crosstabulation A form of table in which the frequencies of cases that combine the values on one or more categorical variables with values on another are laid out in rows and columns.

Curvilinear regression A form of **regression** that is based not on a straight line, but on some mathematically expressible curve.

Data Systematic records made by individuals.

Data analysis The process whereby researchers take the raw data that have been entered into a data matrix to create information and use them to tackle the objectives for which the research was undertaken.

Data display The presentation of data in tables, charts or graphs.

Data dredging Data are explored in every conceivable way to see if any patterns emerge.

Data fusion The merging of the results from two or more separate surveys with different samples into a single database.

Data matrix A record of all the values for all the variables for all the cases laid out in rows and columns.

Data mining A range of techniques for extracting actionable information from large databases, usually stored in a **data warehouse**.

Dataset A record of all the values for all the variables for all the cases in a research project.

Data summary Reducing the data in a distribution of a single **variable** or the relationship between variables to a single statistic that acts as a summary.

Data warehouse A very large database in which data are gathered from disparate sources and converted into a consistent format that can be used to support social policy or management decision making.

Degree of freedom The number of values (and, for some statistics like chi-square, the number of categories) that are free to vary when estimating some kind of statistical parameter.

Demographic properties Relatively fixed characteristics of cases that researchers have chosen to characterize the nature or condition of a case.

Dendrogram A graphical representation of a cluster of cases generated by using **cluster analysis**.

Departure from independence The extent to which observed data depart from what the data would look like if there were no association.

Dependence technique A multivariate analysis technique where one or more variables are identified as **dependent variables** and which are to be predicted or explained by one or more **independent variables**.

Dependent variable A variable that is seen as an outcome or effect of an independent variable.

Derived measurement The use of two or more measures in some combination to generate a total or single score.

Descriptive research Research that is concerned with measuring or estimating the sizes, quantities or frequencies of characteristics.

Descriptive statistics The display and summary of variables in a dataset.

Descriptors Properties that are studied one at a time in order to illustrate or summarize the key features of a set of cases.

Design factor A multiplier used to convert standard errors calculated by methods appropriate to simple random sampling into standard errors appropriate to more complex sample designs.

Direct measurement A one-to-one correspondence between the concept and the variable used to measure. This is made possible either because the characteristic is concretely observable or because the researcher defines the concept in terms of the variable.

Discrete variable A metric variable that records only whole numbers as values and is a result of counting the number of instances involved as a measure of size.

Discriminant analysis A multivariate technique that determines which weightings of metric independent variables best discriminate between two or more groups of cases better than by chance.

Discriminant function The composite of weightings used in **discriminant analysis**.

Discriminant validity The extent to which a measure does not correlate with other measures from which they are meant to be different.

Dummy variable A way of recoding a **nominal variable** into a series of **binary variables**.

Eclectic designs A research design that mixes in different ways any approach or methods that might help to solve a problem.

Editing The scrutiny of returned questionnaires to ensure that as far as possible they are complete, accurate and consistent.

Effect size An objective and standardized measure of the magnitude of the observed effect.

Eigenvalue The proportion of the total variance explained by each **factor** in a **factor analysis**. It is calculated from the sum of the squared **factor loadings**.

Elaboration analysis The process of studying bivariate relationships to see to what extent they may be due to the influence of other variables. It typically involves introducing a third variable into the analysis and assessing its impact on the other two.

Equifinality A situation in which an outcome may be achieved by more than one pathway.

Estimation Using the value of a statistic derived from a sample to estimate the value of the corresponding population parameter.

Eta A statistical measure of association where the independent variable is binary or nominal and the dependent variable is metric.

Euclidean distance A distance measure used in **cluster analysis** which takes the square root of the sum of the squared differences in **values** for each **variable**.

Experiment A research design in which one or more **independent variables** are manipulated by the researcher to examine their effects on one or more **dependent variables**, while controlling for **extraneous variables**.

Explanation A range of procedures or rhetorics that may be used to clarify and make comprehensible the findings of research to an audience.

Exploratory factor analysis A method of **factor analysis** which is used for exploratory purposes to replace a large set of variables with a smaller number of **factors**.

Exploratory research Research aimed at generating ideas, insights or hypotheses.

Extraneous variable A variable whose effect is controlled, minimized or excluded in the design of an experiment.

Factor A term that is commonly used to refer to an **independent variable** in an experimental design. It is also used to refer to a **latent variable** in **factor analysis**.

Factor analysis A multivariate statistical technique used for identifying the underlying structure in a large set of metric variables. It groups together those variables that are highly inter-correlated.

Factor loading The correlation between a **variable** and a **factor** in **factor analysis**.

Frame errors Errors arising from the use of lists of the population to be studied that have various shortcomings.

Frequency The number of times a value occurs in a given distribution for a particular variable.

Frequency table Frequencies of values, usually either categorical or grouped metric, laid out in a column.

Fuzzy set A record of the degree of set membership of a defined category with membership values in the interval between [1] and [0].

Gamma A measure of association for two ordered category variables based on calculating the number of pairs in a crosstabulation having the same rank order of inequality on both variables compared with the number of pairs having the reverse order of inequality.

General linear model (GLM) A combination of **analysis of variance** and **regression** for situations where there is one or more metric dependent variables and a combination of categorical or metric independent variables and **covariates** referred to as **factors**.

Goodness-of-fit test The determination of the extent to which a certain model fits the observed data. The statistic **chi-square** can be used for this purpose.

Hierarchical clustering A clustering procedure used in **cluster analysis** in which either cases are iteratively combined on the basis of some measure on similarity (agglomerative clustering), or the total set of cases is divided into subgroups on a basis specified by the researcher (divisive clustering).

Hierarchical regression A method of multiple regression in which the order in which the independent variables are entered into the regression model is determined by the researcher, based on previous research.

Histogram A graphical display for metric variables in which the width of the bars represents class intervals and the length or height represents the frequency or proportion with which each interval occurs.

Hypothesis A formal statement that the researcher makes about one or more variables relating to a set of cases that are the focus of the research, but which is as yet untested.

Independent samples Samples from different groups of respondents.

Independent variable A variable that is treated as a condition, cause or influence.

Indirect measurement Taking an indicator of a concept as a variable measure of that concept.

Interaction effect The combined effect of two or more independent variables on a dependent variable.

Interdependence technique A technique in multivariate analysis that involves the simultaneous analysis of three or more variables, none being identified as either dependent or independent.

Interval estimate A range of values within which there is a calculated probability that it will contain the true population parameter.

Item analysis A process used in Likert scaling whereby items that do not correlate with the total score are discarded.

Kendall's tau-b A measure of association for two ordinal variables that is similar to **gamma** except that it takes account of those pairs that are tied in both directions.

Kendall's tau-c A measure of association for two ordinal variables based on **pair-by-pair comparisons**, but only when the number of rows and columns is the same.

Kurtosis A measure of the distribution shape based on the extent to which values cluster about the mean compared with a normal distribution.

Lambda A measure of association for two categorical variables based on the **proportional reduction in error**.

Latent variable A variable that cannot be directly observed or measured, but is assumed to be related to several variables that can be measured.

Level of confidence See **Alpha value**.

Likelihood ratio chi-square As distinct from the Pearson chi-square (see **chi-square**) this takes the logarithm of the relationships between observed and expected frequencies.

Likert scale A **summated rating scale** derived from the summation of five- or seven-point ratings of agreement or disagreement with a number of statements relating to an attitude object.

Limited diversity Diversity is limited when many possible configurations do not have an empirical existence (or for large-*n* datasets there are not enough cases in a configuration to be able to draw conclusions). Each configuration that fits into this category is a **logical remainder**.

Line graph A graph that connects a series of data points using continuous lines.

Linear regression See **regression**.

Log likelihood chi-square See **Likelihood ratio chi-square**.

Log-linear analysis An extension of **chi-square** analysis used to detect the patterns of interactions in a set of categorical variables.

Log odds The natural logarithm of the odds ratio, which is the ratio of the odds of an event occurring in one group compared with another. The odds are calculated by taking the probability of an event occurring divided by the probability of that event not occurring.

Logical remainder A row in a truth table without enough cases in it.

Logistic regression A form of **regression** analysis in which the metric **independent variables** are used to predict the probability of a case being in a particular category of a **categorical variable**.

Longitudinal data Measures have been taken at two or more moments of time or the whole study is repeated at a later date. There are three main kinds of non-experimental longitudinal data: repeat cross-sectional, panel studies and cohort studies.

Matrix coverage The number of cases in a subset relationship as a proportion of the total number of cases in the data matrix.

Mean A measure of central tendency for metric variables calculated by totalling all the values in a distribution and dividing by the number of observations made.

Measure The ascertainment of type, order, ranking, number or calibration according to specified standards.

Measurement A process by which the characteristics or properties of cases that are to be used as variables or **set memberships** are specified.

Median A measure of central tendency calculated by taking the value that splits all the observations for a variable into two halves arranged in an ascending or descending series.

Mediating relationship An independent variable is seen as having an effect on a dependent variable indirectly through another independent variable.

Metric table A table used to display metric quantities.

Metric variable A variable that arises from the processes of either calibration or counting.

Missing value An unrecorded value in a data matrix usually from questions in a survey that are unanswered.

Mixed methods Research designs in which either or both research methods and data are combined in various ways.

Mode A measure of central tendency established by taking the most commonly occurring value in a distribution.

Model A simplified description of a system or a structure that is devised to assist the process of making calculations concerning the relationships between variables and making predictions.

Multicollinearity A degree of correlation between three or more metric predictor variables in a regression equation that poses a threat to the validity of the regression analysis.

Multidimensional scaling A concept that consists of two or more dimensions that are not totalled, but kept separate either as a profile or as a point in multidimensional space. Also used to refer to a spatial map in which the perceptions or preferences of consumers are represented in two-dimensional space.

Multiple R^2 The proportion of variation in a dependent variable associated with the variation of the independent variables.

Multiple regression An extension of regression in which the outcome is predicted from a linear combination of two or more predictor variables.

Multiple response question A fixed choice question that allows a respondent to pick more than one response category.

Multi-variable table A table displaying the distributions of two or more variables that are not interlaced.

Multivariate analysis The display, summary or drawing of conclusions from variables taken three or more at a time.

Multivariate analysis of covariance (MANCOVA) A **multivariate analysis of variance** that removes the effects of **covariates** through the use of **regression**-like procedures.

Multivariate analysis of variance (MANOVA) An analysis of variance when there is more than one metric dependent variable.

Multivariate hypothesis A statement about the relationships between three or more properties in a population or sample of cases.

Necessary condition Whenever the outcome is present, the condition is also present. For fuzzy sets, set membership of the outcome is less than or equal to membership of the condition.

Nominal variable A set of values that represent two or more contrasting groups having categories that are exhaustive and mutually exclusive.

Non-hierarchical clustering A technique of clustering used in **cluster analysis**, sometimes also called *k*-means clustering, which involves pre-determining the number of cluster centres and all cases are grouped within a specified threshold from the centre.

Non-parametric statistics Statistics that do not assume that data are metric or that the **mean** or **standard deviation** can be calculated.

Non-probability sample A sample in which the chances of selecting a case from the population of cases is not calculable since the selection is made on a subjective basis.

Non-sampling errors Survey errors that are not a result of the sampling process.

Non-standardized regression coefficient The rate of change in a dependent variable consequent upon a unit change in an independent variable in a multiple regression equation. Also known as a **partial slope**.

Normal curve A curve that describes a **normal distribution.**

Normal distribution A symmetrical bell-shaped distribution that describes the expected probability distribution for random events.

Null hypothesis A statement made about a population of cases in advance of testing it on data.

***n*-way crosstabulation** A bivariate **crosstabulation** that is 'controlled' by a third or further categorical variable by breaking down or 'layering' the original cross-tabulation by categories of further variables.

One-sample *t*-test Calculates the probability of getting a difference between an achieved mean and some specified test value (the hypothesized value) as big or bigger than the one achieved resulting from random sampling fluctuations.

One-way ANOVA An **analysis of variance** in which there is only one basis for categorizing groups of the **independent variable**.

Ordered category variable A set of **values** that represent two or more categories that are exhaustive and mutually exclusive and in addition possess an implied order from high to low.

Outcome coverage The number of cases in a subset relationship as a proportion of the number of cases that are instances of the outcome.

Outliers Values that are substantially different from the general body of values.

Pair-by-pair comparisons A measure of association based on the tendency for all possible combinations of pairs of cases to show similar orderings on both ordinal variables.

Paired comparisons A scaling technique in which respondents are asked to compare objects or images two at a time according to a specified criterion.

Paired samples *t*-test An **analysis of variance** in which two sets of metric values are compared for the same respondents.

Parametric statistics Statistics that assume metric data, and that the calculation of a mean or standard deviation is a legitimate operation.

Partial correlation Used to calculate the original bivariate correlation with the effects of a third metric variable removed.

Partial slope The impact of an independent variable over and above the impacts of all other independent variables in the equation.

Pearson's r See **correlation coefficient**.

Phi A measure of association calculated by taking **chi-square** and dividing by the number of cases and taking the square root.

Pie chart A graphic in which the frequencies or proportion of each category of a categorical variable is represented by a slice of a circle.

Point estimate A single value taken as an estimate of a population parameter.

Population A total set of cases or other units that are the focus of the researcher's attention.

Pre-code Assigning a numerical code to each response category at the questionnaire design stage.

Primary data Data constructed specifically for the research at hand.

Principal components analysis One of several ways of undertaking factor analysis in which both the amount of variance to be accounted for and the number of components to be extracted equal the number of variables.

Probability sample A sample in which the selection of sampling units is made by methods independent of human judgement. Each unit will have a known and non-zero probability of selection.

Properties Characteristics of **cases** that are included in the research population and that the researcher has chosen to observe or measure and then record.

Proportional reduction in error The extent to which it is possible to predict the values of one categorical variable from the values of another categorical variable with which it might be associated.

***p*-value** The probability in a random sample of obtaining a value as extreme or more extreme than the one actually obtained if the **null hypothesis** were true.

Qualitative data Systematic records that consist of words, phrases, text or images.

Quantitative data Systematic records that arise as numbers that result from the systematic capture of classified, ordered, ranked, counted or calibrated characteristics of a sample or population of cases.

Questionnaire Any document that is used as an instrument with which to capture data generated by asking people questions.

Random error Error arising from a random sampling procedure in which there will be chance fluctuations.

Random sample A probability sample in which the selection of sampling units is made by methods independent of human judgement. Each unit will have a known and non-zero probability of selection.

Range The difference between the minimum and maximum value in a distribution of a metric variable.

Ranked variable A set of scale values in which each case has its own rank. There are as many rankings as cases.

Rating An ordered classification of a grade given by a respondent on a survey.

Refusal rate The number of refusals in a survey as a proportion of the number of eligible respondents contacted.

Regression The use of a formula describing a straight line that represents the 'best fit' in a **scattergram** to predict the values of one metric variable from another.

Reliability The extent to which the application of a scale produces consistent results if repeated measures are taken.

Residual analysis A process in **multiple regression** that checks on the distribution of residuals, or errors.

Response rate The number of completed questionnaires as a proportion of the total number of respondents approached.

Row marginal The totals at the end of each row in a **crosstabulation**.

Sample A subset of cases or other units from a total population of cases or units.

Sample design The particular mix of procedures used for the selection of sampling units in a particular piece of research.

Sampling distribution A theoretical distribution of a statistic for all possible samples of a given size that could be drawn from a particular population.

Sampling distribution of chi-square A theoretical distribution for all possible samples of a fixed number of **degrees of freedom** that could be drawn from a particular population.

Sampling error Error that arises from the sampling process. It may be defined more precisely as the difference between the result of a sample and the result that would have been obtained from a census using identical procedures.

Sampling frame A complete list of the population of units or cases from which a sample is to be taken.

Sampling unit Whatever entity is being sampled.

Saturated model A model that perfectly fits the data and therefore has no error. It contains all possible main effects and interactions between variables.

Scale A single-dimensional or multidimensional measure generated from two or more individual measures.

Scattergram A graphical display of the relationship between two metric variables in which each case is represented by a dot that reflects the position of two combined measurements.

Score A total derived from adding up numerical values.

Semantic differential scale A multidimensional scale that represents a profile of characteristics expressed as bipolar opposites like 'sweet–sour' that constitutes a seven-point rating scale.

Sequential mixed design A research design that involves undertaking a research project in two or more stages or phases, each stage or phase acting as an input to the next.

Set membership A property of a case recorded as the extent to which a case is a member of a specified category.

Set-theoretic methods Approaches to analysing social reality through the notion of sets and their relationships. They work with properties as degrees of set membership. They see relations between social phenomena in terms of intersection, union, negation and subset which are interpreted in terms of sufficiency and necessity.

Simple random sample Each sampling unit has an equal chance of being selected from a list.

Skewed set memberships All or nearly all cases hold very high or very low membership in a set – either in the outcome or in a condition. Skewness can arise in two main contexts: very high **matrix coverage** when either all or nearly all the cases manifest the outcome or the condition; or very low **outcome coverage** when the condition never or very seldom occurs.

Skewness A measure of distribution shape based on the difference between the mean and median values in a distribution.

Somers' d A measure of association for two ordinal variables that is similar to **gamma** except that it takes account of those pairs that are tied on one variable but not on the other.

Spearman's rho A measure of correlation between two fully ranked scales.

Spurious relationship A covariation between two variables that is a result of their common relationship with a prior variable.

Stacked bar chart A chart that shows the frequency or percentage of the categories of one variable separated out by the categories of another and placed one on top of the other.

Standard deviation An average of deviations of values about the mean for a given statistic.

Standard error The **standard deviation** of a **sampling distribution**.

Standard error of the estimate The **standard deviation** of the residuals in a **multiple regression** which shows how well spread out the data points are around the regression line.

Standard error of the mean The **standard error** of the **sampling distribution** of the **means** observed for a metric variable.

Standard error of the proportion The **standard error** of the **sampling distribution** of the proportions observed for a **binary variable**.

Statistical inference A process by which sample statistics are used to estimate population parameters or to test statements about a population with a known degree of confidence.

Statistical significance A result that, in a random sample, is unlikely to have arisen by chance with a specified probability.

Stepwise regression A method of **multiple regression** in which the order in which the independent variables are entered into the regression model is based on a statistical criterion.

Structural equation modelling A statistical technique that combines the notions of **correlation** and goodness of fit between a theoretical model and the observed data.

Subset relationship A relationship between two properties such that members of one set are totally contained within a wider set.

Sufficient condition Whenever that **condition** is present, the outcome is also present. For **fuzzy sets**, set membership on the outcome is greater than or equal to membership of the condition across all cases.

Summated rating scale A measure derived from summing together two or more separate rating scales.

Survey The capture of data based on addressing questions to respondents in a formal manner and taking a systematic record of their responses.

Survey population The total set of cases in which the researcher is interested for the purpose of the research.

Symmetrical relationship Changes either of category or of metric value on one variable are mirrored by changes in category or metric value on another variable, that is a degree of **covariation** exists.

Systematic error Error arising from sampling procedures that result in the over- or under-representation of particular kinds of sampling unit mostly in the same direction.

Theory A set of concepts and logically related propositions that work or may work in more than one context.

Three-way analysis A bivariate crosstabulation that is layered by a third variable to examine its impact on the original relationship.

Total survey error The sum of all sources of error, both those arising from the sampling process and non-sampling errors.

Triangulation The use of two or more approaches to research to see if they come to similar conclusions.

Trivial necessary condition The condition always or nearly always occurs, so it readily becomes necessary, whatever the outcome.

Trivial sufficient condition All or nearly all the cases in a dataset manifest the outcome or have high memberships of the outcome. In this situation, all or nearly all **conditions** or **configurations** readily become **sufficient** for the outcome to happen.

Truth table A sorting of cases according to set membership values into one of the 2^k logically possible configurations. Each row of the table is linked to the outcome and can be interpreted as a statement of sufficiency.

***t*-test** An **analysis of variance** in which the categorical **independent variable** is a **binary variable**.

Two-tailed test A statistical test of a non-directional hypothesis.

Unbiased estimate A **standard error** that is adjusted when the sample **standard deviation** is being used as an estimate of the population standard deviation. The sample standard deviation is divided by the square root of the sample size minus one.

Univariate analysis The display, summary or drawing of conclusions from a single variable or set of variables treated one at a time.

Univariate hypothesis A statement about the population **value** of a **variable** or two or more variables that are not combined or related.

Validity The extent to which the application of a scale measures what it is intended to measure.

Value What researchers actually record as a result of the process of assessing **properties**. Such records may relate either to **variables** or to **set memberships**.

Variable Set of values that assess cases relative to one another and arise from one or more of the activities of classifying, ordering, ranking, counting or calibrating the characteristics of cases.

Variance The mean squared deviation of all the values in the distribution of a metric variable.

Varimax A method of factor rotation used in factor analysis that keeps the vectors at right angles so that they are unrelated to one another.

Weighting The application of a multiplying factor to some of the responses given in a survey in order to eliminate or reduce the impact of bias caused by under- or over-representation of respondents with particular characteristics.

Wilks's lambda A procedure used in a **general linear model** to examine the contribution of each main and **interaction effect**. It is a measure of the unexplained variance and is the mirror image of η^2.

XY plot A plot that displays each case's fuzzy membership on a condition or configuration on the X-axis and the membership of the outcome on the Y-axis.

z-score The value of an observation expressed in standard deviation units. The arithmetic mean is subtracted from each observation and divided by the standard deviation.

z-test A test against the null hypothesis carried out on a univariate hypothesis by comparing the result from a sample with a standardized normal distribution.

REFERENCES

Abbott, A. (1992) 'What do cases do? Some notes on activity in sociological analysis', in C. Ragin and H. Becker (eds), *What Is a Case? Exploring the Foundations of Social Inquiry*. Cambridge: Cambridge University Press.

Abbott, A. (1998) 'The causal devolution', *Sociological Methods and Research*, 27 (2): 148–81.

Argyrous, G. (2011) *Statistics for Research with a Guide to SPSS*, 3rd edn. London: Sage.

Assael, H. and Keon, J. (1982) 'Nonsampling versus sampling errors in survey research', *Journal of Marketing*, 46, Spring: 114–23.

Barker, A., Nancarrow, C. and Spackman, N. (2001) 'Informed eclecticism: a research paradigm for the twenty-first century', *International Journal of Market Research*, 43 (1): 3–27.

Berg-Schlosser, D. and Mitchell, J. (2000) *Conditions of Democracy in Europe, 1919–39: Systematic Case Studies*. Basingstoke: Macmillan.

Berg-Schlosser, D. and Mitchell, J. (2003) *Authoritarianism and Democracy in Europe, 1919–39: Comparative Analyses*. Basingstoke: Palgrave Macmillan.

Berger, P. and Luckmann, T. (1966) *The Social Construction of Reality: A Treatise on the Sociology of Knowledge*. New York: Doubleday.

Berk, R. (2004) *Regression Analysis: A Constructive Critique*. Thousand Oaks, CA: Sage.

Boole, G. (1847) *The Mathematical Analysis of Logic: Being an Essay Towards the Calculus of Deductive Reasoning*. Oxford: Basil Blackwell.

Booth, A., Brennan, A., Meir, P., O'Reilly, D., Purshouse, R., Stockwell, T. and Wong, R. (2008) *The independent review of the effects of alcohol pricing and promotion*. Report prepared for the Department of Health. London: The Stationery Office.

Borgna, C. (2013) 'Fuzzy-set analysis: the hidden asymmetries'. COMPASSS Working Papers, www.compasss.org.

Bottomly, P. and Nairn, A. (2004) 'Blinded by science: the managerial consequences of inadequately validated analysis solutions', *International Journal of Market Research*, 46 (2): 171–87.

Byrne, D. (2002) *Interpreting Quantitative Data*. London: Sage.

Campbell, D.T. and Fiske, D.W. (1959) 'Convergent and discriminant validity by the multitrait–multimethod matrix', *Psychological Bulletin*, 56: 85–105.

Caren, N. and Panofsky, A. (2005) 'TQCA: a technique for adding temporality to Qualitative Comparative Analysis', *Sociological Methods and Research*, 43: 147–72.

Cohen, S. and Markowitz, P. (2002) 'Renewing market segmentation: some new tools to correct old problems', ESOMAR Consumer Insight Congress, Barcelona.

Cook, T.D. and Campbell, D.T. (1979) *Quasi-experimentation: Design and Analysis Issues for Field Settings*. Boston, MA: Houghton Mifflin.

Cooper, B. and Glaesser, J. (2011) 'Using case-based approaches to analyse large datasets: a comparison of Ragin's fsQCA and fuzzy cluster analysis', *International Journal of Social Research Methodology*, 14 (1): 31–48.

Cortina, J. (1993) 'What is coefficient alpha? An examination of theory and application', *Journal of Applied Psychology*, 78 (1): 98–104.

Cragun, R. (2013) 'Using SPSS and PASW', Wikibooks. Available at http://en.wikibooks.org/wiki/Using_SPSS_and_PASW.

Creswell, J. (2009) *Research Design: Qualitative, Quantitative, and Mixed Methods Approaches*, 3rd edn. London: Sage.

Creswell, J. and Plano Clark, V. (2011) *Designing and Conducting Mixed Methods Research*, 2nd edn. London: Sage.

De Leeuw, J. (2004) 'Series Editor's Introduction', in R. Berk, *Regression Analysis: A Constructive Critique*. Thousand Oaks, CA: Sage.

Denzin, N. (1978) 'The logic of naturalistic inquiry', in N. Denzin (ed.), *Sociological Methods: A Sourcebook*. New York: McGraw-Hill.

De Vaus, D. (2002) *Analyzing Social Science Data: 50 Key Problems and Data Analysis*. London: Sage.

DeVellis, R.F. (2011) *Scale Development: Theory and Applications*, 3rd edn. London: Sage.

Diamantopoulos, A. and Schlegelmilch, B. (1997) *Taking the Fear Out of Data Analysis*. London: Dryden Press. Republished by Cengage Learning, 2000.

Durand, R. and Lambert, Z. (1988) 'Don't know responses in surveys: analysis and interpretational consequences', *Journal of Business Research*, 16 (2):169–88.

Emmenberger, P. (2010) 'How good are your counterfactuals? Assessing quantitative macro-comparative welfare state research with qualitative criteria', COMPASSS Working Papers, www.compasss.org.

Field, A. (2009) *Discovering Statistics Using SPSS*. London: Sage.

Field, A. (2013) *Discovering Statistics Using IBM SPSS Statistics*, 4th edn. London: Sage.

Fisher, R. (1925) *Statistical Methods for Research Workers*. Edinburgh: Oliver & Boyd.

Goertz, G. and Levy, J. (2007), 'Causal explanation, necessary conditions and case studies', in G. Goertz and J. Levy (eds), *Explaining War and Peace: Case Studies and Necessary Condition Counterfactuals*. Abingdon: Routledge. pp. 9–45.

Gorard, S. (2013) *Research Design: Creating Robust Approaches for the Social Sciences*. London: Sage.

Gordon, R., Harris, F., Mackintosh, A.M. and Moodie, C. (2010a) 'Assessing the cumulative impact of alcohol marketing on young people's drinking: cross-section data findings', *Addiction Research and Theory*, Early Online, 1–10, Informa UK Ltd.

Gordon, R., Mackintosh, A.M. and Moodie, C. (2010b) 'The impact of alcohol marketing on youth drinking behaviour: a two-stage cohort study'. *Alcohol and Alcoholism*, 45 (5): 470–80.

Gordon, W. (1999) *Goodthinking: A Guide to Qualitative Research*, Admap Publications

Green, S. (1991) 'How many subjects does it take to do a regression analysis?', *Multivariate Behavioural Research*, 26: 499–510.

Greene, J., Caracelli, V. and Graham, W. (1989) 'Toward a conceptual framework for mixed-method evaluation designs', *Educational Evaluation and Policy Analysis*, 11: 255–74.

Hacking, I. (1999) *The Social Construction of What?* Cambridge, MA: Harvard University Press.

Hair, J., Black, W., Babin, B. and Anderson, R. (2010) *Multivariate Data Analysis: A Global Perspective*, 7th edn. Upper Saddle River, NJ: Pearson Education.

Hempel, C. (1942) 'Studies in the logic of confirmation', *Mind*, 54: 1–26, 97–121.

Hume, D. (1911) *A Treatise of Human Nature*. New York: E.P. Dutton.

Jong, G., Faulkner, J. and Warland, R. (1976) 'Dimensions of religiosity reconsidered: evidence from cross-cultural study', *Social Forces*, 54 (June): 866–89.

Kent, R. (2007) *Marketing Research: Approaches, Methods and Applications in Europe*. London: Thomson Learning (now Cengage, Andover).

King, G. and Zeng, L. (2007) 'When can history be our guide? The pitfalls of counterfactual inference', *International Studies Quarterly*, 51 (1): 183–210.

Koerner, R. (1980) 'The design factor – an underutilized concept?', *European Research*, November: 243–50.

Latour, B. and Woolgar, S. (1979) *Laboratory Life: The Social Construction of Scientific Facts*. Beverly Hills, CA: Sage.

Lee, N. and Hooley, G. (2005) 'The evolution of "classical mythology" within marketing measure development', *European Journal of Marketing*, 39 (3/4): 365–85.

Lessler, J. and Kalsbeek, W. (1992) *Nonsampling Errors in Surveys*. New York: Wiley.

Likert, R. (1932) 'A technique for the measurement of attitudes', *Archives of Psychology*, 22 (140): 55.

McGivern, Y. (2009) *The Practice of Market Research*. Harlow: Pearson Education.

Miethe, T. and Regoeczi, W. (2004) *Rethinking Homicide: Exploring the Structure and Process Underlying Deadly Situations*. Cambridge: Cambridge University Press.

Morse, J. (1991) 'Approaches to qualitative–quantitative methodological triangulation', *Nursing Research*, 40: 120–3.

Nunnally, J. (1978) *Psychometric Theory*, 2nd edn. New York: McGraw-Hill.

Osgood, C., Suci, G. and Tannenbaum, P. (1957) *The Measurement of Meaning*. Chicago: University of Illinois Press.

Pampel, F. (2000) *Logistic Regression: A Primer*. Thousand Oaks, CA: Sage.

Peterson, R. (1994) 'A meta-analysis of Cronbach's coefficient alpha', *Journal of Consumer Research*, 221 (September): 381–91.

Ragin, C. (1987) *The Comparative Method: Moving beyond qualitative and quantitative strategies*. Berkeley, CA: University of California Press.

Ragin, C. (1992) '"Casing" and the process of social inquiry', in C. Ragin and H. Becker (eds), *What Is a Case? Exploring the Foundations of Social Inquiry*. Cambridge: Cambridge University Press.

Ragin, C. (2000) *Fuzzy Set Social Science*. Chicago: University of Chicago Press.

Ragin, C. (2006) 'Set relations in social research: evaluating their consistency and coverage', *Political Analysis*, 14: 291–310.

Ragin, C. (2008) *Redesigning Social Inquiry: Fuzzy Sets and Beyond*. Chicago: University of Chicago Press.

Ragin, C. and Becker, H. (eds) (1992) *What Is a Case? Exploring the Foundations of Social Inquiry*. Cambridge: Cambridge University Press.

Ragin, C. and Sonnett, J. (2004) 'Between complexity and parsimony: limited diversity, counterfactual cases and comparative analysis', in S. Kropp and M. Minkenberg (eds), *Vergleichen in der Politikwissenschaft*. Wiesbaden: VS Verlag für Sozialwissenschaften. pp. 180–97.

Ragin, C. and Strand, S. (2008) 'Using Qualitative Comparative Analysis to study causal order. Comment on Caren and Panofzky (2005)', *Sociological Methods and Research*, 36: 431–41.

Rihoux, B. and Ragin, C. (eds) (2009) *Configurational Comparative Methods*. Thousand Oaks, CA: Sage.

Schneider, C. and Grofman, B. (2006) 'It might look like regression ... but it's not! An intuitive approach to the presentation of QCA and fs/QCA', COMPASSS Working Paper, No. 32.

Schneider, C. and Wagemann, C. (2012) *Set-Theoretic Methods for the Social Sciences: A Guide to Qualitative Comparative Analysis*. Cambridge: Cambridge University Press.
Schrodt, P. (2006) 'Beyond the linear frequentist orthodoxy', *Political Analysis*, 14: 335–9.
Smith, L. and Foxcroft, D. (2009) 'The effect of alcohol advertising and marketing on drinking behaviour in young people: a systematic review', *BMC Public Health*, 9 (51).
Spicer, J. (2005) *Making Sense of Multivariate Data Analysis*. Thousand Oaks, CA: Sage
Tabachnik, B. and Fidell, L. (2001) *Using Multivariate Statistics*, 4th edn. Boston, MA: Allyn & Bacon.
Tashakkori, A. and Teddlie, C. (1998) *Mixed Methodology: Combining Qualitative and Quantitative Approaches*. Thousand Oaks, CA: Sage.
Uprichard, E. (2009) 'Introducing cluster analysis: what can it teach us about the case?', in D. Byrne and C. Ragin (eds) *The Sage Handbook of Case-Based Methods*. London: Sage.
Yang, K. (2010) *Making Sense of Statistical Methods in Social Research*. London: Sage.

INDEX